THE SYSTEM OF
MINERALOGY

By the late JAMES D. DANA and the late EDWARD S. DANA

THE SYSTEM OF MINERALOGY. *Seventh Edition.*
Rewritten and enlarged by the late Charles Palache,
the late Harry Berman, and Clifford Frondel.
VOL. I. 1944.
VOL. II. 1951.
VOL. III. 1962. Rewritten and enlarged by Clifford Frondel.

By the late JAMES D. DANA

MANUAL OF MINERALOGY. *Seventeenth Edition.*
Revised by Cornelius S. Hurlbut, Jr. 1959.

By the late EDWARD S. DANA

A TEXTBOOK OF MINERALOGY. *Fourth Edition.*
Revised by the late William E. Ford. 1932.

MINERALS AND HOW TO STUDY THEM. *Third Edition.*
Revised by Cornelius S. Hurlbut, Jr. 1949.

THE SYSTEM OF

MINERALOGY

of James Dwight Dana and Edward Salisbury Dana

Yale University 1837-1892

SEVENTH EDITION

Entirely Rewritten and Greatly Enlarged

BY

CLIFFORD FRONDEL

Harvard University

VOLUME III

SILICA MINERALS

JOHN WILEY AND SONS, INC.

NEW YORK AND LONDON

PREFACE

This book describes the mineralogy of the polymorphs of silica. Two other volumes, still in preparation, will deal with the mineralogy of the silicates, to which the silica minerals stand in a broad structural sense as prototypes. The importance of the silica minerals both in scientific and in technological respects, and the historical interest that attaches to certain of them, it is hoped, justify the detail of the treatment here attempted. The properties of silica aside from those of immediate relevance to mineralogy are the subject of a voluminous literature. The bulk of this literature necessarily is left to specialized works in ceramics, chemistry, and other fields. The periodical literature has been examined up to 1960, although a few more recent papers are cited. The description of stishovite was added during publication, and it was not possible to make cross-references or emendations relating thereto throughout the text. The book by K. V. Zinserling, *Artificial Twinning of Quartz*, published by the Academy of Sciences of the USSR (Moscow, 1961), came to my attention too late to be used.

Historically, interest in the descriptive morphology of quartz, although long continued, was predominant in the early part and the middle of the nineteenth century, and the study of physical properties was emphasized in the latter nineteenth and early twentieth centuries. The structural and crystallochemical properties of quartz and of the other polymorphs of silica are now the focal point of interest. This more fundamental research stems from a modern discovery, that of the diffraction of x-rays by crystals.

The familiar term "crystal clear" ultimately derives from the Greek name *crystallos*, applied to transparent rock-crystal. This term, with its connotation of purity, often has subjectively encouraged the selection of colorless, limpid crystals of quartz and of other minerals as standards for the measurement of their physical properties. In some instances the purpose has been more general, as in the selection of Iceland Spar as a datum for the wavelength of x-rays. Only in relatively recent times have analytical methods been developed of sufficient sensitivity to measure the small compositional variations apparently always present in natural quartz. After a century of extremely precise and laborious measurements of the physical properties of quartz, including some of the most precise made on any substance, we now face the realization that few of these measurements were made on quartz of known composition and that none was made on quartz free from solid solution, with the ideal composition SiO_2. The uses to which the compositional variation in quartz and the other polymorphs of silica can be put in genetic mineralogy and in other fields, including those technological in nature, are still largely unexplored.

It is a pleasant duty to acknowledge the aid of others in the preparation of this work. In particular, I am indebted to Mr. Paul E. Desautels of the Division of Mineralogy, United States National Museum, for the preparation of most of the line drawings, and for supplying the photographs of the specimens shown in Figs. 91, 92, 93, 94, 95, and 97. Grateful appreciation also is expressed to Dr. William Parrish, of the Philips Laboratories, for the machine computation of Table 8, giving the interplanar spacings and reflection angles of quartz for various x-radiations, and to Professors Ray Pepinsky and J. Van den Hende, of the Pennsylvania State University, for the machine computation of the angle table (Table 11) for the morphology of quartz. The excellent diffractometer chart for quartz shown in Fig. 10 also was prepared for this book by Dr. Parrish. Dr. Brian Mason, of the American Museum of Natural History, and Dr. Benjamin M. Shaub, Northampton, Massachusetts, very kindly supplied the photographs of specimens shown in Figs. 44 and 45 and Fig. 96, respectively. Acknowledgment is made to the Smithsonian Institution and to the American Museum of Natural History for permission to publish the photographs mentioned. The photograph of the synthetic quartz crystal shown in Fig. 77 is published with the kind permission of the Western Electric Company.

I am also much indebted to my wife, Dr. Judith Weiss Frondel, for assistance in the preparation of the manuscript and in proofreading, and to my colleague, Professor Cornelius S. Hurlbut, Jr., for many discussions.

CLIFFORD FRONDEL

Cambridge, Mass.
July, 1962

CONTENTS

INTRODUCTION

Silica exists in a number of different polymorphs (Table 1). The particular polymorph called low-quartz or simply quartz is by far the most common of these substances and is one of the major constituents of the Earth's outer crust. Quartz crystals and most of the colored and fine-grained varieties of quartz, together with opal, were known as distinct entities over 2000 years ago. In some instances their names have survived with the original meanings to the present day.

The name quartz, first used in the Middle Ages in Saxony for massive vein quartz, did not become an inclusive designation for the colored and fine-grained varieties of this mineral until about the end of the eighteenth century. The mutual identity of these varieties was established partly on morphological grounds, here with reference to amethyst, smoky quartz (morion), and rock-crystal, and partly by qualitative chemical and physical tests. An adequate background for the systematic treatment of the chemistry and crystallography of silica, however, was not available until the first half of the nineteenth century. In this period, the chemical composition of quartz and the chemistry

TABLE 1. POLYMORPHS OF SILICA

Name	Thermal Stability Range at 1 atm.	Symmetry
Quartz	Below 573°	Hexagonal-P; trigonal trapezohedral
High-quartz	573–870°	Hexagonal-P; hexagonal trapezohedral
Low-tridymite	Below 117°	Orthorhombic
Middle-tridymite	117–163°	Hexagonal
High-tridymite	163–1470° (stable 870–1470°)	Hexagonal: dihexagonal dipyramidal
Low-cristobalite	Below 200°	Tetragonal; tetragonal trapezohedral (?)
High-cristobalite	200–1720° (stable 1470–1720°)	Isometric; tetartoidal (?)
Keatite	Metastable at ordinary conditions	Tetragonal; tetragonal trapezohedral
Coesite	Metastable at ordinary conditions	Monoclinic; prismatic
Stishovite	Metastable at ordinary conditions	Tetragonal; ditetragonal dipyramidal

1

of silicon in general were established by Berzelius and others, and the morphological crystallography and optical behavior became known through the work of Herschel, Biot, Weiss, Rose, and Des Cloizeaux.

Tridymite, the first polymorph of silica to be recognized in addition to quartz, was described by vom Rath in 1868, and cristobalite was described by him in 1884. Both substances were originally found in an igneous rock in the Cerro San Cristóbal, Mexico. High-quartz was observed during laboratory experimentation by Le Châtelier in 1889 and was later shown to occur in nature. Tridymite has since been found to include a group of three separate polymorphs closely related in crystal structure, called high-, middle-, and low-tridymite, and cristobalite has been found to include two related polymorphs, called high- and low-cristobalite. The high and low forms of tridymite were first observed by Merian in 1884, and the high and low forms of cristobalite by Mallard in 1890. The stability relations of these polymorphs of silica were comprehensively investigated in the early 1900's by Fenner, who also first recognized the existence of middle-tridymite. Later, three additional polymorphs of silica were obtained by laboratory synthesis: coesite, synthesized in 1953 by L. Coes; keatite, synthesized in 1954 by P. P. Keat; and stishovite, synthesized in 1961 by Stishov and Popova. Two of these polymorphs, coesite and stishovite, have since been identified in nature.

The development of optical microscopy in the latter nineteenth century brought important advances but also some confusion in the interpretation of the fine-grained, fibrous varieties of silica. A number of new polymorphs, later to be discredited, were advanced on the basis of optical characters. Thermal methods of study, developed in this period, resulted in the discovery of high-quartz and have continued to be important in the investigation of the silica polymorphs. Numerous extremely precise measurements of the optical properties and of the density and various other physical properties of quartz, also made during the latter nineteenth and early twentieth centuries, revealed very small but significant variations in these properties in different samples. The significance of the variations was not then understood, because of lack of analytical data of requisite precision, and the quantitative correlation of these variations with compositional changes still remains to be effected.

The development of x-ray diffraction techniques early in the twentieth century brought an understanding of the fundamental structural relations between the various polymorphs and of the crystal chemistry of silica, and provided a powerful new method for the discrimination of these substances. The identification of the so-called amorphous substance opal as a variety of cristobalite and the clarification of the relation of the fibrous types of silica to quartz and cristobalite were achieved primarily by x-ray study. X-ray and thermal studies also have revealed structural variations of the nature of stacking disorder and superstructures extending between tridymite and cristobalite. These variations relate chiefly to the chemical composition and thermal history of particular samples. The current development of relatively sensitive analytical methods, suited to measurements in the range of compositional variation in quartz and the other polymorphs, and adapted to small samples, should contribute further to an understanding of the observed variations in the properties and structural characters of these substances.

PHASE RELATIONS

The polymorph of silica that is thermodynamically stable at ordinary conditions of temperature and pressure is low-quartz. The name quartz as ordinarily applied refers to this particular polymorph. The low forms of tridymite and of cristobalite, and also keatite, stishovite, and coesite, can exist indefinitely in a metastable state at ordinary conditions. Low-quartz is stable under atmospheric pressure at temperatures up to about 573°C., where it undergoes a reversible, rapid (displacive) inversion to high-quartz. High-quartz, closely related in crystal structure to low-quartz, is stable at atmospheric pressure at temperatures from about 573° up to 870°. In some contexts the name quartz is applied as a generic term to both low-quartz and high-quartz. The inversion temperature between the low and high forms varies with the presence of other atoms in solid solution, being materially decreased by Al in substitution for Si, with valence compensation effected by the entrance of alkalies into interstitial positions, and increased by the substitution of Ge for Si. The inversion temperature is in the neighborhood of 573.3° (on heating) for ideal SiO_2. The solid solubility of other atoms is considerably greater in high-quartz than in low-quartz because of the greater openness of the structure and the higher temperatures of formation. Exsolution phenomena have not been observed to accompany the inversion from high- to low-quartz.

Over 870° and up to 1470° the stable polymorph at atmospheric pressure is tridymite (high-tridymite). The inversion between high-quartz and tridymite at 870° is of the sluggish (reconstructive) type. On heating, high-quartz usually persists metastably in the absence of fluxing or mineralizing agents to temperatures over 870° and may ultimately melt. The melting point is not known precisely, but is lower than that of tridymite (about 1670°) or of cristobalite (1723°). The liquid produced by the melting of high-quartz would tend to crystallize to cristobalite at temperatures below 1723°. Tridymite when cooled below the sluggish inversion at 870° may persist metastably to ordinary temperatures, passing through two rapid (displacive) inversions, at about 163° to so-called lower high-tridymite or middle-tridymite, and at about 117° to low-tridymite. The knowledge of the polymorphic behavior and stability ranges of the tridymite-type polymorphs is unsatisfactory because of uncertainties as to the exact chemical composition of the material and because of the presence of stacking disorder and superstructures in the crystal structure. The observed behavior in general depends on the conditions of formation and the thermal history of the particular sample investigated. Most or all natural tridymite probably departs considerably from SiO_2 in composition through the solid solution of Al, alkalies, Ca, and Mg.

Cristobalite (high-cristobalite) is the stable polymorph of silica at atmospheric pressure at temperatures from 1470° up to its melting point at 1723°. The inversion between high-tridymite and cristobalite at 1470° is of the sluggish (reconstructive) type. On heating, tridymite may persist metastably over 1470° in the absence of fluxes up to its melting point at about 1670°. The liquid then formed may crystallize to cristobalite at temperatures below 1723°. On cooling, cristobalite generally persists as a metastable form below the sluggish inversion to tridymite at 1470° down to room temperature, passing through a rapid (displacive) inversion at about 267°, or at a temperature below 267° down to 200° or less, to a structurally related polymorph called low-cristobalite. When heated in the range 573–870° in the presence of a flux, cristobalite may convert first to tridymite as a metastable form, which then converts to high-quartz. As with tridymite, the thermal behavior of cristobalite is complicated by stacking disorder and by compositional variation, and the phenomena observed depend on the prior history of the particular sample. Both cristobalite and tridymite can crystallize directly as metastable forms below 1470° and 860°, respectively.

Liquid silica when cooled below 1720° crystallizes with great difficulty and can be readily supercooled to a glass. The glass is variously known as vitreous silica, silica glass, quartz glass, and lechatelierite. On prolonged heating at temperatures over about 1000° silica glass slowly devitrifies and forms cristobalite as a stable or metastable phase. Silica formed by chemical methods at essentially ordinary temperature and pressure in water solution generally affords an essentially structureless gel or sol that on standing or slight warming ages into a poorly crystallized cristobalite-type phase. The differences in free energy between the three main polymorphs of silica are very small, and the stability relations apparently can be considerably altered by the entrance of H_2O or of cations into interstitial positions. The metastable formation of the relatively open, highly symmetric structure of cristobalite apparently is favored in this way over tridymite and quartz.

In addition to the polymorphs mentioned above, three additional polymorphs have been synthesized at high temperatures and pressures. Keatite is a metastable form that has been obtained over the investigated range from 380° to 585° and from 5000 to 18,000 p.s.i. It is converted to the other known polymorphs of silica when heated with a flux in their temperature stability ranges. Coesite and stishovite are high-pressure polymorphs that can exist metastably at ordinary temperatures and pressures. The stability field of coesite and of other silica polymorphs is indicated in Fig. 1.

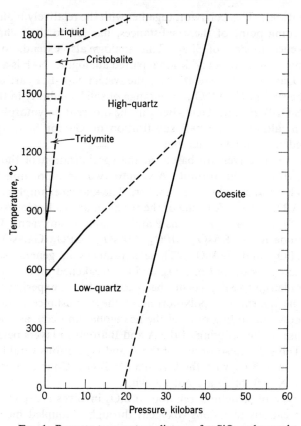

FIG. 1. Pressure-temperature diagram for SiO_2 polymorphs.

STRUCTURALLY RELATED SUBSTANCES

A number of different substances are related in their crystal structures to polymorphs of silica. These substances can be arranged into three categories on the basis of the structural relationship: as simple analogues, coupled derivatives, and stuffed derivatives.

The simple analogues conform to the formula AX_2. In them the atomic arrangement is identical with that in SiO_2, but either the A position is wholly occupied by one other kind of atom in place of Si, with the X position occupied by oxygen, or, in addition, the X position is wholly occupied by a different kind of atom in place of oxygen. These simple analogues, which include GeO_2 and BeF_2, show the same general polymorphic behavior as SiO_2 itself. Since the anion-cation bond in GeO_2 and BeF_2 is weaker than that between Si and O, the stability ranges of the several polymorphs are displaced toward lower temperatures. This behavior, together with the relatively high solubility and lower melting point of these substances, has led to their characterization as weakened models of SiO_2. This analogy also extends to polycomponent systems containing an additional paired analogue. NaF is a weakened model of CaO, and the phase relations in the system $NaF-BeF_2$ are analogous[1] to those in the system $CaO-SiO_2$. The range of solid solubility in the systems SiO_2-GeO_2, SiO_2-BeF_2, and GeO_2-BeF_2 in the different polymorphs involved is not known, although a small substitution of Ge for Si in quartz has been observed in synthetic material.

The coupled derivatives are based on the specialization of the A (or Si) structural position in the formula AX_2 into two or more non-equivalent positions of identical coordination number. The known examples conform to the formula ABX_4, where A is one of the trivalent atoms B, Ga, Al, Mn, and Fe, B is one of the pentavalent atoms P and As, and X is oxygen. The known examples include BPO_4, $BAsO_4$, $AlPO_4$, $AlAsO_4$, $GaPO_4$, $GaAsO_4$, $MnPO_4$, $MnAsO_4$, $FePO_4$, and $FeAsO_4$. These substances in general show polymorphism analogous to that of SiO_2, and are weakened models, although a complete parallelism to SiO_2 has not been demonstrated experimentally in all instances. The quartz-type polymorphs of these substances, so far as investigated, show a doubling of c_0 of the hexagonal unit cell, as compared to quartz, because of the ordering of the A and B atoms in layers perpendicular to this axis (Table 2). Disordering of the A and B positions would produce a simple analogue of SiO_2 with the formula $(A, B)O_2$. Coupled derivatives of SiO_2 in which X is fluorine are not known.

The third type of structural relation to SiO_2 involves a departure from a 1 : 2 ratio of cations to anions (oxygen) through a coupled mechanism of substitution for Si that introduces ions of appropriate size into vacant interstitial positions in the SiO_2 structure-type. These structures, called stuffed derivatives of silica, chiefly comprise aluminosilicates. One common type, which illustrates the principles involved, conforms to the formula $(V_{1-x}A_x)$ $(Si_{1-x}B_x)O_2$, where V denotes an interstitial vacancy, A is a monovalent ion of appropriate size, usually Li, Na, or K, and B is a trivalent ion of appropriate size, usually Al. In other types, the interstitial positions may be systematically occupied by two kinds of monovalent ions such as Na and K, or by a divalent ion such as Ba, Sr, or Pb, and the Si may be replaced in part or entirely by Al, Fe, Mg, or Ca. Some examples are listed in Table 3. These all represent stoichiometric end-compositions. The mechanism of derivation, however, is continuous in that it can produce a serial variation in composition, Instances are known in which a partial solid solution series extends by a mechanism of this general type either from a stoichiometric silicate end-composition toward a silica polymorph or, conversely, from a silica polymorph toward a stoichiometric silicate end-composition.

Most of the stuffed derivatives of silica are based on the relatively open

TABLE 2. SOME STRUCTURAL ANALOGUES OF QUARTZ

Formula	a_0 in Å	c_0 in Å	Ref.
SiO_2	4.9029	5.3936	...
GeO_2	4.972	5.648	2
BeF_2	4.750	5.225 (at 200°)	1
$AlPO_4$	4.975	10.84	3
BPO_4	4.470	9.926	4
$BAsO_4$	4.562	10.33	4

TABLE 3. SOME STUFFED DERIVATIVES OF SILICA[5]

Basic Structure	Derivative Structures
High-cristobalite	$NaAlSiO_4$ (high-carnegieite)
	Na_2CaSiO_4
	$K_2Al_2O_4$
	$K_2Fe_2O_4$
	$Na_2Al_2O_4$
High-tridymite	$KAlSiO_4$ (kalsilite)
	$KAlSiO_4$ (kaliophilite)
	$KNa_3Al_4Si_4O_{16}$ (nepheline)
	$K_2MgSi_3O_8$ (alpha form)
	$BaAl_2O_4$
	$SrAl_2O_4$
	$BaFe_2O_4$
	$PbFe_2O_4$
	$KLiSO_4$
High-quartz	$LiAlSiO_4$ (high-eucryptite)
Keatite	$LiAlSi_2O_6$ (beta-spodumene)[17]

structures of the high-temperature polymorphs. Stuffed derivatives based on low-quartz are not known, primarily because this substance is relatively close packed and restricts the entrance of the interstitial atom needed for valence compensation. Low-quartz, however, does show a very small range of compositional variation by this general mechanism that is significant with regard to the observed variation in the properties of this polymorph. The formation of stuffed derivatives is in general accompanied by more or less distortion of the basic silica framework. Stuffed derivatives in general do not exhibit polymorphism that closely parallels that of silica itself, primarily because of changes in the structural stability caused by the introduction of the stuffing atoms and by alteration of the strength of the Si-O bond.

ILL-DEFINED POLYMORPHS OF SiO₂

A considerable number of additional polymorphs of an ill-defined or speculative nature have been put forth on various grounds. Many have been based on optical evidence. These include two biaxial forms indicated by transitory optical phenomena at the 573° inversion between high- and low-quartz,[6] a monoclinic form based on the anomalous dichroism of amethyst,[7] and forms based on the optical elongation and other characters of various types of fibrous silica (lussatite, quartzine, pseudochalcedonite, cubosilicite, etc.).[8] A low-temperature non-piezoelectric modification of quartz has been reported[9] but not verified, and a metastable modification of tridymite with an inversion near 440° has been reported by dilatometric study.[10] Several unidentified phases, one probably identical with keatite, have been reported[11] by heating silicic acid. Prismatic crystals of what might be a new modification have been found in a silicate melt.[12] Vestan,[13] a supposed triclinic form found as phenocrysts in an igneous rock, and the asmanite[14] of meteorites are identical with quartz and tridymite, respectively. Royite[15] doubtless is quartz. A fibrous orthorhombic polymorph apparently isostructural with SiS₂ has been reported[16] by oxidation of SiO at high temperatures. Hypothetical polymorphs can be obtained by extrapolation from silicates in which coupled mechanisms of solid solution exist that theoretically lead to SiO₂ as an end-composition.

References

1. Roy, Roy, and Osborne: *J. Am. Ceram. Soc.*, **36**, 185 (1953).
2. Zachariasen: *Zs. Kr.*, **67**, 226 (1928).
3. Brill and De Bretteville: *Am. Min.*, **33**, 750 (1948); Winkhaus: *Jb. Min.*, Abh. **83**, 1 (1951).
4. Dachille and Glasser: *Acta Cryst.*, **12**, 820 (1959).
5. Buerger: *Am. Min.*, **39**, 600 (1954).
6. Steinwehr: *Zs. Kr.*, **99**, 292 (1938) and *Naturwiss.*, **25**, 348 (1937).
7. Pancharatnam: *Proc. Indian Ac. Sci.*, **40A**, 196 (1954); Trommsdorf: *Jb. Min.*, Beil. Bd. **72**, 464 (1937).
8. Michel-Lévy and Munier-Chalmas: *Bull. soc. min.*, **15**, 161 (1892) and *C. R.*, **110**, 649 (1890); Wallerant: *Bull. soc. min.*, **20**, 52 (1897); Lacroix: *C. R.*, **130**, 430 (1900); Bombicci: *Mem. accad. sci. ist. Bologna*, **8**, 67 (1899) and *Jb. Min.*, **1**, 189 (1901); Riva: *Zs. Kr.*, **31**, 406 (1899); Lengyel: *Föld. Közl.*, **66**, 278 (1936).
9. Osterberg: *Phys. Rev.*, **49**, 552 (1936); see also Balamuth, Rose, and Quimby: *Phys. Rev.*, **49**, 703 (1936) and Pavlovic and Pepinsky: *J. Appl. Phys.*, **25**, 1344 (1954).
10. Travers and Goloubinoff: *Rev. métall.*, **23**, 27, 100 (1926).
11. Endell: *Kolloid.-Zs.*, **111**, 19 (1948).
12. Fouqué and Lévy: *Synth. des min. et des roches*, Paris, 1882, p. 88; Morozewicz: *Min. Mitt.*, **18**, 158 (1898).
13. Jenzsch: *Ann. Phys.*, **105**, 320 (1878).
14. Maskelyne: *Phil. Trans.*, **161**, 361 (1871).
15. Sharma: *Proc. Indian Ac. Sci.*, **12B**, 215 (1940).
16. Weiss and Weiss: *Zs. anorg. Chem.*, **276**, 95 (1954).
17. Skinner and Evans: *Am. J. Sci.*, **258A**, 312 (1960).
18. Boyd and England: *J. Geophys. Res.*, **65**, 749 (1960).

QUARTZ

QUARTZ. Crystallos, κρύσταλλos *Greek*. Crystallus *Latin*; *Pliny* (*Nat. Hist.*, **37**, 9, 10, A.D. 77). Mormorion [smoky quartz] *Pliny* (**37**, 63); Morion. Quarz, Kiesel *Germ.* Qvarts *Swed.* Quarzo *Ital.* Cuarzo *Span.* Quartz *Fr.*
Haytorite *Tripe* (*Phil. Mag.*, **1**, 40, 1827). Cotterite *Harkness* (*Min. Mag.*, **2**, 82, 1878). Royite *Sharma* (*Proc. Indian Ac. Sci.*, **12B**, 215, 1940).

SYMMETRY

Quartz crystallizes in the trigonal trapezohedral class (3 2) of the rhombohedral subsystem. The lattice type is hexagonal. The trigonal trapezohedral class is characterized by one axis of three-fold symmetry and three polar axes of two-fold symmetry perpendicular thereto and separated by angles of 120°. There is no center of symmetry and no plane of symmetry. The general and special forms of quartz are listed below.

CRYSTAL FORMS IN THE TRIGONAL TRAPEZOHEDRAL CLASS

Pinacoid	$\{0001\}$	
Hexagonal prism	$\{10\bar{1}0\}$	
Trigonal prisms	(L) $\{2\bar{1}\bar{1}0\}$	(R) $\{11\bar{2}0\}$
(right and left)		
Ditrigonal prisms	(L) $\{i\bar{k}h0\}$	(R) $\{hki0\}$
(right and left)		
Rhombohedra	(+) $\{h0\bar{h}l\}$	(−) $\{0h\bar{h}l\}$
(positive and negative)		
Trigonal dipyramids	(L) $\{2h.\bar{h}.\bar{h}.l\}$	(R) $\{h.h.\overline{2h}.l\}$
(right and left)		
Trigonal trapezohedra	(+L) $\{i\bar{k}hl\}$	(+R) $\{hki l\}$
(positive and negative;	(−L) $\{kh\bar{i}l\}$	(−R) $\{ki\bar{h}l\}$
right and left)		

Three different sets of axes of reference have been employed to describe the geometry of quartz and of crystals based on hexagonal and rhombohedral lattices in general in addition to the Bravais set of axes used in this and in virtually all other crystallographic publications. These axial sets are described below and are illustrated in Figs. 2, 3, 4, and 5. Weiss (1816–17)[1] placed the positive ends of the three horizontal *a*-axes at 60° to each other, with the *c*-axis perpendicular to their plane

9

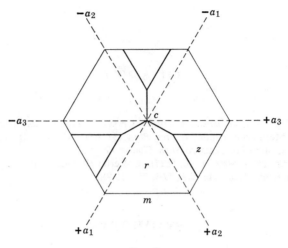

FIG. 2

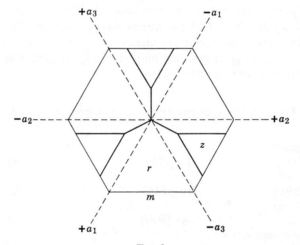

FIG. 3

(Fig. 2). Bravais (1851)[2] placed the positive ends of the a-axes at 120° to each other (Fig. 3), with c again perpendicular to their plane, yielding an adequate treatment of rhombohedral lattices. The four index symbol $\{hkil\}$ is now known as the Bravais symbol. Miller (1839)[3] used three equivalent axes not in a plane but parallel to the polar edges of a rhombohedron (Fig. 4). The interaxial angle α is the defining constant instead of the axial ratio $a:c$ as in the Bravais set. The Miller axes have certain advantages in the description of crystals based on a rhombohedral lattice, but are objectionable for hexagonal lattices. Schrauf (1861)[4] employed an orthogonal, pseudohexagonal axial set (Fig. 5), with $b:a = 1:\sqrt{3}$. In this, as in the Miller axes, any form of a hexagonal crystal cannot be represented by a single set of face indices that are permutations of each other.

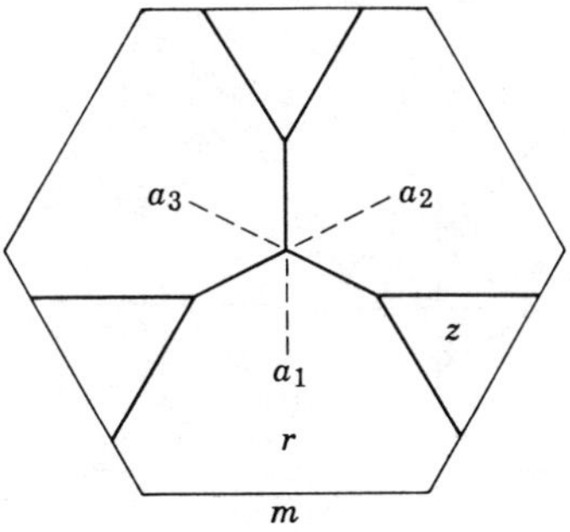

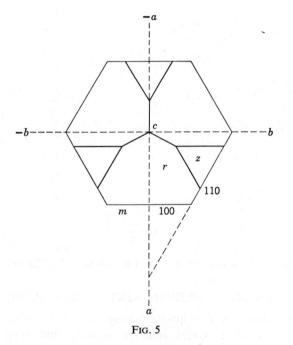

FIG. 4

FIG. 5

FIGS. 2–5. Crystallographic axes of reference of Weiss (1816) (Fig. 2), Bravais (1851) (Fig. 3), Miller (1839) (Fig. 4), and Schrauf (1861) (Fig. 5).

In addition to the so-called crystallographic or geometrical axes of reference described above, an additional set of orthogonal axes is often used in reference to piezoelectric and other physical properties of quartz. These axes also have been variously selected and labeled. A recommended[5] set for the description of the elastic and piezoelectric properties of right and left quartz is shown in Fig. 6. Some authors have used a right-handed system for the description of both right and left quartz.

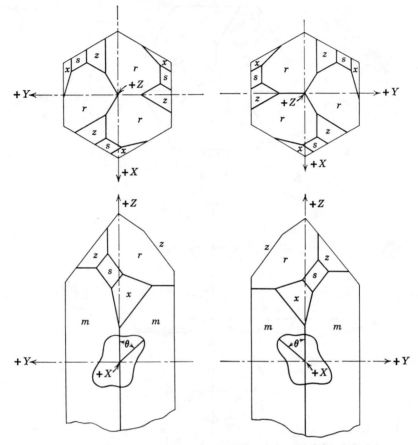

FIG 6. Piezoelectric axes of reference in right-handed and left-handed quartz.

ENANTIOMORPHISM; STRUCTURAL HAND

The crystal class to which quartz belongs has the quality of enantio-morphism, a name (from Greek *enantios, opposite,* and *morphosis, form*) describing the relation between two identical but non-superposable objects or mirror-images. Structurally, quartz belongs to one or the other of two enantiomorphous space groups for which, according to conventions and terminology described beyond, the notation $P3_12$ is appropriate for *structurally* right-handed crystals and $P3_22$ for *structurally* left-handed crystals.

The hand of quartz, however, always has been described in terms of morphological, optical, and other structure-dependent characters of quartz that have been put into a self-consistent set by means of conventions. The particular set of conventions in general use and here followed is that consistent with the morphological criteria of Weiss (1836).[6]   In this usage, the seeming hand of the morphological and other characters, which may be termed the conventional hand, is *opposite* to the hand of the crystal structure. A morphologically right-handed crystal thus is structurally left-handed, and conversely. The unqualified terms right and left as used in this book and in the literature refer to the conventional hand.

The structural feature that expresses the enantiomorphism of quartz appears as a helical arrangement of the SiO_4 tetrahedra or Si atoms along the three-fold axes as shown in the portion of the crystal structure illustrated in Fig. 9, page 18. A fundamental convention may now be mentioned concerning the sense or so-called hand of a helix or screw. Perhaps it is conveniently best described by a corkscrew. The helix of a corkscrew is said to be right-handed if it enters the cork as it is thrust in away from the observer and the handle turned clockwise. When viewed directly, with the axis of the helix held vertically and perpendicular to the line of sight, the helix or thread of a right-handed corkscrew slopes upward to the right. Virtually all corkscrews, wood screws, and the like are made with a right-handed helix or thread, and are called right-handed, primarily because most people are right handed and the musculature of the right forearm is such that a clockwise turn in driving in is easier and more forceful. It may be noted that the term clockwise, with reference to the direction of motion of the hands of a clock viewed face on, and the term right-handed are taken as synonymous.

The quartz structure shown in Fig. 9 represents a left-handed helix, and structurally this quartz is described as left-handed. It would at the same time, however, be described as right-handed on the basis of certain long-standing conventions based on crystal morphology and on the direction of rotation of the plane of polarized light as it passes through the crystal toward the observer. This circumstance arose in the fact that it is impossible except under certain conditions to distinguish by x-ray means between enantiomorphous structures because of the inherent centrosymmetry of x-ray diffraction effects (Friedel's Law). Very recently, however, it has been possible to establish the structural hand of quartz through the anomalous scattering produced when an atom in the crystal (Si) has an absorption edge just on the long-wavelength side of the monochromatic radiation used for the diffraction. In accordance with theoretical considerations as to the optical properties, quartz that is structurally left-handed on the basis of its helical arrangement is found to be right-handed on the basis of the accepted morphological and optical conventions. The relation between the hand of the crystal structure and the hand of the morphology and the rotary polarization was predicted by Niggli[8] in 1926. The latter conventions may now be described.

MORPHOLOGICAL HAND

With regard to the morphology, crystallographers in general have accepted the convention of Weiss (1816), who described enantiomorphism in quartz independently of Herschel. This convention has since been adopted for all

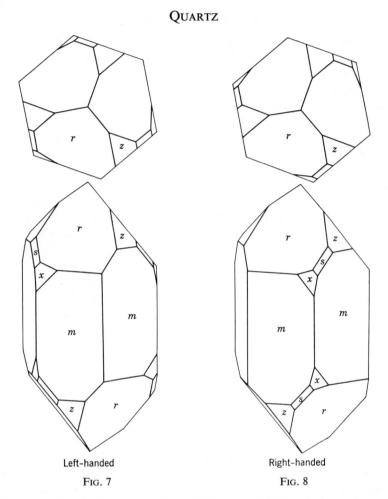

Left-handed   Right-handed

FIG. 7   FIG. 8

similar crystals. In the convention of Weiss, a right-handed quartz crystal has the trigonal pyramid s in the upper right corner of the prism face which is below the positive rhombohedron $r\{10\bar{1}1\}$, and the striations, if any, on s slope upward to the right. The positive right trigonal trapezohedron x is similarly situated. The right trigonal prism $a\{11\bar{2}0\}$, a very rare form in quartz, would be situated along the prism edge below the s and x faces. Opposite relations obtain in a left-handed quartz crystal. A right-handed crystal is illustrated in Fig. 8, and a left-handed crystal in Fig. 7. The drawings themselves are not mirror-images, although the crystals are, because of the use of the clinographic projection. Weiss did not use the term hand, but characterized a right-twisted (*rechtsgewundenen*) crystal as one in which the s and x faces, x usually being somewhat elongate, sloped upward to the right as in a right-handed screw. The positive and negative unit rhombohedra r and z can be confused; r reflects x-rays more strongly than

z, in about the ratio 100 : 75, and r in general tends to be larger in size and brighter in luster than z. Etching criteria also are useful (see *Etching Phenomena*). Historically, Brewster (1823)[9] believed from the opposing optical rotation of (Brazil-twinned) amethyst that two types of quartz existed, and Herschel (1821)[10] established this fact. The relation between morphology and optical hand was further described by Dove (1836).[11] Biot (1814)[12] earlier believed that the optical rotation was a property of the ultimate particles—molecules—of silica and was independent of their mode of aggregation. The crystals called morphologically left-handed by Herschel on the basis of his convention for the optical rotation are morphologically right-handed by the Weiss convention. Haüy[13] figured crystals of both right- and left-handed quartz, although not with the same forms. His trapezohedral crystals are all left-handed; one with the trigonal pyramid is right-handed, but others show this form repeated on all corners because of Dauphiné twinning (not then recognized). Cappeler (1723) also figured crystals with s and x faces, one apparently a Dauphiné twin.

OPTICAL HAND

With regard to the optical phenomena, the most widely used convention to describe hand is that of Biot (1814). This convention is identical in result with the morphological convention of Weiss. Biot discovered that quartz rotates the plane of polarization of light traveling parallel to the optic axis either to the right or to the left. (Rotatory polarization itself had been discovered earlier by Arago in 1811, in quartz.) In Biot's convention, the quartz is termed right-handed if the plane of polarization is found to be rotated clockwise when the observer looks through the analyzing Nicol toward the source of the light. The actual test involves a clockwise rotation of the analyzing Nicol; the (imaginary) helix generated by the clockwise rotation of the plane of polarization as it advances through the crystal toward the observer is left-handed, as is the structural hand. Opposite relations obtain in an optically left-handed crystal.

If a plate of right-handed quartz cut at right angles to the optic axis is employed, a clockwise rotation of the analyzing Nicol will cause the polarization colors of a parallel beam of white light to change in the order of their increasing refringence, or descend in Newton's scale, from red to violet. In a left-handed plate of quartz, the Nicol again being rotated clockwise and the observer looking toward the source of light, the order of change of the colors is the opposite, from violet to red. In monochromatic light the Nicol must be rotated to the right or to the left to produce extinction in a right-handed or a left-handed crystal, respectively. In strongly convergent white light or monochromatic light, the ring systems move inward in a right-handed plate and outward in a left-handed plate as the analyzer is rotated clockwise. Biot did not correlate his optical observations with the crystal morphology, and in fact was not aware of the existence of the two morphologically enantiomorphous types of quartz.

Herschel (1821) used an opposite convention for the description of the rotatory polarization. He described the optical effects produced by the rotation of the analyzing Nicol as if seen by an observer looking from the source of light in the direction of travel. The direction of rotation of the analyzing Nicol required to produce the same optical effects as in Biot's convention is now opposite in sense, reversing the hand. Biot's optical convention should be used, since it has priority and gives a description consistent with the morphological convention and with the conventions now generally adopted to relate etching, pyroelectric, and piezoelectric phenomena to the hand of quartz.

The conventions employed to describe the various enantiomorphic properties of quartz in the scientific and technical literature frequently are not explicitly stated or are used inconsistently with each other. The Weiss and Herschel conventions often are used separately for the morphological and optical characters without making the difference clear. Particular confusion exists in the literature relating to the elastic and piezoelectric properties of crystals. An added complication exists here in the equations for the transformations of the elastic and piezoelectric constants in right and left quartz with reference to a single orthogonal set of reference axes, a clockwise

TABLE 4. RELATIVE FREQUENCY OF RIGHT AND LEFT QUARTZ

Number of Crystals Examined	Per Cent Left	Per Cent Right	Locality	Ref.
4442[a]	50.05	49.95	Brazil	17
2415	50.68	49.32	Brazil, Colombia	18
1811	50.6	49.4	Switzerland	21
6404[b]	50.61	49.39	Russia	14
298 (untwinned)	50.7	49.3	U.S., Alaska	19
383 (Dauphiné)	50.1	49.9	U.S., Alaska	19
214 (untwinned)	52.3	47.7	Austria	20
840 (Dauphiné)	50.7	49.3	Austria	20
Total 16,807 Avg. (weighted)	50.5	49.5		

[a] Data probably of low precision through failure to recognize the true extent of Brazil twinning.

[b] Figure obtained after eliminating 931 crystals from localities which gave proportions differing by more than 5 per cent from the average of all other work. If these data are included, the proportion of left crystals in the total of 7335 crystals is 51.15 per cent.

rotation being variously called positive or negative. A collation of usage is available.[5] The advantage of describing these properties in terms of a right-handed set of orthogonal axes in right quartz and a left-handed set in left quartz (Fig. 6) is that the signs of the elastic and piezoelectric constants are thereby left unchanged.

RELATIVE FREQUENCY OF RIGHT AND LEFT QUARTZ

Right- and left-handed quartz have virtually the same frequency of occurrence in nature. There appears to be a slight preponderance of left quartz, of about 1 per cent (Table 4). Possibly significant variations in the proportion of right and left quartz at different localities have been reported.[14] A few widely variant studies[15] based on extremely small samples have been omitted from Table 4. Several statements based on general experience have been made in the literature that right and left quartz are of (essentially) equal abundance; also that left is more abundant than right.[22] It has been observed that right and left quartz cause the catalytic cleavage of certain racemates into the corresponding right and left forms.[16]

References

1. Weiss: *Abh. Ak. Wiss. Berlin*, 318, 1816–17.
2. Bravais: *J. école polytechn.*, **20**, 117 (1851).
3. Miller: *Treatise on Cryst.*, 1839.
4. Schrauf: *Ann. Phys.*, **114**, 221 (1861).
5. Cady and Van Dyke: *Proc. Inst. Radio Eng.*, **30**, 495 (1942).
6. Weiss: *Abh. Ak. Wiss. Berlin, Phys. Kl.*, 187, 1836.
7. de Vries: *Nature*, **181**, 1193 (1958).
8. Niggli: *Zs. Kr.*, **63**, 295 (1926).
9. Brewster: *Trans. Roy. Soc. Edinburgh*, **9**, 139 (1823) (paper presented in 1819).
10. Herschel: *Trans. Cambridge Phil. Soc.*, **1**, 43 (1821).
11. Dove: *Ann. Phys.*, **40**, 607 (1837); *Jb. Min.*, 550, 1838; *Mitt. Ges. Naturforsch. Freunde Berlin*, 37, 1836.
12. Biot: *Mem. cl. sci. math. phys., inst. imp. France*, **1**, 241–243, etc. (1814).
13. Haüy: *Traité de minéralogie*, Paris, first ed., 1801; second ed., 1822.
14. Vistelius: *Mém. soc. russe min.*, **79**, 191 (1950) and Laemmlein: *Mém. soc. russe min.*, **73**, 94 (1944), citing and including earlier work by others, including Kokkoros (1935).
15. Bindrich: *Zs. Kr.*, **59**, 113 (1924); Thompson: *Rocks and Min.*, **12**, 38 (1937); Osann: *Jb. Min.*, **1**, 108 (1891); Mügge: *Jb. Min.*, **1**, 1 (1892); Heritsch: *Min. Mitt.*, **3**, 115 (1953).
16. Schwab and Rudolph: *Naturwiss.*, **20**, 363 (1932).
17. Trommsdorff: *Jb. Min.*, Beil. Bd. **72**, 464 (1937).
18. Hurlbut: *Am. Min.*, **31**, 443 (1946).
19. Gault: *Am. Min.*, **34**, 142 (1949).
20. Brandenstein and Heritsch: *Min. Mitt.*, **2**, no. 4, 424 (1951).
21. Friedlaender: *Beitr. Geol. Schweiz, Geotech. Ser.*, Lfg. **29**, 29, 87 (1951).
22. Van Dyke: *Proc. Inst. Radio Eng.*, **28**, 399 (1940).

STRUCTURAL CRYSTALLOGRAPHY

The crystal structure of quartz is based on a hexagonal lattice. The unit cell has the nominal dimensions 4.913 Å, c_0 5.405 Å, and contains Si_3O_6. The atoms are arranged according to the enantiomorphous pair of space groups $D_3^4 = C3_12$ and $D_3^6 = C3_22$. The atomic positions are (in $D_3^4 = C3_12$):

Si: (a) $\bar{u}, \bar{u}, \frac{1}{3}$; $u\,0\,0$; $0, u, \frac{2}{3}$.

O: (c) xyz; $y - x, \bar{x}, z + \frac{1}{3}$; $\bar{y}, x - y, z + \frac{2}{3}$;

$\quad\quad x - y, \bar{y}, \bar{z}$; $y, x, \frac{2}{3} - z$; $\bar{x}, y - x, \frac{1}{3} - z$.

Parameters[1]: $u = 0.465$; $x = 0.415$; $y = 0.272$; $z = 0.120$ (all ± 0.003).

The Si atoms are in four-coordination with oxygen, and constitute the (SiO_4) tetrahedron found as the basic unit of structure in all the known polymorphs of SiO_2 and in silicates in general. In quartz each oxygen is shared with two Si atoms, the (SiO_4) tetrahedra thus being linked by sharing of each of the corner oxygen atoms to form a three-dimensional network. A similar situation obtains in the other polymorphs of SiO_2, but the geometry of the linkage differs. The (SiO_4) tetrahedron in quartz is almost symmetrical, with an Si-O distance of 1.61 Å. Each oxygen has in addition to two Si atoms six adjacent oxygen atoms at distances ranging between 2.60 and 2.67 Å. The Si-O bond is of intermediate type, roughly half covalent and half ionic.

The details of the atomic arrangement are indicated in Fig. 9. The enantiomorphism of quartz arises in the helical arrangement of the Si atoms or

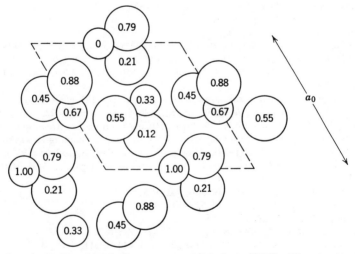

FIG. 9. Atomic arrangement in low-quartz, projected on (0001). The structure is left-handed. Small circles silicon, large circles oxygen.

(SiO$_4$) tetrahedra on the three-fold axes of the space group, a structurally right-handed helix obtaining in morphologically left-handed quartz, and a structurally left-handed helix obtaining in morphologically right-handed quartz. The structural hand cannot be established by x-ray techniques, except in Cr$K\alpha$ radiation, because of the centrosymmetry of the diffraction effects themselves. In Cr$K\alpha$ radiation, however, which has a wavelength near the resonance level of Si, however, a difference of intensity may be observed between $hk\bar{\imath}l$ and $\overline{hk\imath}l$ reflections which then permits a direct identification of hand.[2]

Historically, the first analysis of the crystal structure of quartz, by Gibbs[3] in 1926, followed from the determination of the simpler structure of high-quartz by Bragg and Gibbs[4] and by Wyckoff.[5] The structure of low-quartz and its relation to that of high-quartz were established through a postulated mechanism of shift of the Si atoms from their position in high-quartz during the displacive inversion of these polymorphs at 573°. This shift, which does not break Si-O-Si bonds or change the hand of the structure, proceeds in such a way from high-quartz as to destroy one set of two-fold axes and to convert the six-fold axes into three-fold axes. The existence of the (SiO$_4$) tetrahedron as the basic unit of structure in cristobalite and tridymite as well as in high- and low-quartz also was established[6] at this time. The atomic parameters in low-quartz were refined in later studies.[7] The lattice type and unit cell dimensions of quartz were first determined[8] by W. H. Bragg in 1914, and the (enantiomorphous) space group was later established[9] from his x-ray data. The centrosymmetric point group (Laue symmetry) of quartz was determined from Laue photographs in 1915 by Rinne.[10] The first x-ray powder diffraction observations on SiO$_2$ were made[11] in 1917 on quartz, cristobalite, silica glass, and silica gel. The structural identity of chalcedony with quartz and of calcined chalcedony and quartz with cristobalite was established[12] by this method in 1922 by Hull.

Before the advent of x-ray diffraction techniques, the structure of quartz had been theorized in connection with the optical properties,[13] the piezoelectric and pyroelectric behavior,[14] and the geometrical relations to other SiO$_2$ polymorphs.[15]

References

1. Brill, Hermann, and Peters: *Ann. Phys.*, **41**, 233 (1942) and *Naturwiss.*, **27**, 676 (1939).
2. de Vries: *Nature*, **181**, 1193 (1958).
3. Gibbs: *Proc. Roy. Soc. London*, **110A**, 443 (1926).
4. W. H. Bragg and Gibbs: *Proc. Roy. Soc. London*, **109A**, 405 (1925).
5. Wyckoff: *Am. J. Sci.*, **11**, 101 (1926) and *Zs. Kr.*, **63**, 507 (1926).
6. Gibbs: *Proc. Roy. Soc. London*, **113A**, 351 (1926); Wyckoff: *Am. J. Sci.*, **9**, 448 (1925) and *Zs. Kr.*, **62**, 189 (1925).
7. Wei: *Zs. Kr.*, **92**, 355 (1935); Machatschki: *Zs. Kr.*, **94**, 222 (1936); Brill, Hermann, and Peters (1942).

TABLE 5. UNIT CELL DIMENSIONS OF QUARTZ (VALUES OF HIGHER PRECISION)
Dimensions in angstrom units at 25°C. calculated from original kX values at 18°
by factor Å = kX(1.00202) and thermal coefficients of Jay[18]

a_0 in Å at 25°	c_0 in Å at 25°	a_0 in kX at 18°	c_0 in kX at 18°	c_0/a_0	$d(hkil)$ in kX at 18°	Other Data	Material	Ref.
	5.404852		5.393616				Not described	6
4.91361	5.40495	4.90320	5.39371	1.10004			Not described	7
					3.33666 $(10\bar{1}1)$		Not described	8
					1.177628 (1340)		Not described	9
					3.336274 $(10\bar{1}1)$		Not described	10
4.913303		4.902890					Not described	11
4.912996	5.404567	4.902585	5.393333	1.10010		Inv. temp.	Brazil, colorless	12
4.913631	5.405053	4.903219	5.393818	1.100056			Synthetic, 290°	12
4.913250	5.404765	4.902839	5.393530	1.100083			Synthetic, 380°	12
4.913144	5.404736	4.902735	5.393500	1.10010			Synthetic, 390°	12
4.91329	5.40486	4.90288	5.39362	1.10009		d	Not described	13
4.91350	5.40507	4.90309	5.39383	1.100088		Anal., d, n, inv. temp.	Arkansas, colorless	2

8. W. H. Bragg: *Proc. Roy. Soc. London*, **89**, 575 (1914); W. H. Bragg and W. L. Bragg: *X-rays and Crystal Structure*, London, 1915.
9. Gibbs: *Proc. Roy. Soc. London*, **107A**, 561 (1925); Huggins: *Phys. Rev.*, **19**, 363 (1922); W. H. Bragg and Gibbs (1925).
10. Rinne: *Sitzber. Sachs. Ak. Wiss., Math. Phys. Kl., Leipzig*, **67**, 303 (1915).
11. Kyropoulos: *Zs. anorg. Chem.*, **99**, 197 (1917).
12. Hull in Washburn and Navias: *Proc. Nat. Ac. Sci.*, **8**, 1 (1922).
13. Sohncke: *Mathem. Ann.*, **9**, 504 (1876), *Ann. Phys.*, Erg. Bd. **8**, 16 (1878), and *Zs. Kr.*, **13**, 229 (1887); Wulff: *Zs. Kr.*, **17**, 629 (1890); Beckenkamp: *Zs. anorg. Chem.*, **110**, 290 (1920).
14. Kelvin: *Phil. Mag.*, **36**, 331 (1893); Beckenkamp: *Zs. Kr.*, **32**, 18 (1899).
15. Beckenkamp: *Zs. Kr.*, **34**, 569, 588 (1901) and **36**, 483 (1902); Sosman: *Properties of Silica*, New York, 1927, Ch. 15, 16.

UNIT CELL DIMENSIONS

The unit cell dimensions of natural quartz vary significantly, but neither the range of variation nor the dimensions for the pure end-composition are clearly established. The following values probably are close to the average for colorless natural quartz crystals:

$$a_0\ 4.90290 \pm 0.0003,\ c_0\ 5.39365 \pm 0.0003,\ \text{in kX units at } 18°C.$$

The deviation stated probably includes the range of variation in such material. An indexed tabulation of d-spacings and 2θ angles calculated from these dimensions is given in Table 8. The experimental measurements on which these values are based, given beyond, were obtained on euhedral quartz crystals formed hydrothermally at low to moderate temperatures. Precision measurements are lacking on quartz grains from igneous rocks or other high-temperature aluminous environments in which a greater range of variation might be expected. The unit cell dimensions of pure quartz, which are lower than those stated, are not known.

Table 5 includes reported measurements of the unit cell dimensions of quartz of relatively high precision, about 1 part in 30,000 or better. Table 6 includes measurements known or believed to be of lower precision, but superior to about 1 part in 5000. Numerous measurements of still lower precision or of unknown or uncertain precision have been reported. Table 7 lists measurements of the interplanar spacing of $(23\bar{5}4)$ made on certain samples of natural and synthetic quartz for which the high-low inversion temperature has been determined. Unit cell dimensions cannot be calculated from these spacings because the axial ratios of the various samples are not known. Data on analyzed quartz crystals from Swiss localities are given in Table 7A.

It has long been known that there is a small but significant variation in the chemical composition of quartz. This has been recognized largely through the observed variation in certain properties, such as the indices of refraction

TABLE 6. UNIT CELL DIMENSIONS OF QUARTZ (VALUES OF LOWER PRECISION)
Dimensions in angstrom units at 25° and in kX units at 18°
related by factor 1.00202 and thermal coefficients of Jay[18]

a_0 in Å at 25°	c_0 in Å at 25°	a_0 in kX at 18°	c_0 in kX at 18°	c_0/a_0	$d(hkil)$ in kX at 18°	Other Data	Material	Ref.
4.913	5.404	4.903	5.393	1.0999			Not described	14
4.91329	5.40451	4.90288	5.39327	1.10002			Not described?	15
(4.91329)		(4.90288)			4.24605 ($10\bar{1}0$)		Cornwall, colorless	16
(4.91333)		(4.90292)			4.24595 ($10\bar{1}0$)		St. Gotthard, colorless	16
(4.91321)		(4.90280)			4.24630 ($10\bar{1}0$)		Durham, milky	16
(4.91362)		(4.90321)			4.24620 ($10\bar{1}0$)		Bohemia, rose	16
(4.91350)		(4.90309)						16
4.91289	5.40442	4.90248	5.39319	1.10009		Anal., d, n	Brazil, dark smoky	2
4.9134	5.4049	4.9030	5.3937	1.10008			Carrara, colorless	2
4.9134	5.4049	4.9030	5.3937	1.10008			Synthetic, 800°, NaWO$_3$	2
4.9141	5.4055	4.90334	5.3943	1.10005			Synthetic, 600°, high Al	2
4.9144	5.4059	4.9037	5.3947	1.10006			Synthetic, 890°, very high Al	2
4.91375	5.40533	4.9040	5.39410	1.10009		Anal.	Brazil, rose	1
4.9134	5.4049	4.9030	5.3937	1.10003		Spec. anal.	Synthetic, colorless	1
4.9133	5.4051	4.9029	5.3939	1.10009		Spec. anal.	Synthetic, colorless	1
4.9132	5.4050	4.9028	5.3938	1.10011		Spec. anal.	California, smoky	1
4.9132	5.4052	4.9028	5.3940	1.10014		Spec. anal.	Synthetic, colorless	1
4.9139	5.4052	4.9035	5.3940	1.09997		Spec. anal.	Synthetic, colorless	1
4.9134	5.4047	4.9030	5.3935	1.09999		Spec. anal.	Brazil, amethyst	1
4.9131	5.4047	4.9027	5.3935	1.10006		Spec. anal.	Amethyst, "greened"	1
4.9134	5.4050	4.9030	5.3938	1.10005		Spec. anal.	Herkimer, colorless	1
4.9138	5.4047	4.9034	5.3935	1.09990		Spec. anal.	Volcanic quartz	1
4.9141	5.4052	4.9037	5.3940	1.09993		Spec. anal.	Rose	1
4.9131	5.4051	4.9027	5.3939	1.10014		Spec. anal.	Citrine	1

TABLE 7. DATA ON THE INVERSION TEMPERATURE AND INTERPLANAR SPACING OF CERTAIN QUARTZ SAMPLES[17]

a_0 in Å	c_0 in Å	c_0/a_0	d (23$\bar{5}$4) in Å at Room Temperature	Heating	Cooling	Occurrence
4.9131	5.4036	1.0998	0.79120	573.3°	573.2°	Vein quartz
(4.9027 kX at 18°)	(5.3924)		0.79122	572.4	572.7	Phenocryst, rhyolite
			0.79122	573.7	573.8	Novaculite
			0.79122	573.6	573.5	Cavity in dolomite
			0.79123	573.6	574.0	Geode in limestone
			0.79123	573.7	574.3	Novaculite
			0.79123	573.6	573.4	Vein quartz Spec. anal.
			0.79123	...	...	Synthetic
			0.79125	572.8	572.5	Rhyolite tuff
			0.79125	573.5	573.5	Cavity in limestone
			0.79126	571.8	572.0	Phenocryst, rhyolite
			0.79126	570.3	573.2	Spec. anal.
			0.79126	570	574	Cavity in sulfide vein
			0.79130	571.8	571.9	Zoned pegmatite Spec. anal.
			0.79133	573 ± 2	573 ± 2	Chert
			0.79137	571.7	574.2	...
			0.79164	...	...	Synthetic, Ge-bearing
			0.79167	623(?)	...	Synthetic, Ge-bearing
			0.79170	545	564	Pegmatite
4.9172	5.4069	1.0996	0.79179	536	...	Fibrous
(4.9068 kX at 18°)	(5.3957)					Spec. anal.
			0.79187	592	591	Synthetic, Ge-bearing

TABLE 7A. UNIT CELL DIMENSIONS OF ANALYZED ALPINE QUARTZ[22] AT 25°

a_0 in Å (± 0.0001)	c_0 in Å (± 0.0001)	Values in Atoms per 10^6 Si				Color
		Al	Li	Na	H	
4.9129	5.4049	13	2.5	2.5	3	Smoky
4.9133	5.4045	115	66	30	18	Colorless
4.9145	5.4055	2350	1250	25	930	Zoned

and density, that can be measured with very high precision, or through measurement of properties, such as electric resistivity, that are relatively sensitive to compositional variation. Only in recent years, however, has direct evidence been afforded of variation in the unit cell dimensions. This is primarily due to the difficulty of experimentally measuring the cell edges to the precision required. Also, unfortunately, many of the precision measurements of the cell dimensions that have been made were not obtained on samples that had been chemically analyzed, and characterized in other regards, including their density and their irradiation response, so that the significance of the observed variations is unknown.

The nature of the mechanisms of compositional variations in quartz, their extent, and the effect that they have on the unit cell dimensions and axial ratio are not clearly understood. The principal mechanisms involve the entrance of Al in substitution for Si and of Li, Na, or Al into interstitial positions (see further under *Chemical Composition*). The general effect is to increase the cell volume and to decrease the axial ratio. The substitution of Al for Si increases both a_0 and c_0, although not equally, while the interstitial type of solid solution mainly expands a_0[1]. Increase in the content of Al tends to lower the high-low inversion temperature. The cell dimensions of quartz also are increased by the substitution of Ge for Si, and the inversion temperature is raised thereby (Table 7). The substitution of Ti, Mn, B, or P for Si would in all cases be expected to increase the cell volume.

The numerical relation between the amount of Al in solid solution and the cell dimensions is not known. The maximum variation also is not known. The data cited in Tables 5 and 6 indicate a range in colorless quartz crystals of at least 0.0006 in a_0 and of 0.0005 in c_0. The range of variation in synthetic quartz grown from solutions containing Al and alkalies is larger. A sample[2] grown hydrothermally at about 890° with LiAlSiO$_4$ as a coexisting phase had a_0 and c_0 both about 0.0014 kX higher than the lowest precision values reported for (presumably nearly pure) quartz. This sample had a very sluggish high-low inversion at about 556° (rising temperature), among the lowest values yet reported.[3] The first-formed phase in this experiment presumably was high-quartz, in which the solubility of Al and Li is relatively large.

The value of a_0 has been found to decrease and the high-low inversion temperature to increase with increasing content of sodium carbonate or silicate in solutions from which quartz is prepared hydrothermally from silica gel.[4]

Rose quartz apparently has slightly higher cell dimensions than either pure quartz or the average values for colorless quartz. Quartz that is rendered smoky by x-ray irradiation also should have higher values than pure quartz because the color centers responsible for the effects are known to be substitutional Al. Natural smoky quartz should have larger cell dimensions than pure quartz for the same reason, but a direct comparison with colorless quartz of the same bulk composition is not possible unless the mechanisms by which the elements present in solid solution are housed are identified. The change in cell dimensions produced by the thermal bleaching of smoky quartz apparently is extremely small.[5] When Al ions in substitution for Si in quartz are made to migrate in an electric field, the region cleared of Al acquires smaller cell dimensions.[5]

Precision determinations of the cell dimensions of the microcrystalline and fibrous, chalcedonic varieties of quartz are lacking. Relatively large variations have been reported in some work of low precision. The available evidence indicates that any variation in these varieties is less than 0.002 in both a_0 and c_0. Preferred orientation and particle size effects become important in the x-ray photography of these materials.

X-RAY POWDER DIFFRACTION DATA

Quartz gives an excellent x-ray powder diffraction pattern with low-background, high-intensity, sharp lines and well-resolved $\alpha_1\alpha_2$ doublets in the high-angle region. The axial coefficients of thermal expansion are fairly low and are known over a wide range of temperature.[18] For these reasons quartz is useful as an internal or calibration standard in x-ray work. Since the unit cell dimensions of quartz vary significantly, a randomly selected natural colorless quartz crystal can be employed for calibration purposes only to a precision set by the limits of variation. The calculated 2θ angles of Table 8 are based on a probable average value of the cell dimensions of such material, and will afford a precision of less than about 1 part in 10,000. If a higher

TABLE 8. INTERPLANAR SPACINGS AND REFLECTION ANGLES FOR QUARTZ
Data based on the cell dimensions a_0 4.91331 Å and c_0 5.40488 Å at 25°C. The wavelengths employed are the Cauchois-Hulubei values multiplied by 1.00202 to convert kX to Å, as follows:

	$K\alpha_1$	$K\alpha_2$	$K\alpha$	$K\beta$
Cu	1.54051	1.54433	1.54178	1.39217
Co	1.78892	1.79278	1.79021	1.62075
Fe	1.93597	1.93991	1.93728	1.75653
Cr	2.28962	2.29351	2.29092	2.08480

TABLE 8 (cont'd)

hkil	d		Cu	Co	Fe	Cr
10Ī0	4.25505	α_1	20.859	24.269	26.299	31.215
		α_2	20.911	24.323	26.353	31.270
		α	20.876	24.287	26.317	31.233
		β	18.831	21.958	23.824	28.361
10Ī1	3.34331	α_1	26.640	31.036	33.660	40.049
		α_2	26.707	31.104	33.730	40.119
		α	26.662	31.058	33.683	40.072
		β	24.034	28.055	30.460	36.334
11Ī0	2.45666	α_1	36.545	42.704	46.410	55.550
		α_2	36.639	42.801	46.510	55.653
		α	36.576	42.736	46.444	55.585
		β	32.920	38.522	41.894	50.215
10Ī2	2.28124	α_1	39.467	46.170	50.216	60.243
		α_2	39.569	46.275	50.325	60.356
		α	39.501	46.205	50.252	60.281
		β	35.533	41.616	45.287	54.380
11Ī1	2.23647	α_1	40.291	47.149	51.293	61.579
		α_2	40.395	47.257	51.405	61.695
		α	40.326	47.185	51.330	61.617
		β	36.268	42.489	46.245	55.562
20Ī0	2.12753	α_1	42.451	49.723	54.128	65.108
		α_2	42.562	49.837	54.247	65.232
		α	42.488	49.761	54.167	65.150
		β	38.195	44.779	48.763	58.676
20Ī1	1.97968	α_1	45.795	53.721	58.545	70.660
		α_2	45.915	53.846	58.675	70.798
		α	45.835	53.763	58.588	70.706
		β	41.172	48.328	52.673	63.545
11Ī2	1.81781	α_1	50.140	58.951	64.349	78.067
		α_2	50.273	59.091	64.496	78.225
		α	50.184	58.998	64.398	78.120
		β	45.030	52.949	57.782	69.980

TABLE 8   (cont'd)

hkil	d		Cu	Co	Fe	Cr
0003	1.80163	α_1	50.622	59.534	64.998	78.904
		α_2	50.757	59.675	65.146	79.064
		α	50.667	59.581	65.047	78.958
		β	45.457	53.462	58.351	70.703
20$\bar{2}$2	1.67166	α_1	54.874	64.698	70.769	86.445
		σ_2	55.022	64.855	70.934	86.629
		α	54.923	64.750	70.824	86.507
		β	49.216	57.995	63.389	77.155
10$\bar{1}$3	1.65904	α_1	55.327	65.251	71.389	87.267
		α_2	55.476	65.409	71.556	87.453
		α	55.376	65.303	71.444	87.329
		β	49.615	58.479	63.927	77.852
21$\bar{3}$0	1.60826	α_1	57.232	67.582	74.010	90.768
		α_2	57.387	67.748	74.186	90.966
		α	57.284	67.638	74.068	90.834
		β	51.293	60.515	66.199	80.805
21$\bar{3}$1	1.54146	α_1	59.959	70.939	77.800	95.920
		α_2	60.123	71.115	77.989	96.136
		α	60.014	70.997	77.863	95.992
		β	53.689	63.433	69.467	85.101
11$\bar{2}$3	1.45282	α_1	64.035	76.002	83.561	103.997
		σ_2	64.213	76.195	83.770	104.246
		α	64.094	76.066	83.631	104.080
		β	57.257	67.807	74.389	91.697
30$\bar{3}$0	1.41835	α_1	65.785	78.194	86.074	197.635
		α_2	65.969	78.395	86.292	107.902
		α	65.846	78.261	86.147	107.724
		β	58.783	69.689	76.518	94.604
21$\bar{3}$2	1.38204	α_1	67.743	80.662	88.919	111.859
		α_2	67.934	80.872	89.148	112.147
		α	67.806	80.732	88.995	111.955
		β	60.485	71.798	78.912	97.919
20$\bar{2}$3	1.37489	α_1	68.144	81.169	89.505	112.745
		α_2	68.336	81.381	89.737	113.039
		α	68.208	81.240	89.582	112.843
		β	60.833	72.231	79.403	98.607

TABLE 8 (cont'd)

2θ

hkil	d		Cu	Co	Fe	Cr
30$\bar{3}$1	1.37190	α_1	68.312	81.383	89.753	113.122
		α_2	68.505	81.596	89.985	113.417
		α	68.377	81.454	89.830	113.220
		β	60.980	72.413	79.611	98.897
10$\bar{1}$4	1.28784	α_1	73.467	87.981	97.464	125.479
		α_2	73.680	88.220	97.730	125.858
		α	73.538	88.061	97.552	125.606
		β	65.436	77.989	85.995	108.077
30$\bar{3}$2	1.25589	α_1	75.659	90.831	100.844	131.442
		α_2	75.880	91.082	101.127	131.876
		α	75.733	90.915	100.938	131.587
		β	67.319	80.371	88.745	112.200
22$\bar{4}$0	1.22833	α_1	77.669	93.470	104.008	137.500
		α_2	77.898	93.733	104.307	138.003
		α	77.746	93.558	104.107	137.667
		β	69.040	82.560	91.288	116.127
21$\bar{3}$3	1.19977	α_1	79.882	96.408	107.570	145.181
		α_2	80.121	96.685	107.889	145.807
		α	79.962	96.501	107.676	145.389
		β	70.927	84.977	94.113	120.646
22$\bar{4}$1	1.19779	α_1	80.042	96.621	107.830	145.792
		α_2	80.280	96.899	108.151	146.430
		α	80.121	96.714	107.937	146.004
		β	71.062	85.151	94.318	120.981
11$\bar{2}$4	1.18395	α_1	81.171	98.136	109.689	150.454
		α_2	81.415	98.422	110.020	151.201
		α	81.252	98.232	109.799	150.701
		β	72.021	86.388	95.772	123.392
31$\bar{4}$0	1.18014	α_1	81.488	98.564	110.216	151.890
		α_2	81.733	98.851	110.551	152.679
		α	81.570	98.660	110.327	152.151
		β	72.290	86.735	96.182	124.083
31$\bar{4}$1	1.15297	α_1	83.835	101.752	114.186	166.355
		α_2	84.090	102.057	114.547	168.092
		α	83.920	101.854	114.306	166.910
		β	74.275	89.313	99.235	129.404

TABLE 8 (*cont'd*)

hkīl	*d*		2θ Cu	Co	Fe	Cr
$20\bar{2}4$	1.14062	α_1	84.954	103.291	116.130	
		α_2	85.215	103.604	116.505	
		α	85.041	103.396	116.255	
		β	75.218	90.546	100.706	132.097
$22\bar{4}2$	1.11824	α_1	87.072	106.238	119.911	
		α_2	87.343	106.568	120.315	
		α	87.162	106.348	120.045	
		β	76.996	92.886	103.516	137.555
$30\bar{3}3$	1.11444	α_1	87.444	106.760	120.590	
		α_2	87.716	107.094	121.000	
		α	87.534	106.872	120.726	
		β	77.307	93.298	104.013	138.573
$31\bar{4}2$	1.08151	α_1	90.829	111.593	127.024	
		α_2	91.117	111.957	127.494	
		α	90.925	111.714	127.180	
		β	80.125	97.059	108.598	149.088
$40\bar{4}0$	1.06376	α_1	92.786	114.459	131.001	
		α_2	93.084	114.844	131.515	
		α	92.885	114.587	131.171	
		β	81.742	99.246	111.303	156.996
$10\bar{1}5$	1.04770	α_1	94.646	117.242	135.011	
		α_2	94.955	117.648	135.578	
		α	94.749	117.377	135.199	
		β	83.272	101.335	113.918	168.473
$40\bar{4}1$	1.04374	α_1	95.119	117.957	136.072	
		α_2	95.430	118.370	136.654	
		α	95.222	118.095	136.265	
		β	83.659	101.867	114.589	174.194
$21\bar{3}4$	1.03455	α_1	96.238	119.672	138.671	
		α_2	96.556	120.099	139.293	
		α	96.344	119.815	138.877	
		β	84.573	103.130	116.193	
$22\bar{4}3$	1.01489	α_1	98.744	123.608	145.025	
		α_2	99.076	124.071	145.772	
		α	98.854	123.762	145.271	
		β	86.608	105.971	119.852	

TABLE 8   (cont'd)

hkil	d		Cu	Co	Fe	Cr
				2θ		
40$\bar{4}$2	0.989838	α_1	102.185	129.282	155.876	
		α_2	102.538	129.806	156.993	
		α	102.302	129.457	156.242	
		β	89.374	109.909	125.067	
11$\bar{2}$5	0.989426	α_1	102.244	129.383	156.100	
		α_2	102.597	129.908	157.229	
		α	102.362	129.558	156.470	
		β	89.421	109.977	125.159	
31$\bar{4}$3	0.987200	α_1	102.566	129.932	157.354	
		α_2	102.921	130.464	158.550	
		α	102.683	130.109	157.745	
		β	89.677	110.346	125.659	
30$\bar{3}$4	0.978328	α_1	103.871	132.206	163.322	
		α_2	104.235	132.767	164.997	
		α	103.992	132.392	163.860	
		β	90.715	111.854	127.720	
32$\bar{5}$0	0.976176	α_1	104.194	132.779	165.145	
		α_2	104.560	133.348	167.056	
		α	104.316	132.969	165.752	
		β	90.971	112.229	128.237	
20$\bar{2}$5	0.963715	α_1	106.118	136.293		
		α_2	106.497	136.913		
		α	106.244	136.499		
		β	92.488	114.468		131.380
32$\bar{5}$1	0.960634	α_1	106.608	137.219		
		α_2	106.991	137.854		
		α	106.735	137.430		
		β	92.873	115.041		132.201
41$\bar{5}$0	0.928528	α_1	112.104	148.863		
		α_2	112.527	149.763		
		α	112.244	149.161		
		β	97.123	121.560		142.123
32$\bar{5}$2	0.918114	α_1	114.060	153.932		
		α_2	114.499	155.023		
		α	114.205	154.291		
		β	98.606	123.928		146.115

TABLE 8   (cont'd)

hkil	d		Cu	Co	Fe	Cr
40$\bar{4}$3	0.916007	α_1	114.467	155.096		
		α_2	114.910	156.242		
		α	114.614	155.473		
		β	98.913	124.425	146.991	
41$\bar{5}$1	0.915122	α_1	114.639	155.603		
		α_2	115.084	156.774		
		α	114.787	155.988		
		β	99.043	124.635	147.367	
22$\bar{4}$4	0.908907	α_1	115.871	159.542		
		α_2	116.326	160.960		
		α	116.022	160.005		
		β	99.965	126.148	150.160	
0006	0.900813	α_1	117.534	166.382		
		α_2	118.004	168.639		
		α	117.690	167.093		
		β	101.199	128.212	154.307	
21$\bar{3}$5	0.897153	α_1	118.309	171.119		
		α_2	118.787	175.273		
		α	118.468	172.255		
		β	101.770	129.184	156.444	
31$\bar{4}$4	0.888855	α_1	120.125			
		α_2	120.620			
		α	120.289			
		β	103.095	131.485	162.294	
10$\bar{1}$6	0.881281	α_1	121.858			
		α_2	122.371			
		α	122.028			
		β	104.344	133.719	170.517	
41$\bar{5}$2	0.878140	α_1	122.599			
		α_2	123.121			
		α	122.772			
		β	104.874	134.687		
30$\bar{3}$5	0.859747	α_1	127.251			
		α_2	127.827			
		α	127.442			
		β	108.121	140.977		

TABLE 8 (cont'd)

hkil	d		Cu	Co	Fe	Cr
32$\bar{5}$3	0.858285	α_1	127.646			
		α_2	128.227			
		α	127.838			
		β	108.391	141.532		
50$\bar{5}$0	0.851010	α_1	129.675			
		α_2	130.284			
		α	129.877			
		β	109.760	144.445		
11$\bar{2}$6	0.845748	α_1	131.215			
		α_2	131.846			
		α	131.424			
		β	110.781	146.741		
50$\bar{5}$1	0.840654	α_1	132.770			
		α_2	133.424			
		α	132.986			
		β	111.793	149.151		
40$\bar{4}$4	0.835828	α_1	134.307			
		α_2	134.986			
		α	134.531			
		β	112.777	151.648		
20$\bar{2}$6	0.829521	α_1	136.421			
		α_2	137.138			
		α	136.658			
		β	114.099	155.332		
41$\bar{5}$3	0.825360	α_1	137.890			
		α_2	138.634			
		α	138.136			
		β	114.996	158.131		
33$\bar{6}$0	0.818885	α_1	140.310			
		α_2	141.105			
		α	140.573			
		β	116.432	163.465		
50$\bar{5}$2	0.811715	α_1	143.217			
		α_2	144.082			
		α	143.502			
		β	118.085	173.415		

TABLE 8 (cont'd)

hkil	d		2θ			
			Cu	Co	Fe	Cr
22$\bar{4}$5	0.811488	α_1	143.314			
		α_2	144.181			
		α	143.600			
		β	118.138	173.999		
33$\bar{6}$1	0.809645	α_1	144.109			
		α_2	144.997			
		α	144.401			
		β	118.575			
42$\bar{6}$0	0.804129	α_1	146.621			
		α_2	147.582			
		α	146.937			
		β	119.911			
31$\bar{4}$5	0.797121	α_1	150.164			
		α_2	151.250			
		α	150.521			
		β	121.677			
42$\bar{6}$1	0.795374	α_1	151.124			
		α_2	152.249			
		α	151.493			
		β	122.129			
32$\bar{5}$4	0.791283	α_1	153.523			
		α_2	154.759			
		α	153.927			
		β	123.210			
21$\bar{3}$6	0.785926	α_1	157.078			
		α_2	158.525			
		α	157.549			
		β	124.673			
33$\bar{6}$2	0.783696	α_1	158.747			
		α_2	160.318			
		α	159.256			
		β	125.298			
42$\bar{6}$2	0.770732	α_1	175.967			
		β	129.150			

TABLE 8   (*cont'd*)

hkil	d		Cu	Co	Fe	Cr
50$\bar{5}$3	0.769485	β	129.542			
41$\bar{5}$4	0.765261	β	130.901			
51$\bar{6}$0	0.764230	β	131.241			
30$\bar{3}$6	0.760412	β	132.527			
10$\bar{1}$7	0.759719	β	132.765			
40$\bar{4}$5	0.758206	β	133.290			
51$\bar{6}$1	0.756703	β	133.821			
33$\bar{6}$3	0.745491	β	138.047			
11$\bar{2}$7	0.736600	β	141.817			
51$\bar{6}$2	0.735390	β	142.365			
42$\bar{6}$3	0.734306	β	142.865			
22$\bar{4}$6	0.726409	β	146.773			
20$\bar{2}$7	0.725805	β	147.094			
32$\bar{5}$5	0.724486	β	147.808			
50$\bar{5}$4	0.720095	β	150.326			
31$\bar{4}$6	0.716050	β	152.877			
60$\bar{6}$0	0.709175	β	157.949			
41$\bar{5}$5	0.704354	β	162.424			
51$\bar{6}$3	0.703550	β	163.293			
60$\bar{6}$1	0.703148	β	163.744			
33$\bar{6}$4	0.700317	β	167.396			
43$\bar{7}$0	0.699526	β	168.629			

precision is desired, a particular quartz sample must be employed for which the unit cell dimensions have been independently determined to the precision desired. Two other precise tabulations of 2θ angles and d-spacings are available but are of less value for this purpose, since they happen to be based in one case on cell dimensions near the lower limit of variation,[3] and in the other on dimensions near the maximum observed values[19] (see comparison below.)

Ref.:	3	Present Work	19
a_0	4.902585	4.90290	4.90320
c_0	5.393333	5.39365	5.39371
$a_0:c_0$	1.10010	1.10009	1.10004

All values in kX units at 18°.

An indexed x-ray powder diffraction pattern of quartz[20] recorded by the diffractometer (chart) method[21] is shown in Fig. 10.

References

1. Cohen and Sumner: *Am. Min.*, **43**, 58 (1958).
2. Frondel and Hurlbut: *J. Chem. Phys.*, **23**, 1215 (1955) and U.S. Signal Corps Eng. Labs., Ft. Monmouth, N.J., *Contract Rpt. DA* 36-039 *SC*-15350, 1953.
3. Keith: *Am. Min.*, **40**, 530 (1955).
4. Sabatier and Wyart: *C.R.*, **239**, 1053 (1954).
5. Hammond, Chi, and Stanley: U.S. Signal Corps Eng. Labs., Ft. Monmouth, N.J., *Eng. Rpt. E*-1162, November 1955.
6. Elg: *Zs. Phys.*, **106**, 315 (1937).
7. Wilson and Lipson: *Proc. Phys. Soc.*, **53**, 245 (1941).
8. Kunzl and Köppel: *C.R.*, **196**, 787 (1933).
9. Brogren: *Ark. Mat., Astron., Fys.*, **36B**, no. 6 (1949).
10. Brogren and Friskopp: *Ark. Mat., Astron., Fys.*, **36B**, no. 4 (1949).
11. Brogren and Haeggblom: *Ark. Fys.*, **2**, 1 (1950).
12. Keith: *Proc. Phys. Soc.*, **63B**, 208, 1034 (1950).
13. Brogren: *Ark. Fys.*, **7**, 47 (1953).
14. Harrington: *Am. J. Sci.*, **13**, 467 (1927).
15. Bergqvist: *Zs. Phys.*, **66**, 494 (1930).
16. Bradley and Jay: *Proc. Phys. Soc.*, **45**, 407 (1933).
17. Keith and Tuttle: *Am. J. Sci.*, Bowen Vol., 203 (1952).
18. Jay: *Proc. Roy. Soc. London*, **142A**, 237 (1933); Wilson and Lipson (1941).
19. Parrish: Philips Labs. (Irvington-on-Hudson), *Res. Lab. Tech. Rpt.* **68**, 1953.
20. A tabulation of older powder data is given by Swanson, Fuyat, and Ugrinic: *Nat. Bur. Stds. Circ.* **539**, III, 24, 1954.
21. Parrish: *Seventh Nat. Conf. Clays and Clay Minerals*, Pergamon, New York, 1960, p. 230; Parrish, Hamacher, and Lowitzsch: *Philips Tech. Rev.*, **16**, 123 (1954); Parrish and Kohler: *Rev. Sci. Instr.*, **27**, 795 (1956).
22. Bambauer: *Schweiz. min. pet. Mitt.*, **41**, 335 (1961).

MORPHOLOGICAL CRYSTALLOGRAPHY

The morphological characters of quartz are described in this section. The varieties of quartz that are based on characters other than outward form, such as the color and the mode of aggregation, are described in other sections (see *Varieties of Quartz*).

Historically, the morphology of quartz is of interest in that it was in crystals of this species that Steno recognized in 1669 the fundamental law of the constancy of interfacial angles. Enantiomorphism also was first recognized morphologically in this species. Quartz crystals were known to Pliny, who in his *Historia naturalis* (*ca.* A.D. 77) noted the hexagonal shape of the prism and the fact that the pyramidal termination does not always have the same appearance. Recognizable descriptions and illustrations of quartz crystals are found in many works extending in time up to the emergence of crystallography as a formal science toward the end of the eighteenth century. The first angular measurement on quartz was made in 1708 by J. J. Scheuchzer, who determined the angle $r \wedge z$ over the termination as 75° [actually 76°26′], and systematic measurements were made by Romé de l'Isle (1783), Malus (1817), W. Phillips (1817), and Haüy (1801, 1822). Haüy took the rhombohedron as the primitive form, with the angle $(10\bar{1}1)(\bar{1}011)$ as 85°36′; he listed 8 forms in 1801, including the trapezohedron x, and 13 forms in 1822. The forms $m\,r\,z$ and s (left) and unidentified steep rhombohedra were earlier recognized by de l'Isle. The interpolar angle r to z was first measured to high precision by Kupffer in 1823, and his axial ratio [1.09997 from 46°15′51.6″] has been accepted by most crystallographers to the present time. The morphology of quartz has since received more study than that of any other mineral species with the exception of calcite. Particular reference may be made to the monographic studies of Des Cloizeaux (1858)[1] and of G. Rose (1846)[2] and to the summary account of the morphology by V. Goldschmidt (1922).[3] Among later workers G. vom Rath in the period 1870–1885 contributed a number of rare and new forms. Very numerous studies of material from particular localities or of individual crystals have been published.[4] The present work lists 112 well-established forms, for which an angle table is provided, and 423 rare, questionable, or discredited forms.

The influence of the crystal structure on the morphology of quartz was discussed by Niggli,[5] who related the facial development in its zonal relations to the packing density of directions in the structure. The morphology clearly is not responsive to the "law of Bravais," which relates the relative size and frequency of occurrence of the crystal forms to the reticular density of planes of the translation lattice, or to a modification[6] of this theory, in which an added effect due to screw axes and glide planes is postulated. The main difficulties here are the virtual absence on quartz of forms such as the trigonal and ditrigonal prisms and the basal pinacoid and the unequal occurrence of positive and negative forms. The relative importance of the forms of quartz also has been discussed on purely theoretical grounds in terms of zonal relations[7] and by V. Goldschmidt (1897)[8] from the point of view of number-series (law of complication).

References

1. Des Cloizeaux: *Mém. sur la cristallisation et la structure interieure du quartz*, Mém. Ac. Sci. France, **15**, 1858, 211 pp., 123 figs.
2. Rose: Über das Krystallisationssystem des Quarzes, *Abh. Ak. Wiss. Berlin*, 1846, 58 pp., 50 figs.
3. V. Goldschmidt: *Atlas der Krystallformen*, **8**, text and tables, Heidelberg, 1922.
4. See summaries of older literature in Goldschmidt (1922) and Hintze: *Handbuch der Mineralogie*, **1**, Lfg. 9, 1353, Berlin, 1905, and also Gallitelli: *Per. min. Rome*, **6**, 105 (1935); Fagnani and Weber: *Bull. soc. fribourg. sci. nat.*, **39**, 67 (1950); Friedlaender: *Beitr. Geol. Schweiz, Geotech. Ser.*, no. 29 (1951); Buttgenbach: *Les minéraux de Belge et du Congo Belge*, Liège, 1947, p. 102. The complete literature later than 1922 cannot be cited here; see *Mineral Abstracts* and similar works.
5. *Niggli: Zs. Kr.*, **63**, 295 (1926).
6. Donnay and Harker: *Amer. Min.*, **22**, 446 (1937).
7. E. Weiss: *Abh. naturfor. Ges. Halle*, **5**, 51 (1860).
8. Goldschmidt: *Zs. Kr.*, **28**, 32 (1897).

AXIAL RATIO AND ANGLE TABLE

Axial Ratio. The unit cell used to describe the morphology of quartz was originally selected by Haüy (1801) and has been employed by almost all later workers. It conforms to the cell found by x-ray study. A nominal value for the axial ratio, $a:c = 1:1.10009$, is here used for purposes of morphological description. The nominal value $a:c = 1:1.0999$ was earlier used by Dana (1892),[1] Hintze (1904),[2] Goldschmidt (1922),[3] and others.

The best morphological measurements of the axial ratio of quartz afford values of c differing by as much as 0.002. The determinations of the ratio reported in the literature whose precision appears to be of the order of $30''$ of arc or better for the angular measurements are cited in Table 9. A significant variation in the axial ratio has been shown to exist by precision x-ray measurements (see *Structural Crystallography*), accompanying variation in the chemical composition. The maximum difference thus found is about 0.0002, and might prove to be less than 0.0001 if the precision of the measurements could be evaluated with certainty. A variation of 0.0001 in the ratio $a:c = 1:1.1000$ corresponds to a change in the rho angle to r or z of $9.1''$ and in the interpolar angle r to z of $6.1''$. These angular differences are about at the limits of accuracy of optical goniometry, as set both by the instrument and by the quality of the

QUARTZ

TABLE 9. AXIAL RATIO OF QUARTZ AT ROOM TEMPERATURE FROM PRECISE
MORPHOLOGICAL MEASUREMENTS

$a:c$	Interpolar Angle r to z	Locality	Ref.
1:1.09997	46°15′51.6″	St. Gotthard	5
1:1.1002	46°16′3.7″	St. Gotthard	6
1:1.1002	46°16′4.8″	Herkimer and Marmoros	6
1:1.1006	46°15′30″	Urals	7
1:1.0998	46°15′42″	Riesengrund	8
1:1.1012	46°17′5″	Rosenlaui	8
1:1.1018	46°17′43.8″	North Carolina	8
1:1.1000	46°15′54″	Herkimer	9
1:1.0997	(85°44′33.5″)	Binnenthal	10
1:1.0998	(103°33′55″)	Binnenthal	10

finest crystals, and are considerably less than the accuracy of conventional measurements. The latter generally are not more precise than a few minutes of arc on a quartz crystal of good quality, as tested by the variation of the phi and rho angles of all the reflections from the terminal r and z faces. It hence appears that a serial variation in the axial ratio of quartz accompanying compositional variation cannot be recognized by morphological measurements. A nominal value of the morphological axial ratio is taken for this reason.

The observed variation in the morphological axial ratio (Table 9) is largely due to differences in the gross physical quality of the measured crystals. The main sources of error here are the presence of vicinal hillocks on the terminal rhombohedral faces, and the presence of through-going internal boundaries, such as the composition planes of twins, and lineage structures, across which there is a small angular divergence. X-ray goniometric measurements made on crystal plates cut across the boundaries of natural Dauphiné twins frequently show divergences of 5′ of arc or more. The r and z faces are the best in quality of the inclined forms on quartz, and the optical goniometric measurements leading to the axial ratio always are based thereon. These faces, however, usually are seen on close examination to be replaced by a three-sided vicinal pyramid with interfacial angles of about 1° (see *Vicinal Surfaces*). The faces of this vicinal pyramid may be unequal in size, and sometimes only one face may be developed that completely replaces, in a sense, the true position of r or z. Although the optical reflections from these vicinal faces may be of the highest quality, they will lead to slightly erroneous measurements[11] of the interpolar angles of r and z. Variations in the physical quality of the r and z faces early led to the belief that there were significant differences in their polar angles, and that the axial ratios of smoky quartz, amethyst, and colorless quartz were different.

The axial ratio decreases at an increasing rate with increasing temperature up to the inversion point at 573°, where it decreases abruptly to a slightly lower value and then remains essentially constant up to at least 1250° in the high-quartz region.[4] The most probable values of the ratio for various temperatures are cited in Table 10. The temperature coefficient at 0° of the ratio is 6.14 × 10⁻⁶. The variation in the rho angle of r or z from 0° to about 570° is expressed by the following exponential formula (where T is the temperature in degrees Centigrade):

$$\rho = 51°47.4' - [0.01132T + 0.01335 (e^{0.01T} - 1)]$$

In the neighborhood of room temperature the ratio varies by about 0.000006 per degree C., corresponding to a variation in the rho angle of r or z of about 0.55″ per degree.

TABLE 10. VALUES OF THE AXIAL RATIO OF QUARTZ AND
HIGH-QUARTZ AT DIFFERENT TEMPERATURES

Temp.	c/a	Temp.	c/a
−250°	1.1015	500°	1.0956
−200	1.1014	550	1.0946
−100	1.1009	573	{1.0940 / 1.0922}
0	1.1003		
100	1.0996	600	1.0921
200	1.0988	800	1.0916
300	1.0979	1000	1.0915
400	1.0969	1200	1.0916

Angle Table. All the forms that have been reported for quartz in the literature are collected in Tables 11 and, 12. Table 11 comprises an angle table for the well-established forms, with the angles given in degrees, minutes, and seconds. The reference angles are as follows: ϕ, the azimuthal angle to $(11\bar{2}0)$; ρ, the interfacial angle to (0001); M, the interfacial angle to $(1\bar{1}10)$; A_2, the interfacial angle to $(\bar{1}2\bar{1}0)$; R, the interfacial angle to $(10\bar{1}1)$; Z, the interfacial angle to $(01\bar{1}1)$. The form designation and angles are shown in Fig. 11. The last column to the right in Table 11 indicates the relative frequency of occurrence of these forms: VVC denotes very, very common or universal forms; VC, very common forms; C, common forms, noted by at least 10 observers; LC, less common forms, noted by at least 6 observers; and R, rare forms. Table 12 lists questionable and discredited forms:[12] PV denotes probably valid forms; D, doubtful or uncertain forms; and X,

TABLE 11. ANGLE TABLE FOR QUARTZ FORMS

Hexagonal—P; trigonal trapezohedral—3 2

$a : c = 1 : 1.100090;\ \alpha = 93°56'10";\ p_0 : r_0 = 1.27026 : 1;\ \lambda = 85°45'50"$

	h k i l	Miller	φ	ρ	M	A_2	R	Z	Frequency
c	0 0 0 1	111	...	00°00'00"	90°00'00"	90°00'00"	51°47'28"	51°47'28"	R
m	1 0 Ī 0	2ĪĪ	30°00'00"	90°00'00"	60°00'00"	90°00'00"	38°12'32"	66°51'58"	VVC
a	1 1 2̄ 0	10Ī	00°00'00"	90°00'00"	90°00'00"	60°00'00"	47°07'02"	47°07'02"	R
'a	2 Ī Ī 0	1Ī0	60°00'00"	90°00'00"	30°00'00"	120°00'00"	47°07'02"	90°00'00"	R
K	2 1 3̄ 0	5Ī4	10°53'36"	90°00'00"	79°06'24"	70°53'36"	42°04'24"	53°33'37"	R
k	5 1 6̄ 0	11.4̄.7	21°03'06"	90°00'00"	68°56'54"	81°03'06"	39°05'10"	60°23'58"	R
ω	1 0 Ī 3	522	30°00'00"	22°56'56"	78°45'28"	90°00'00"	28°50'24"	43°42'54"	R
'ω	0 1 Ī 3	441	−30°00'00"	22°56'56"	101°14'31"	70°15'53"	43°42'54"	28°50'24"	R
D	2 0 2̄ 5	311	30°00'00"	26°56'08"	76°54'33"	90°00'00"	24°51'12"	43°09'44"	R
π	1 0 Ī 2	411	30°00'00"	32°25'16"	74°27'02"	90°00'00"	19°22'04"	42°52'49"	LC
'π	0 1 Ī 2	110	−30°00'00"	32°25'16"	105°32'57"	62°20'04"	42°52'49"	19°22'04"	C
r	1 0 Ī 1	100	30°00'00"	51°47'20"	66°52'00"	90°00'00"	00°00'00"	46°15'58"	VVC
z	0 1 Ī 1	22Ī	−30°00'00"	51°47'20"	113°07'59"	47°07'10"	46°15'58"	00°00'00"	VVC
q	11 0 ĪĪ 10	32.Ī.Ī	30°00'00"	54°24'36"	66°00'30"	90°00'00"	02°37'15"	47°11'51"	R
Ξ	9 0 9̄ 8	26.Ī.Ī	30°00'00"	55°01'01"	65°48'59"	90°00'00"	03°13'40"	47°25'38"	R
Π	6 0 6̄ 5	17.Ī.Ī	30°00'00"	56°44'02"	65°17'14"	90°00'00"	04°56'41"	48°06'12"	LC
λ	5 0 5̄ 4	14.Ī.Ī	30°00'00"	57°47'52"	64°58'13"	90°00'00"	06°00'31"	48°32'31"	R
'λ	0 5 5̄ 4	332	−30°00'00"	57°47'52"	115°01'46"	42°52'39"	48°32'31"	06°00'31"	R
C	13 0 Ī3 10	12.Ī.Ī	30°00'00"	58°48'08"	64°40'43"	90°00'00"	07°00'47"	48°58'09"	R
t	4 0 4̄ 3	11.Ī.Ī	30°00'00"	59°26'29"	64°29'50"	90°00'00"	70°39'08"	49°14'51"	R
J	7 0 7̄ 5	19.2.2̄	30°00'00"	60°39'07"	64°09'44"	90°00'00"	08°51'41"	49°47'16"	R

Table 11 (cont'd)

hkil	Miller	ϕ	ρ	M	A_2	R	Z	Frequency
g 10 0 $\overline{10}$ 7	9$\overline{1}\overline{1}$	30° 00′ 00″	61° 08′ 33″	64° 01′ 45″	90° 00′ 00″	09° 21′ 12″	50° 00′ 45″	R
j 3 0 $\overline{3}$ 2	8$\overline{1}\overline{1}$	30° 00′ 00″	62° 18′ 30″	63° 43′ 18″	90° 00′ 00″	10° 31′ 09″	50° 33′ 22″	LC
j′ 0 3 $\overline{3}$ 2	554	−30° 00′ 00″	62° 18′ 30″	116° 16′ 41″	39° 55′ 46″	50° 33′ 22″	10° 31′ 09″	C
U 8 0 $\overline{8}$ 5	7$\overline{1}\overline{1}$	30° 00′ 00″	63° 48′ 06″	63° 20′ 31″	90° 00′ 00″	12° 00′ 45″	51° 16′ 28″	R
i 5 0 $\overline{5}$ 3	13.$\overline{2}$.$\overline{5}$	30° 00′ 00″	64° 43′ 00″	63° 07′ 16″	90° 00′ 00″	12° 55′ 39″	51° 43′ 35″	C
i′ 0 5 $\overline{5}$ 3	887	−30° 00′ 00″	64° 43′ 00″	116° 52′ 43″	38° 27′ 27″	51° 43′ 35″	12° 55′ 39″	C
I 0 13 $\overline{13}$ 7	20.20.$\overline{19}$	−30° 00′ 00″	67° 01′ 41″	117° 24′ 34″	37° 07′ 20″	52° 54′ 21″	15° 14′ 20″	R
l 2 0 $\overline{2}$ 1	5$\overline{1}\overline{1}$	30° 00′ 00″	68° 30′ 52″	62° 16′ 23″	90° 00′ 00″	16° 43′ 31″	53° 41′ 31″	VC
l′ 0 2 $\overline{2}$ 1	1$\overline{1}\overline{1}$	−30° 00′ 00″	68° 30′ 52″	117° 43′ 36″	36° 18′ 28″	53° 41′ 31″	16° 43′ 31″	VC
χ 0 13 $\overline{13}$ 6	19.19.$\overline{20}$	−30° 00′ 00″	70° 01′ 55″	118° 01′ 49″	35° 30′ 54″	54° 30′ 54″	18° 14′ 34″	R
S 7 0 $\overline{7}$ 3	17.$\overline{4}$.4	30° 00′ 00″	71° 21′ 22″	61° 43′ 16″	90° 00′ 00″	19° 34′ 01″	55° 15′ 00″	R
μ 5 0 $\overline{5}$ 2	4$\overline{1}\overline{1}$	30° 00′ 00″	72° 31′ 16″	61° 30′ 57″	90° 00′ 00″	20° 43′ 55″	55° 54′ 30″	R
′μ 0 5 $\overline{5}$ 2	778	−30° 00′ 00″	72° 31′ 16″	118° 29′ 02″	34° 18′ 21″	55° 54′ 30″	28° 43′ 55″	R
w 0 11 $\overline{11}$ 4	556	30° 00′ 00″	74° 01′ 31″	118° 43′ 50″	33° 38′ 06″	56° 46′ 28″	22° 14′ 10″	R
M 3 0 $\overline{3}$ 1	722	30° 00′ 00″	75° 17′ 47″	61° 04′ 39″	90° 00′ 00″	23° 30′ 26″	57° 31′ 12″	VC
′M 0 3 $\overline{3}$ 1	445	−30° 00′ 00″	75° 17′ 47″	118° 55′ 21″	33° 06′ 20″	57° 31′ 12″	23° 30′ 26″	VC
ν 22 0 $\overline{2}$2 7	17.$\overline{5}$.5	30° 00′ 00″	75° 56′ 15″	60° 59′ 11″	90° 00′ 00″	24° 08′ 54″	57° 54′ 01″	R
′σ 0 10 $\overline{10}$ 3	13.13.$\overline{17}$	−30° 00′ 00″	76° 42′ 43″	119° 07′ 05″	32° 33′ 31″	58° 21′ 49″	24° 55′ 22″	R
h 7 0 $\overline{7}$ 2	16.$\overline{5}$.5	30° 00′ 00″	77° 19′ 25″	60° 48′ 11″	90° 00′ 00″	25° 32′ 04″	58° 43′ 57″	C
′h 0 7 $\overline{7}$ 2	334	−30° 00′ 00″	77° 19′ 25″	119° 11′ 49″	32° 20′ 13″	58° 43′ 57″	25° 32′ 04″	C
Y 4 0 $\overline{4}$ 1	3$\overline{1}\overline{1}$	30° 00′ 00″	78° 51′ 57″	60° 37′ 14″	90° 00′ 00″	27° 04′ 36″	59° 40′ 25″	VC
Υ 0 4 $\overline{4}$ 1	557	−30° 00′ 00″	78° 51′ 57″	119° 22′ 45″	31° 49′ 05″	59° 40′ 25″	27° 04′ 36″	VC
θ 13 0 $\overline{13}$ 3	29.$\overline{10}$.$\overline{10}$	30° 00′ 00″	79° 42′ 12″	60° 31′ 52″	90° 00′ 00″	27° 54′ 57″	60° 11′ 28″	R
Δ 14 0 $\overline{14}$ 3	31.$\overline{11}$.$\overline{11}$	30° 00′ 00″	80° 25′ 29″	60° 27′ 35″	90° 00′ 00″	28° 38′ 08″	60° 38′ 25″	R

TABLE 11 (cont'd)

h k i l	Miller	φ	ρ	M	A_2	R	Z	Frequency
'Δ 0 14 14̄ 3	17.17.25	−30°00'00"	80°25'29"	119°32'25"	31°21'17"	60°38'25"	28°38'08"	R
e 5 0 5̄ 1	11.4̄.4	30°00'00"	81°03'08"	60°24'06"	90°00'00"	29°15'47"	61°02'01"	C
'e 0 5 5̄ 1	223	−30°00'00"	81°03'08"	119°35'53"	31°11'11"	61°02'01"	29°15'47"	C
d 11 0 11̄ 2	833̄	30°00'00"	81°51'15"	60°19'59"	90°00'00"	30°03'54"	61°32'23"	C
'd 0 11 11̄ 2	13.13.20̄	−30°00'00"	81°51'15"	119°40'01"	30°59'11"	61°32'22"	30°03'54"	LC
ζ 6 0 6̄ 1	13.5.5	30°00'00"	82°31'30"	60°16'50"	90°00'00"	30°44'09"	61°57'56"	VC
'ζ 0 6 6̄ 1	7.7.11̄	−30°00'00"	82°31'30"	119°43'09"	30°49'58"	61°57'56"	30°44'09"	VC
γ 13 0 13̄ 2	28.11̄.11	30°00'00"	83°05'39"	60°14'22"	90°00'00"	31°18'18"	62°19'44"	R
φ 7 0 7̄ 1	522̄	30°00'00"	83°35'00"	60°14'29"	90°00'00"	31°47'39"	62°38'33"	VC
'φ 0 7 7̄ 1	8.8.13̄	−30°00'00"	83°35'00"	119°47'32"	30°36'57"	62°38'33"	31°47'39"	VC
B 8 0 8̄ 1	17.7.7	30°00'00"	84°22'47"	60°09'31"	90°00'00"	32°35'26"	63°09'22"	LC
'B 0 8 8̄ 1	335̄	−30°00'00"	84°22'47"	119°50'28"	30°28'25"	63°09'22"	32°35'26"	R
A 9 0 9̄ 1	19.8̄.8	30°00'00"	85°00'03"	60°07'37"	90°00'00"	33°12'42"	63°33'32"	LC
T 10 0 10̄ 1	733̄	30°00'00"	85°29'55"	60°06'07"	90°00'00"	33°42'34"	63°52'59"	LC
'T 0 10 10̄ 1	11.11.19̄	30°00'00"	85°29'55"	119°53'52"	30°18'16"	63°52'59"	33°42'34"	LC
Ψ 11 0 11̄ 1	23.10̄.10	30°00'00"	85°54'23"	60°05'03"	90°00'00"	34°07'02"	64°08'58"	VC
'Ψ 0 11 11̄ 1	447̄	−30°00'00"	85°54'23"	119°54'56"	30°15'07"	64°08'58"	34°07'02"	VC
X 16 0 16̄ 1	11.5.5̄	30°00'00"	87°10'59"	60°02'23"	90°00'00"	35°23'38"	64°59'20"	LC
Q 0 17 17̄ 1	6.6.11̄	−30°00'00"	87°20'55"	119°57'05"	30°06'21"	65°05'53"	35°33'34"	R
Y 18 0 18̄ 1	37.17̄.17	30°00'00"	87°29'44"	60°01'53"	90°00'00"	35°42'23"	65°11'44"	R
ξ 1 1 2 2	52̄1	00°00'00"	47°43'43"	90°00'00"	68°17'07"	23°07'59"	23°07'59"	C
'ξ 2 1̄ 1 1	51̄2	60°00'00"	47°43'43"	50°08'46"	111°42'52"	23°07'59"	65°24'48"	LC
s 1 1 2̄ 1	412	00°00'00"	65°33'28"	90°00'00"	62°55'22"	24°54'14"	28°54'14"	VVC
's 2 1̄ 1̄ 1	42̄1	60°00'00"	65°33'28"	37°57'45"	117°04'37"	28°54'14"	75°10'13"	VC

TABLE 11 (cont'd)

	h	k	i	l	Miller	φ	ρ	M	A₂	R	Z	Frequency
κ	11	12	$\overline{23}$	11	$15.4.\overline{8}$	−01° 26′ 16″	66° 30′ 35″	91° 19′ 07″	61° 25′ 20″	30° 31′ 35″	28° 25′ 14″	R
η	7	6	$\overline{13}$	7	$9\overline{2}4$	02° 32′ 34″	63° 56′ 30″	87° 42′ 56″	65° 31′ 48″	26° 05′ 36″	29° 54′ 56″	R
Ω	7	9	$\overline{16}$	7	$10.3.\overline{6}$	−04° 07′ 40″	68° 21′ 48″	93° 50′ 11″	58° 33′ 58″	33° 37′ 33″	27° 43′ 07″	R
P	3	4	$\overline{7}$	3	$13.4.\overline{8}$	−04° 42′ 54″	68° 46′ 50″	94° 23′ 40″	57° 56′ 05″	34° 18′ 49″	77° 36′ 13″	R
V	7	5	$\overline{12}$	5	$\overline{8}14$	05° 29′ 46″	69° 20′ 34″	84° 51′ 28″	67° 09′ 52″	27° 28′ 27″	35° 27′ 43″	R
E	5	7	$\overline{12}$	5	$22.7.\overline{14}$	−05° 29′ 46″	69° 20′ 34″	96° 08′ 31″	57° 05′ 24″	35° 27′ 43″	27° 28′ 27″	R
Z	2	3	$\overline{5}$	5	$4\overline{2}1$	−06° 35′ 12″	47° 55′ 02″	94° 53′ 01″	63° 44′ 41″	28° 01′ 00″	18° 14′ 58″	R
α	3	2	$\overline{5}$	3	$11.2.\overline{4}$	06° 35′ 12″	61° 33′ 02″	84° 12′ 41″	69° 33′ 03″	21° 46′ 34″	31° 51′ 15″	VC
P	2	3	$\overline{5}$	2	$3\overline{1}2$	−06° 35′ 12″	70° 35′ 12″	96° 11′ 36″	55° 54′ 12″	36° 31′ 53″	27° 20′ 22″	R
O	3	5	$\overline{8}$	3	$14.5.\overline{10}$	−08° 12′ 47″	71° 21′ 22″	97° 46′ 46″	54° 07′ 01″	38° 29′ 25″	27° 14′ 36″	C
H	2	1	$\overline{3}$	5	$10.4.\overline{1}$	10° 53′ 34″	33° 54′ 27″	83° 56′ 54″	79° 28′ 43″	21° 54′ 40″	32° 21′ 32″	R
'H	1	2	$\overline{3}$	5	320	−10° 53′ 34″	33° 54′ 27″	96° 03′ 05″	68° 52′ 47″	32° 21′ 32″	21° 54′ 40″	R
R	2	1	$\overline{3}$	3	$8\overline{2}1$	10° 53′ 36″	48° 14′ 28″	81° 53′ 43″	75° 51′ 57″	15° 01′ 42″	31° 14′ 17″	LC
'R	1	2	$\overline{3}$	3	$7\overline{4}2$	−10° 53′ 36″	48° 14′ 28″	98° 06′ 17″	60° 45′ 56″	31° 14′ 17″	15° 01′ 42″	LC
L	2	1	$\overline{3}$	2	$7\overline{1}2$	10° 53′ 36″	59° 14′ 37″	80° 39′ 13″	73° 39′ 47″	17° 22′ 43″	34° 14′ 03″	R
'L	1	2	$\overline{3}$	2	$2\overline{1}1$	−10° 53′ 36″	59° 14′ 36″	99° 20′ 46″	65° 45′ 57″	34° 14′ 02″	17° 22′ 43″	VC
ρ	2	1	$\overline{3}$	1	$20\overline{1}$	10° 53′ 36″	73° 25′ 47″	79° 33′ 51″	71° 42′ 56″	27° 22′ 25″	41° 46′ 49″	LC
'ρ	1	2	$\overline{3}$	1	$5\overline{2}4$	−10° 53′ 36″	73° 25′ 47″	100° 26′ 08″	51° 08′ 11″	41° 46′ 49″	27° 22′ 25″	C
Σ	10	5	$\overline{15}$	2	$9\overline{1}6$	10° 53′ 36″	83° 12′ 45″	79° 11′ 01″	71° 01′ 56″	35° 52′ 09″	48° 28′ 12″	R
W	3	7	$\overline{10}$	3	$16.7.\overline{14}$	−13° 00′ 14″	75° 07′ 10″	102° 33′ 38″	48° 45′ 46″	44° 25′ 22″	27° 44′ 24″	LC
ψ	2	5	$\overline{7}$	7	$16.10.\overline{5}$	−13° 53′ 52″	48° 34′ 28″	100° 22′ 32″	58° 40′ 26″	33° 30′ 31″	12° 45′ 27″	R
G	3	8	$\overline{11}$	3	$17.8.\overline{16}$	−14° 42′ 16″	76° 30′ 55″	104° 17′ 25″	46° 50′ 17″	46° 34′ 59″	28° 12′ 22″	C
N	3	1	$\overline{4}$	4	$11.2.\overline{1}$	16° 06′ 07″	48° 52′ 03″	77° 56′ 32″	79° 34′ 37″	11° 04′ 31″	35° 11′ 27″	LC
'N	1	3	$\overline{4}$	4	$3\overline{2}1$	−16° 06′ 07″	48° 52′ 03″	102° 03′ 28″	27° 07′ 48″	35° 11′ 27″	11° 04′ 31″	LC

TABLE 11 (cont'd)

h k i l	Miller	ϕ	ρ	M	A_2	R	Z	Frequency
u 3 1 4̄ 1	81̄4	16° 06' 07"	77° 41' 00"	74° 16' 41"	76° 25' 41"	28° 42' 07"	48° 22' 39"	VC
'u 1 3 4̄ 1	21̄2	−16° 06' 07"	77° 41' 00"	105° 43' 19"	45° 15' 06"	48° 22' 39"	28° 42' 07"	VC
o 2 7 9̄ 16	970	−17° 47' 01"	33° 01' 04"	99° 34' 48"	66° 11' 56"	36° 15' 33"	20° 25' 38"	R
β 2 7 9̄ 9	20.14.7	−17° 47' 01"	49° 07' 16"	103° 21' 06"	55° 56' 45"	36° 29' 05"	09° 46' 53"	C
f 4 1 5̄ 5	14.2.1̄	19° 06' 23"	49° 20' 21"	75° 37' 22"	81° 45' 27"	08° 45' 21"	37° 30' 37"	LC
τ 1 4 5̄ 4	10.7.5̄	−19° 06' 23"	55° 30' 18"	105° 39' 03"	51° 27' 47"	39° 15' 44"	09° 31' 11"	C
y 4 1 5̄ 1	10.2̄.5̄	19° 06' 23"	80° 15' 08"	71° 10' 46"	79° 15' 56"	30° 05' 52"	52° 17' 18"	VC
'y 1 4 5̄ 1	748	−19° 06' 23"	80° 15' 08"	108° 49' 13"	41° 50' 22"	52° 17' 18"	30° 05' 52"	VC
ι 2 9 1̄1̄ 2	536	−20° 10' 24"	81° 10' 53"	109° 55' 29"	40° 37' 54"	53° 41' 33"	30° 41' 45"	R
x 5 1 6̄ 1	41̄2	21° 03' 06"	81° 57' 08"	69° 09' 55"	81° 08' 26"	31° 13' 33"	54° 51' 14"	VVC
'x 6 1̄ 5̄ 1	42̄1	38° 56' 53"	81° 57' 08"	51° 30' 22"	98° 51' 35"	31° 13' 33"	68° 31' 37"	LC
−'x 1 5 6̄ 1	8.5.1̄0	−21° 03' 06"	81° 57' 08"	110° 50' 04"	39° 38' 28"	51° 51' 14"	31° 13' 33"	VC
−x 1̄ 6 5̄ 1	5.8.1̄0	−38° 56' 53"	81° 57' 08"	128° 29' 30"	22° 28' 30"	51° 51' 14"	31° 13' 33"	R
ε 7 1 8̄ 7	22.1.2̄	23° 24' 47"	53° 52' 27"	71° 16' 44"	84° 41' 01"	05° 38' 45"	47° 00' 57"	R
v 7 1 8̄ 1	16.5.8	23° 24' 47"	84° 02' 49"	66° 43' 13"	83° 26' 56"	32° 48' 27"	57° 59' 50"	R
b 9 1 1̄0 9	28.1.2̄	24° 42' 28"	53° 23' 52"	70° 19' 43"	85° 49' 13"	04° 26' 18"	42° 54' 12"	R
'b 1̄ 9 1̄0 9	20.17.1̄0	−24° 47' 28"	53° 23' 52"	109° 40' 15"	49° 00' 34"	42° 54' 12"	04° 26' 18"	R
n 12 1 1̄3 1	26.1̄0.13	26° 02' 12"	86° 24' 17"	64° 01' 05"	86° 02' 40"	34° 48' 16"	61° 31' 04"	LC
'n 1 12 1̄3 1	54̄8	−26° 02' 12"	86° 24' 17"	115° 58' 55"	34° 07' 48"	61° 31' 04"	34° 48' 16"	C

TABLE 12. LIST OF QUESTIONABLE AND DISCREDITED FORMS FOR QUARTZ

PV = probably valid; D = doubtful; X = discredited

D	$43\overline{7}0$	PV	$02\overline{2}5$	PV	$70\overline{7}4$	D	$19.0.\overline{19}.2$
D	$11.8.\overline{19}.0$	X	$0.13.\overline{13}.32$	D	$90\overline{9}5$	D	$12.0.\overline{12}.1$
D	$10.7.\overline{17}.0$	X	$0.7.\overline{7}.17$	D	$09\overline{9}5$	D	$0.12.\overline{12}.1$
D	$32\overline{5}0$	X	$0.5.\overline{5}.12$	D	$0.29.\overline{29}.16$	PV	$13.0.\overline{13}.1$
D	$8.5.\overline{13}.0$	X	$0.8.\overline{8}.19$	D	$13.0.\overline{13}.7$	D	$0.13.\overline{13}.1$
D	$7.4.\overline{11}.0$	X	$10.0.\overline{10}.21$	D	$39.0.\overline{39}.02$	D	$14.0.\overline{14}.1$
D	$25.11.\overline{36}.0$	D	$03\overline{3}5$	D	$0.23.\overline{23}.11$	D	$15.0.\overline{15}.1$
D	$52\overline{7}0$	D	$20\overline{2}3$	D	$0.21.\overline{21}.10$	D	$0.15.\overline{15}.1$
D	$31\overline{4}0$	X	$02\overline{2}3$	PV	$17.0.\overline{17}.8$	PV	$0.16.\overline{16}.1$
D	$41\overline{5}0$	X	$19.0.\overline{19}.28$	D	$07\overline{7}3$	D	$0.18.\overline{18}.1$
D	$12.1.\overline{13}.0$	X	$13.0.\overline{13}.19$	X	$26.0.\overline{26}.11$	D	$19.0.\overline{19}.1$
D	$0.1.\overline{1}.32$	PV	$03\overline{3}4$	D	$0.17.\overline{17}.7$	D	$0.20.\overline{20}.1$
D	$1.0.\overline{1}.30$	PV	$40\overline{4}5$	D	$13.0.\overline{13}.5$	D	$0.22.\overline{22}.1$
D	$0.1.\overline{1}.23$	PV	$04\overline{4}5$	D	$80\overline{8}3$	X	$23.0.\overline{23}.1$
D	$1.0.\overline{1}.20$	PV	$21.0.\overline{21}.20$	D	$08\overline{8}3$	D	$28.0.\overline{28}.1$
D	$1.0.\overline{1}.10$	D	$0.12.\overline{12}.11$	D	$19.0.\overline{19}.7$	X	$0.30.\overline{30}.1$
D	$10\overline{1}9$	D	$0.11.\overline{11}.10$	X	$45.0.\overline{45}.16$	X	$31.0.\overline{31}.1$
D	$01\overline{1}8$	PV	$09\overline{9}8$	D	$0.20.\overline{20}.7$	D	$0.35.\overline{35}.1$
X	$0.4.\overline{4}.29$	D	$80\overline{8}7$	PV	$0.22.\overline{22}.7$	X	$0.41.\overline{41}.1$
D	$10\overline{1}7$	D	$70\overline{7}6$	X	$0.13.\overline{13}.4$	D	$46.0.\overline{46}.1$
D	$0.2.\overline{2}.13$	D	$0.27.\overline{27}.23$	X	$0.23.\overline{23}.7$	X	$50.0.\overline{50}.1$
X	$0.5.\overline{5}.31$	D	$06\overline{6}5$	PV	$10.0.\overline{10}.3$	X	$3.3.\overline{6}.20$
X	$01\overline{1}6$	PV	$16.0.\overline{16}.13$	X	$0.27.\overline{27}.8$	PV	$11\overline{2}3$
X	$0.4.\overline{4}.23$	PV	$90\overline{9}7$	PV	$0.17.\overline{17}.5$	X	$7.7.\overline{14}.16$
X	$2.0.\overline{2}.11$	PV	$09\overline{9}7$	X	$41.0.\overline{41}.11$	X	$44\overline{8}9$
D	$10\overline{1}5$	D	$04\overline{4}3$	PV	$15.0.\overline{15}.4$	D	$22\overline{4}3$
X	$20\overline{2}9$	D	$11.0.\overline{11}.8$	PV	$90\overline{9}2$	X	$33\overline{6}2$
X	$0.7.\overline{7}.29$	D	$0.11.\overline{11}.8$	PV	$19.0.\overline{19}.4$	X	$23.24.\overline{47}.23$
X	$10\overline{1}4$	D	$18.0.\overline{18}.13$	X	$24.0.\overline{24}.5$	X	$19.20.\overline{39}.19$
X	$0.4.\overline{4}.15$	D	$07\overline{7}5$	PV	$16.0.\overline{16}.3$	D	$15.16.\overline{31}.2$
D	$0.7.\overline{7}.23$	D	$0.23.\overline{23}.16$	PV	$17.0.\overline{17}.3$	X	$15.14.\overline{29}.15$
X	$0.5.\overline{5}.16$	D	$13.0.\overline{13}.9$	D	$23.0.\overline{23}.4$	X	$14.15.\overline{29}.14$
X	$0.11.\overline{11}.34$	X	$47.0.\overline{47}.30$	PV	$15.0.\overline{15}.2$	X	$99.92.\overline{191}.61$
X	$0.19.\overline{19}.56$	D	$0.11.\overline{11}.7$	X	$0.23.\overline{23}.3$	PV	$13.12.\overline{25}.25$
X	$0.7.\overline{7}.20$	D	$17.0.\overline{17}.10$	D	$17.0.\overline{17}.2$	D	$12.13.\overline{25}.2$
X	$03\overline{3}8$	D	$19.0.\overline{19}.11$	D	$09\overline{9}1$	D	$10.11.\overline{21}.10$

TABLE 12   (*cont'd*)

D	11.10.$\overline{21}$.11	D	13.9.$\overline{22}$.9	D	3.6.$\overline{9}$.13	D	7.22.$\overline{29}$.7
X	8.9.$\overline{17}$.26	D	13.19.$\overline{32}$.32	PV	21$\overline{3}$4	X	5.16.$\overline{21}$.21
PV	8.9.$\overline{17}$.8	X	22.15.$\overline{37}$.22	D	42$\overline{6}$1	X	4.13.$\overline{17}$.17
D	45.40.$\overline{85}$.29	PV	32$\overline{5}$8	X	17.8.$\overline{25}$.42	D	3.10.$\overline{13}$.11
D	17.15.$\overline{32}$.62	D	23$\overline{5}$8	D	32.15.$\overline{47}$.47	D	10.3.$\overline{13}$.3
X	8.7.$\overline{15}$.22	PV	23$\overline{5}$7	X	7.15.$\overline{22}$.37	D	3.10.$\overline{13}$.3
D	15.13.$\overline{28}$.15	PV	32$\overline{5}$5	X	15.7.$\overline{22}$.22	D	5.17.$\overline{22}$.18
X	6.7.$\overline{13}$.20	PV	32$\overline{5}$1	D	7.15.$\overline{22}$.7	D	72$\overline{9}$8
X	7.6.$\overline{13}$.19	X	23.15.$\overline{38}$.38	D	13.6.$\overline{19}$.6	D	27$\overline{9}$8
PV	6.5.$\overline{11}$.6	D	19.12.$\overline{31}$.12	X	6.13.$\overline{19}$.6	D	72$\overline{9}$7
D	5.6.$\overline{11}$.6	D	27.17.$\overline{44}$.1	X	5.11.$\overline{16}$.27	D	72$\overline{9}$3
PV	6.5.$\overline{11}$.5	D	8.5.$\overline{13}$.21	D	31.14.$\overline{45}$.45	D	27$\overline{9}$2
D	5.6.$\overline{11}$.5	D	5.8.$\overline{13}$.18	D	9.4.$\overline{13}$.6	D	3.11.$\overline{14}$.14
PV	9.11.$\overline{20}$.9	PV	8.5.$\overline{13}$.8	D	4.9.$\overline{13}$.4	X	11.3.$\overline{14}$.3
D	5.4.$\overline{9}$.14	X	8.5.$\overline{13}$.5	D	3.7.$\overline{10}$.17	D	4.15.$\overline{19}$.17
D	4.5.$\overline{9}$.14	X	11.18.$\overline{29}$.11	D	7.3.$\overline{10}$.14	PV	15.4.$\overline{19}$.6
D	5.4.$\overline{9}$.13	D	23.14.$\overline{37}$.23	D	3.7.$\overline{10}$.10	D	4.15.$\overline{19}$.4
D	45$\overline{9}$9	D	53$\overline{8}$8	D	5.2.$\overline{7}$.12	D	5.19.$\overline{24}$.24
PV	45$\overline{9}$5	D	35$\overline{8}$8	D	25$\overline{7}$5	PV	14$\overline{5}$9
PV	45$\overline{9}$4	D	53$\overline{8}$5	D	25$\overline{7}$4	PV	41$\overline{5}$6
D	19.15.$\overline{34}$.19	D	17.10.$\overline{27}$.7	D	25$\overline{7}$3	D	2.8.$\overline{10}$.9
PV	15.19.$\overline{34}$.5	D	7.12.$\overline{19}$.13	D	25$\overline{7}$2	D	41$\overline{5}$3
D	9.7.$\overline{16}$.10	D	7.12.$\overline{19}$.12	X	5.13.$\overline{18}$.31	PV	41$\overline{5}$2
PV	3.4.$\overline{7}$.11	D	26.15.$\overline{41}$.26	PV	8.3.$\overline{11}$.11	D	12.3.$\overline{15}$.4
D	43$\overline{7}$7	D	7.4.$\overline{11}$.18	PV	8.3.$\overline{11}$.3	X	6.25.$\overline{31}$.28
PV	34$\overline{7}$7	D	4.7.$\overline{11}$.18	X	4.11.$\overline{15}$.26	X	25.6.$\overline{31}$.6
PV	43$\overline{7}$4	D	7.4.$\overline{11}$.11	D	36.13.$\overline{49}$.25	D	21.5.$\overline{26}$.7
D	12.9.$\overline{21}$.7	D	5.9.$\overline{14}$.9	X	5.14.$\overline{19}$.33	D	5.21.$\overline{26}$.5
X	41.30.$\overline{71}$.41	D	9.5.$\overline{14}$.5	D	13$\overline{4}$9	D	17.4.$\overline{21}$.9
X	11.8.$\overline{19}$.27	X	6.11.$\overline{17}$.28	PV	31$\overline{4}$7	D	13.3.$\overline{16}$.16
D	8.11.$\overline{19}$.12	D	11.6.$\overline{17}$.11	D	9.3.$\overline{12}$.17	D	13.3.$\overline{16}$.7
D	8.11.$\overline{19}$.2	D	8.15.$\overline{23}$.23	D	31$\overline{4}$5	D	13.3.$\overline{16}$.4
X	5.7.$\overline{12}$.19	D	15.8.$\overline{23}$.15	D	31$\overline{4}$3	D	22.5.$\overline{27}$.5
D	5.7.$\overline{12}$.7	D	8.15.$\overline{23}$.15	D	13$\overline{4}$3	D	2.9.$\overline{11}$.20
D	10.7.$\overline{17}$.10	D	9.17.$\overline{26}$.1	D	31$\overline{4}$2	PV	2.9.$\overline{11}$.11
D	9.13.$\overline{22}$.21	D	6.3.$\overline{9}$.13	D	62$\overline{8}$3	PV	9.2.$\overline{11}$.3
D	14.3.$\overline{17}$.17	D	66.10.$\overline{76}$.11	X	1.10.$\overline{11}$.11	D	19.1.$\overline{20}$.15
PV	14.3.$\overline{17}$.3	D	5.33.$\overline{38}$.5	X	10.1.$\overline{11}$.2	D	1.19.$\overline{20}$.1
D	4.20.$\overline{24}$.21	D	47.7.$\overline{54}$.19	PV	10.1.$\overline{11}$.1	X	39.2.$\overline{41}$.31

Table 12 (*cont'd*)

PV	$51\bar{6}6$	X	$7.1.\bar{8}.15$	D	$11.1.\overline{12}.12$	D	$20.1.\overline{21}.18$
D	$15\bar{6}6$	D	$1.7.\bar{8}.13$	D	$11.1.\overline{12}.11$	D	$1.20.\overline{21}.1$
PV	$15\bar{6}5$	D	$71\bar{8}8$	PV	$1.11.\overline{12}.11$	D	$61.3.\overline{64}.4$
PV	$51\bar{6}3$	PV	$17\bar{8}7$	D	$11.1.\overline{12}.8$	X	$6.125.\overline{131}.16$
D	$51\bar{6}2$	D	$71\bar{8}4$	D	$11.1.\overline{12}.7$	PV	$21.1.\overline{22}.17$
D	$10.2.\overline{12}.3$	PV	$71\bar{8}3$	D	$11.1.\overline{12}.4$	D	$21.1.\overline{22}.2$
X	$5.26.\overline{31}.5$	D	$21.3.\overline{24}.8$	D	$1.11.\overline{12}.1$	D	$1.21.\overline{22}.1$
D	$16.3.\overline{19}.3$	D	$17\bar{8}1$	PV	$12.1.\overline{13}.13$	D	$22.1.\overline{23}.22$
X	$33.6.\overline{39}.7$	D	$22.3.\overline{25}.22$	PV	$1.12.\overline{13}.13$	D	$1.22.\overline{23}.1$
D	$11.2.\overline{13}.2$	D	$15.2.\overline{17}.17$	PV	$1.12.\overline{13}.10$	PV	$23.1.\overline{24}.1$
PV	$2.11.\overline{13}.2$	D	$2.15.\overline{17}.17$	D	$12.1.\overline{13}.9$	D	$24.1.\overline{25}.14$
D	$28.5.\overline{33}.8$	D	$15.2.\overline{17}.9$	D	$37.3.\overline{40}.31$	D	$73.3.\overline{76}.67$
D	$17.3.\overline{20}.3$	D	$2.15.\overline{17}.2$	PV	$3.38.\overline{41}.1$	D	$1.27.\overline{28}.28$
X	$5.29.\overline{34}.5$	X	$23.3.\overline{26}.14$	PV	$13.1.\overline{14}.14$	D	$1.27.\overline{28}.1$
X	$47.8.\overline{55}.9$	D	$3.23.\overline{26}.11$	D	$1.13.\overline{14}.1$	D	$1.29.\overline{30}29$
PV	$1.6.\bar{7}.13$	D	$8.1.\bar{9}.10$	PV	$1.14.\overline{15}.14$	D	$3.92.\overline{95}.1$
PV	$61\bar{7}9$	D	$81\bar{9}8$	D	$14.1.\overline{15}.4$	D	$35.1.\overline{36}.1$
PV	$16\bar{7}7$	X	$81\bar{9}7$	D	$5.70.\overline{75}.1$	D	$37.1.\overline{38}.9$
D	$61\bar{7}6$	D	$81\bar{9}4$	X	$6.85.\overline{91}.11$	D	$41.1.\overline{42}.37$
PV	$16\bar{7}6$	D	$18\bar{9}2$	X	$99.7.\overline{106}.11$	D	$44.1.\overline{45}.9$
PV	$2.12.\overline{14}.11$	D	$81\bar{9}1$	D	$29.2.\overline{31}.41$	D	$1.51.\overline{52}.52$
PV	$61\bar{7}3$	D	$18\bar{9}1$	D	$1.15.\overline{16}.16$	X	$1.117.\overline{118}.1$
D	$61\bar{7}2$	D	$5.42.\overline{47}.5$	X	$3.45.\overline{48}.1$		
PV	$61\bar{7}1$	D	$17.2.\overline{19}.11$	D	$3.47.\overline{50}.1$		
PV	$16\bar{7}1$	D	$1.9.\overline{10}.11$	D	$1.16.\overline{17}.1$		
D	$5.31.\overline{36}.5$	PV	$1.9.\overline{10}.10$	X	$162.10.\overline{172}.17$		
X	$50.8.\overline{58}.9$	X	$9.1.\overline{10}.8$	PV	$1.17.\overline{18}.17$		
D	$19.3.\overline{22}.22$	D	$9.1.\overline{10}.6$	PV	$17.1.\overline{18}.1$		
X	$2.13.\overline{15}.28$	D	$18.2.\overline{20}.9$	D	$35.2.\overline{37}.27$		
D	$2.13.\overline{15}.13$	PV	$1.9.\overline{10}.1$	D	$18.1.\overline{19}.2$		
D	$13.2.\overline{15}.7$	D	$19.2.\overline{21}.13$	D	$1.18.\overline{19}.1$		
D	$13.2.\overline{15}.6$	D	$2.19.\overline{21}.2$	X	$37.2.\overline{39}.29$		
PV	$2.13.\overline{15}.2$	D	$1.10.\overline{11}.21$	D	$3.56.\overline{59}.1$		

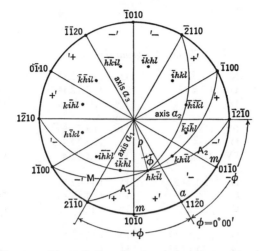

FIG. 11. Stereographic projection showing designation and angles in quartz.

discredited forms. It should be noted in this connection that the distinction between right and left and between positive and negative forms is not always clearly made in published morphological descriptions, or is doubtful because the absence of twinning has not been demonstrated. Furthermore, some of the small modifying forms that have been described, usually vicinal or rounded, may be solution rather than growth surfaces. The relative frequency of occurrence of at least the less common dimensionally equivalent forms hence is uncertain. The absence of a unique, readily produced cleavage has been a handicap.

The typical habits taken by quartz crystals and the relative frequencies of the common forms are discussed in detail in the following section on *Crystal Habit.*

References

1. Dana: *System of Mineralogy,* New York, sixth ed., 1892, p. 183.
2. Hintze: *Handbuch der Mineralogie,* Berlin, **1,** Lfg. 8, 1904 p. 1266.
3. Goldschmidt, V.: *Atlas der Krystallformen,* **7,** Heidelberg, 1922, p. 60.
4. Sosman: *Properties of Silica,* New York, 1927, p. 373; Kôzu and Takané: *Sci. Rpts. Tohoku Univ.,* Ser. III, **3,** no. 3, 239 (1929).
5. Kupffer: *Ak. Wiss. Berlin, Preisschrift. phys. Kl.,* 61, 1825.
6. Dauber: *Ann. Phys.,* **103,** 114 (1858).
7. Koksharov: *Mater. Min. Russlands, St. Petersburg,* **8,** 127 (1878).
8. Gill: *Zs. Kr.,* **22,** 115 (1893).
9. Wright: *J. Wash. Ac. Sci.,* **3,** 485 (1913).
10. Kalb: *Zs. Kr.,* **86,** 458 (1933), on same crystal.
11. See Kalb (1933).
12. Based on an unpublished critical survey by the late Professor Charles Palache, Harvard University, 1945.

CRYSTAL HABIT

Form Frequency. The prism $m\{10\bar{1}0\}$ and the positive and negative rhombohedra $r\{10\bar{1}1\}$ and $z\{01\bar{1}1\}$ are virtually universal forms on quartz crystals, usually as large faces, and in general these three forms dominate the crystal habit of this species. The positive rhombohedron r is somewhat more important than the negative rhombohedron z both in frequency of occurrence and in relative size of development. When both rhombohedra are present in combination, as is usually the case, the faces of r generally predominate. Although r often is present to the exclusion of z, the reverse situation is unusual (but could easily go unrecognized). The forms r and z are sometimes referred to as the major and minor rhombohedra, respectively.

In addition to m, r, and z, approximately 535 established, doubtful, or discredited forms have been reported on quartz (Tables 11 and 12). All these additional forms are found with only occasional exceptions as relatively small modifying faces. The trigonal pyramid $s\{11\bar{2}1\}$ and the positive trigonal trapezohedron x $\{51\bar{6}1\}$ rank immediately after but considerably below m, r, and z in frequency of occurrence. The trigonal pyramid $\{11\bar{2}2\}$ is much less common than s $\{11\bar{2}1\}$. The right and left equivalents of the trigonal trapezohedra and trigonal pyramids apparently are of equal importance in size and frequency, but morphological surveys sufficiently extensive to establish this point with certainty have not been made.

The relative frequency of the few remaining forms of common occurrence on quartz is uncertain. The positive right trigonal trapezohedra $\{32\bar{5}3\}$, $\{12\bar{3}2\}$, $\{21\bar{3}1\}$, $\{31\bar{4}1\}$, and $\{41\bar{5}1\}$ and their positive left equivalents are often noted as small faces. These forms usually occur in combination with x $\{51\bar{6}1\}$, which is by far the most common trigonal trapezohedron; $\{31\bar{4}1\}$ and $\{41\bar{5}1\}$ follow $\{51\bar{6}1\}$ in frequency of occurrence. These and other trigonal trapezohedra, including many of a vicinal nature, often occur together in a striated or rounded zone in which the assigning of indices may be difficult or arbitrary. The trigonal pyramid $\{11\bar{2}1\}$ generally is also present. Occasionally, at particular localities, quite large faces of otherwise rare trapezohedral forms may be present with s, x, and other common modifying forms either very small in size or entirely absent. The positive forms of the trigonal trapezohedra are much more common than the negative ones, and usually are larger in size. The occurrence of negative right trapezohedra on left-handed quartz has been noted,[1] and vice versa. This matter involves the notation employed for the trigonal trapezohedra; some authors reverse the hand of the negative trapezohedra from that given in Fig. 11. Confusion also can be caused by concealed twinning, and the suspected simultaneous occurrence of general forms of opposite hand or of opposite sign should be confirmed by etching the crystal.

A number of established modifying forms also occur in the zone of positive rhombohedra with $r\{10\bar{1}1\}$, including $i\{50\bar{5}3\}$, $l\{20\bar{2}1\}$, $M\{30\bar{3}1\}$, $\Upsilon\{40\bar{4}1\}$, $e\{50\bar{5}1\}$, $\zeta\{60\bar{6}1\}$, $\phi\{70\bar{7}1\}$, $h\{70\bar{7}2\}$, $\psi\{11.0.\bar{1}\bar{1}.1\}$. Of these, $\{30\bar{3}1\}$ and $\{40\bar{4}1\}$ are the most frequently observed, and both apparently are more important forms than $u\{31\bar{4}1\}$; $\{70\bar{7}1\}$ is fairly frequent. The modifying positive and negative rhombohedra when present tend to occur in oscillatory combination, together with r and m and narrow intervening faces of a vicinal nature. Occasionally a positive rhombohedron other than r and not necessarily of relatively low indices may develop large faces and dominate the habit of the crystal. The negative rhombohedra, including $z\{01\bar{1}1\}$ are in general less frequent forms than their positive equivalents; $\{07\bar{7}1\}$, $\{03\bar{3}1\}$, $\{04\bar{4}1\}$, and $\{06\bar{6}1\}$ are the most common. On individual crystals the positive zone usually predominates. Obtuse rhombohedra such as $\{01\bar{1}2\}$ and $\{10\bar{1}2\}$ are relatively uncommon.

With regard to the other forms, the right and left ditrigonal and trigonal prisms are extremely rare or are of doubtful occurrence either alone or in combination. The basal pinacoid $\{0001\}$ also is very rare; when observed, it generally has an uneven or rough, coplanar surface.

Reference

1. MacCarthy: *Zs. Kr.*, **67**, 29 (1928); Aminoff: *Ark. Kemi, Min., Geol.*, **7**, 1 (1919); Groth: *Ann. Phys.*, **137**, 435 (1869).

Form Development. The habit of euhedral quartz crystals is quite varied in detail, when account is taken of the varied combinations of modifying forms, but the main habits as determined by the common forms of large facial development are few in number.

(*a*) *Prismatic Habit.* The characteristic habit of quartz is short prismatic by extension along the c-axis, with the prism $m\{10\bar{1}0\}$ in combination with the rhombohedra $r\{10\bar{1}1\}$ and $z\{01\bar{1}1\}$ as dominant forms. The ratio of width to length of the crystals as measured along the a-axis and c-axis directions generally falls within the range from 2:3 to 1:4. Long prismatic crystals with an elongation of more than about 1:6 are uncommon, and acicular habits are very rare. The simple habit $m\ r\ z$ is extremely common (Figs. 12-15). Small modifying faces of general forms, usually $s\{11\bar{2}1\}$ and $x\{51\bar{6}1\}$, are often present and are useful in the determination of the hand of the crystal. In a right-handed crystal, the right trigonal pyramid s lies to the upper right of the prism face which is below the positive rhombohedron r, and the striations, if any, on s slope upward to the right; the positive right trigonal trapezohedron x is similarly situated. Opposite relations are found in left-handed crystals. The distribution of the faces of the positive and negative and right and left trigonal trapezohedra is indicated in Fig. 16. The separate identity of r and z can be established by etch-figures; r usually is relatively large and has a bright luster, whereas z often is dull. The faces of the modifying

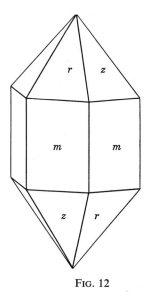

FIG. 12

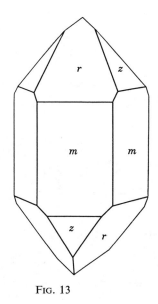

FIG. 13

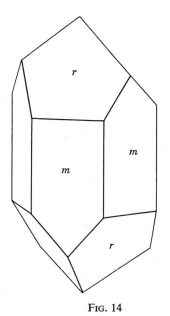

FIG. 14

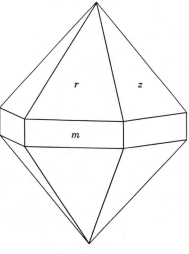

FIG. 15

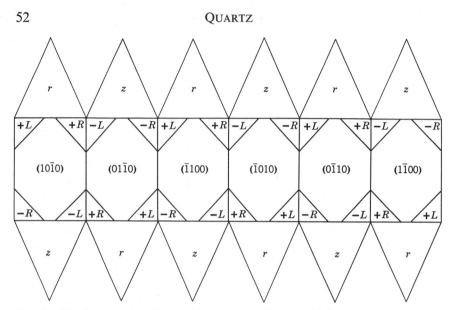

FIG. 16. Distribution of positive and negative and of right and left trigonal trapezohedra.

forms s and x (and of other general forms) usually are unequally developed in size and rarely are present in full complement. Sometimes one face of s is developed to relatively large size, while the other faces of this form are small or absent.

Variants of the prismatic habit appear through the development of the trigonal trapezohedral and rhombohedral zones (Figs. 17–27). Forms in the trapezohedral zone when present in number or in large facial develement may give the crystal a tapering or screw-like appearance. Crystals from Diamond Hill, Rhode Island, show the prism terminated by s with a series of trigonal trapezohedra, with rhombohedral faces virtually lacking. The distribution of the faces of the modifying general forms often is complicated by twinning on laws maintaining parallelism of the crystal axes (see *Twinning*), and identification of the morphology then is facilitated by optical and etching tests to reveal twin boundaries and hand. Natural etch markings are of aid when present. Untwinned crystals showing the full complement of modifying forms at both terminations, and twinned crystals similarly showing the full repetition of these faces, are almost never found in nature. With increasing development of steep rhombohedra such as $\{30\bar{3}1\}$, $\{40\bar{4}1\}$, $\{50\bar{5}1\}$, and their negative equivalents the crystal habit passes into the acute rhombohedral type described beyond. These steep forms usually are present only as narrow modifying faces on crystals of the common prismatic habit and then but rarely.

The prism faces almost always are horizontally striated or grooved, and this feature may aid in the proper orientation of misshapen crystals. The striations are bounded by r or z surfaces along their length, and by s or x faces

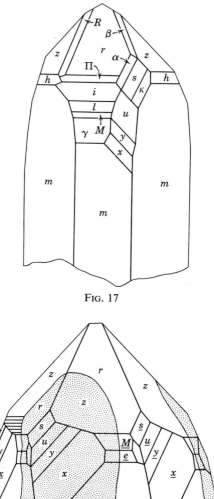

FIG. 17

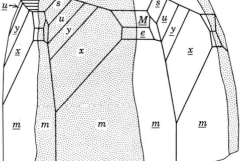

FIG. 18

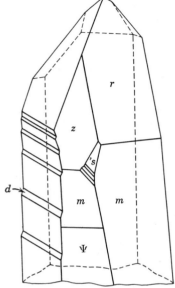

FIG. 19

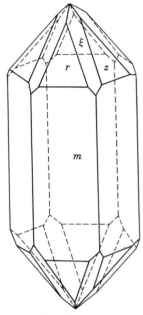

FIG. 20

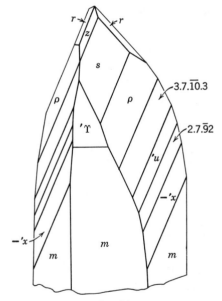

FIG. 21

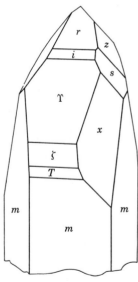

FIG. 22

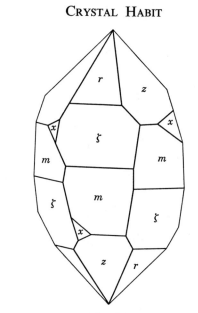

FIG. 23

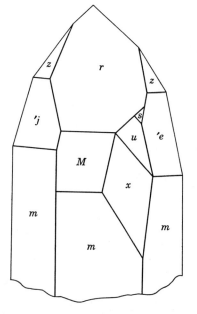

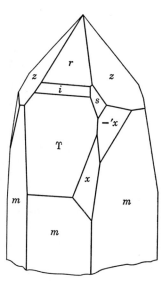

FIG. 24   FIG. 25

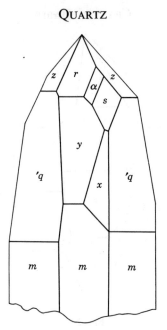

FIG. 26

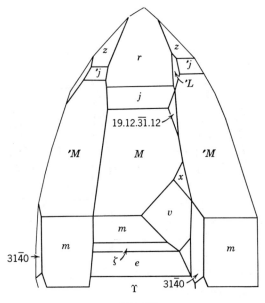

FIG. 27

at their ends.[1] In twinned crystals the striations do not continue uninterrupted across twin boundaries that appear on the prism surfaces. The faces of r generally are more lustrous than those of z, as noted; the faces of s also are generally lustrous, although sometimes striated. Of the trigonal trapezohedra, x often is relatively smooth and lustrous, whereas u and the negative trigonal trapezohedra in general tend to be dull.

(b) *Equant Habits.* Equant crystals are generally produced by the large development of one or both of the unit rhombohedra with the absence or near absence of the prism $m\{10\overline{1}0\}$, as in Figs. 28 and 29. The term dipyramidal habit (a misnomer) or quartzoid is applied to crystals with an equal or nearly equal development of both r and z, this simulating a hexagonal pyramid, with m absent or present only as very narrow faces. This habit is typical of high-quartz, with the (hexagonal) form $\{10\overline{1}1\}$, but it is frequently observed also in undoubted low-quartz. All gradations are found between this habit and the normal prismatic habit. An equant pseudocubic[2] habit is produced when the faces of r are developed to the exclusion of m and z, since the interfacial angle rr is $85°46'$. Small facets of m and z are often present. Pseudocubic crystals composed of z also are known. Modifying faces of general forms are extremely rare on crystals of these habits.

(c) *Acute Rhombohedral Habits.* This category includes crystals elongated along c with a three-sided tapering appearance caused by the large development of faces of a steep positive rhombohedron or, less frequently, of a steep negative rhombohedron; the prism faces appearing on alternate edges generally are quite small or may be lacking. The apex usually is terminated by quite small faces of r and z, and modifying faces of s and x, if present, also tend to be relatively small. The taper of the crystals rarely is uniform, because of oscillatory combination of a number of steep or vicinal rhombohedra, or with m or the unit rhombohedra, and the surfaces may be deeply grooved or striated horizontally or show abrupt changes in slope. Crystals of an acute rhombohedral habit are relatively rare. Probably the most common type is that based on the positive rhombohedron $\{30\overline{3}1\}$. This form may occur

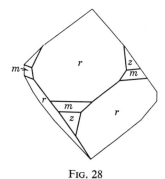

FIG. 28   FIG. 29

alone, except for very small terminal faces of r and z, or it may also be modified by $\{10\bar{1}0\}$. Usually only the top half of the crystal is developed, giving the appearance of a three-sided pyramid. The faces of $\{30\bar{3}1\}$ usually are horizontally grooved and striated by oscillatory combination with m and r. Crystals of this habit that weigh several pounds have been observed. Crystals of the so-called Tessin habit, from Val Maggia and other localities in Ticino (Tessin), Switzerland,[3] are characterized by the predominance of $\{30\bar{3}1\}$ or $\{50\bar{5}3\}$; the crystals sometimes are pseudohemimorphic along c by the development of r and z at one termination and of a series of acute positive or of positive and negative rhombohedra at the other. Pseudohemimorphic crystals also have been observed from Dauphiné and other localities. The unusual tapering smoky crystals from North Carolina, described by vom Rath[4] and others, also are characterized by the large development of $\{30\bar{3}1\}$, with $\{0\bar{3}31\}$, the negative trapezohedron $\{1\bar{2}\bar{3}2\}$, and a variety of other forms, including many very steep rhombohedra. The combination of $\{30\bar{3}2\}$ and $\{0\bar{3}32\}$ together with r and z and modifying forms also has been observed.

Crystals composed wholly of a single steep rhombohedron have been noted but are quite unusual. Doubly terminated needle-like to rod-like crystals up to 1.2 in. long and 0.12 in. across composed wholly of $\{18.0.\bar{18}.1\}$ or of this form combined with small terminal faces of r and z have been described[5] (see Fig. 30). The combination of $\{0.11.\bar{11}.1\}$ and r also has been noted.[6] Small doubly terminated crystals composed of only $\{60\bar{6}1\}$ and $\{10\bar{1}1\}$, sometimes as interpenetration twins on the Dauphiné Law, occur with stibnite and siderite at Kisbánya, Romania[7] (Fig. 31). The combination of $\{70\bar{7}1\}$ with $\{13.0.\bar{13}.7\}$ and $\{20\bar{2}1\}$ without r and z has been described;[8] also $\{30\bar{3}1\}$ and $\{51\bar{6}1\}$. Other steep rhombohedra that have been observed[9] as the dominant form include $\{40\bar{4}1\}$, $\{60\bar{6}1\}$, $\{70\bar{7}1\}$, $\{70\bar{7}2\}$, $\{70\bar{7}5\}$, and $\{34.0.\bar{34}.1\}$. The precise indexing of the very steep forms is uncertain, as is, in many cases, the identification of these rhombohedra as positive or negative. The positive rhombohedra generally tend to be smooth and lustrous, as well as predominant in size, and the negative rhombohedra to be striated or grooved and dull. Crystals with an acute appearance due to the large development of the trigonal pyramid $s\{11\bar{2}1\}$ have been observed. More commonly, a single face of $\{11\bar{2}1\}$ may be developed to very large size as a modification on a crystal otherwise of normal prismatic habit.

Tapering crystals with a hexagonal rather than a rhombohedral aspect also are found, because of the more or less equal development of steep positive and negative equivalent rhombohedra, such as $\{30\bar{3}1\}$ with $\{0\bar{3}31\}$ or $\{40\bar{4}1\}$ with $\{0\bar{4}41\}$ (Fig. 32). Tapering shapes also may arise as coplanar surfaces resulting from the oscillatory combination of m with r or rz.

The crystal habit of quartz is responsive to the environmental conditions obtaining during crystallization, but little is known of the matter either experimentally or in natural occurrences. It is often observed that the crystals from a particular locality have a characteristic habit or combination

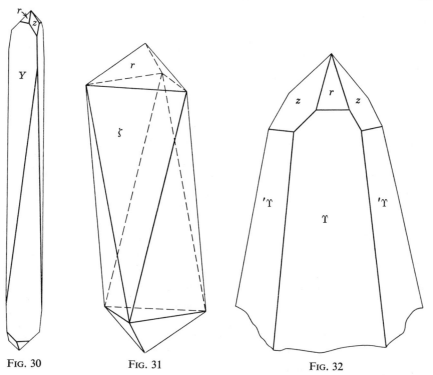

FIG. 30   FIG. 31   FIG. 32

of modifying forms. The s and x forms apparently are more common on quartz crystals formed at relatively high temperatures. Both the pseudocubic and dipyramidal habits of quartz appear from their manner of occurrence, as in gypsum beds and in hematite deposits formed in the meteoritic circulation, to be typical of a low temperature of formation.

References

1. Seager: *Min. Mag.*, **30**, 12 (1953); van Praagh and Willis: *Nature*, **169**, 623 (1952).
2. Tarr: *Am. Min.*, **14**, 19 (1929); Tarr and Lonsdale: *Am. Min.*, **14**, 50 (1929); Witteborg: *Cbl. Min.*, 1933A, 289; Websky: *Jb. Min.*, 113, 1874.
3. Cavinato: *Mem. reale accad. Lincei, Cl. sci. fis., mat., nat.*, ser. 6, **1**, 323 (1925); Bianchi: *Mem. reale accad. Lincei, Cl. sci. fis., mat., nat.*, ser. 5, **14**, 57 (1923) and **32**, 243 (1923); Casasopra: *Schweiz. min. pet. Mitt.*, **19**, 261 (1939).
4. vom Rath: *Zs. Kr.*, **10**, 156, 487 (1885) and **12**, 453, 535 (1887); Gill: *Zs. Kr.*, **22**, 99 (1893); Pogue and Goldschmidt: *Am. J. Sci.*, **34**, 414 (1912).
5. Palache: *Am. Min.*, **13**, 297 (1928).
6. Lacroix: *Min. de France*, **3**, 1901, pp. 81, 86.
7. Drugman: *Min. Mag.*, **25**, 259 (1939).
8. Rosický: *Věstník Stát. geol. ústavu Českoslov. rep.*, **5**, 155 (1929).
9. Bindrich: *Zs. Kr.*, **59**, 113 (1924); Heddle: *Min. Scotland*, **1**, 1923, 43; Rosický (1929); Goldschmidt: *Atlas der Krystallformen*, **7**, text and tables, Heidelberg, 1922.

Vicinal Surfaces. The term vicinal face is applied to planar, crystallographically irrational surfaces that approach very closely in angle and position to forms of simple indices. They may either modify or wholly replace such forms. Vicinal planes may arise through either crystal growth or solution, in the latter case often being termed etch, prerosion, or solution faces. In many descriptions their origin is not indicated or known. Vicinal surfaces due to both growth and solution are very common on quartz. The principal vicinal growth surfaces on quartz appear on the terminal unit rhombohedra. On these faces the vicinal development often appears as an extremely flat three-sided pyramid with interfacial angles up to 1° or so. The apex of the pyramid generally is near the center of the face but may be greatly displaced laterally. In some instances an entire face of r or z may be replaced by one vicinal surface. Vicinal faces are an important factor in limiting the precision of morphological measurements on quartz. Two main types of vicinal pyramids have been recognized:[1] one with the two lower edges inclined at an angle of about 90° (Type I) and another with the two lower edges inclined at roughly 160° (Type II), as seen when the r or z face is viewed perpendicularly. The attitude of the upper edge, which terminates at the apex of the crystal, is useful in determining the hand of crystal and the separate identities of the r and z faces, as shown in Fig. 33. Vicinal faces also may be of aid in recognizing and identifying twinning.[2] In many cases the vicinal development on r and z is irregular or has a rounded outline, or consists of a number of overlapping individual pyramids. A terraced appearance is very common.

Vicinal growth surfaces also may appear on the prism faces, usually as rectangular pyramids showing features associated with the development of s

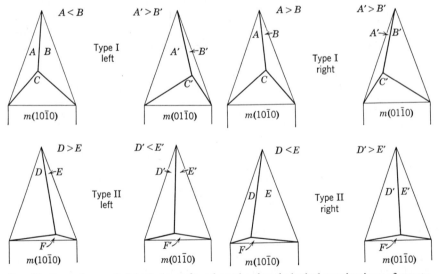

FIG. 33. Some types of vicinal planes found on the rhombohedral terminations of quartz, and their relation to the hand of the crystal.

and x faces and hence useful in the identification of hand. Those areas of alternate prism faces adjacent to the r and z faces may show significant differences in the details of vicinal development and other surface features. Vicinal faces also appear throughout the rhombohedral and trigonal trapezohedral zones, and many of the high index forms reported for quartz are of this general nature. The vicinal faces associated with $s\{11\bar{2}1\}$ are asymmetrical four-sided pyramids elongated toward the edge of s with r. Vicinal surfaces on quartz also occur asociated with re-entrant angles due to parallel growth or to contact twinning. They may bound striations and grooves on prism and other planes. Vicinal faces associated with healed cracks have been observed on amethyst.[3] Others of the reported vicinal forms on quartz are the result of solution. These surfaces tend to be gently rounded, or to be slightly pitted or grooved, and may grade into planar vicinal surfaces; this feature is particularly common in the trigonal trapezohedral zone. The vicinal surfaces produced by artificial and natural solvents on quartz crystals have been frequently described.[4]

Growth Hillocks; Spiral Dislocations. Other surface features that are closely related to vicinal growth surfaces occur on natural and synthetic quartz crystals. The most common are low, extremely broad conical hillocks, having a ratio of height to width ranging up to 1:100 or more, and ranging in dimension up to several centimeters. The hillocks vary from more or less equant or oval shapes to elongate forms with the apex located toward one end. Small isolated hillocks may occur on a single face, sometimes in such number that they overlap to give a pebbled, shingled, or undulating appearance, or a single large hillock may occupy most or all of a face. The hillocks are relatively flat on the r faces, and are more numerous, usually overlap, and are steeper on $\{0001\}$. Under magnification the hillocks may show spiral terraces that terminate at the apex in a spiral dislocation extending into the quartz.[5] The growth spirals on hillocks present on $\{0001\}$ have the appearance of a rounded equilateral triangle, the sense of the spiral not being related to the hand of the quartz. Study of the hillocks by multiple-beam interferometry[6] also reveals a stepped structure, presumably of a spiral nature, formed by the spreading of the growth layers. The adsorption of radioactive ions has been applied to the study of surface imperfections in quartz.[7]

References

1. Kalb and Witteborg: *Jb. Min.*, Beil. Bd. **56**, 334 (1927); Kalb: *Cbl. Min.*, 1927A, 279, and *Zs. Kr.*, **86**, 439 (1933); Witteborg: *Cbl. Min.*, 1933A, 289; Tokody: *Mat. Termés. Ért.*, **55**, 1001 (1937); Virovlanski: *Mém. soc. russe min.*, **67**, 425, 446 (1938); Bond and Andrus: *Am. Min.*, **37**, 622 (1952).
2. Kalb (1933); Padurow: *Jb. Min., Monatsh.*, Abt. A, 106, 1948.
3. Laemmlein: *Zs. Kr.*, **88**, 470 (1934) and *C. R. ac. sci. USSR*, 555, 1934.
4. Cf. Gill: *Zs. Kr.*, **22**, 99 (1893); Molengraaff: *Zs. Kr.*, **14**, 201 (1888) and **17**, 138 (1889); Gonnard: *Bull. soc. min.*, **38**, 78 (1915); Cavinato: *Mem. reale accad, Lincei, Cl. sci.*, **1**, 246 (1925).
5. Augustine and Hale: *J. Phys. Chem. Solids*, **13**, 344 (1960) with literature; Butuzov

and Ikornikova: *C. R. ac. sci. USSR,* **97,** 89 (1954); Zimonyi: *Acta Phys. Ac. Sci. Hung.,* **8,** 119 (1957); Keymeulen: *Naturwiss.,* **44,** 489 (1957).
6. Tolansky: *Multiple-Beam Interferometry,* Oxford, 1948; see also Seager: *Min. Mag.,* **30,** 1 (1953).
7. Antkiw, Waesche, and Senftle: *Am. Min.,* **41,** 363 (1956).

Misshapen Crystals.   Natural crystals of quartz often depart considerably from an ideal shape through the unequal development of equivalent faces of one or more of the forms present on the crystal, or by extension along one or several crystallographic directions that are part of a larger equivalent set. The distortion may be symmetrical, affording crystals that simulate orthorhombic or monoclinic symmetry, but it generally is asymmetrical. The prism faces in general tend to be equally developed, so that the crystals have a hexagonal cross-section, but occasionally crystals are observed that are more or less tabular by flattening on a pair of prism faces, or in which a pair of opposing prism faces are relatively small, giving a diamond-shaped cross-section to the prism zone. Some crystals have alternate faces of $\{10\bar{1}0\}$ relatively small in size, so that the prism zone appears as if trigonal prisms were present. This feature usually is associated with the development of a very steep vicinal positive or negative rhombohedron in oscillatory combination with alternate prism faces.

Distortion associated with the terminal rhombohedral faces of r and z also is common (Fig. 34). The faces of these forms often are asymmetrically developed; one face of r may be developed to large size, giving the termination a slanting appearance, or the faces of r and z may develop unequally in such a way as to give the termination an orthorhombic or monoclinic aspect. Crystals simulating orthorhombic appearance and somewhat resembling topaz have been described;[1] they have a diamond-shaped cross-section in the prism zone, because of the reduced size or the absence of a parallel pair of prism faces, and are terminated by facing and equally developed pairs of faces of a series of positive and negative rhombohedra, thus resembling an orthorhombic prism and pyramids. An approximation to orthorhombic symmetry also is shown by twisted crystals that are distorted by elongation parallel to an a-axis (see *Twisted Crystals*) when these crystals are oriented with this axis vertical. Crystals also are sometimes found that are greatly flattened on a parallel pair of r faces with the prism faces nearly or entirely lacking.

Quartz crystals also occur misshapen by elongation along an a-axis accompanied by flattening on the pair of prism faces parallel to this axis, resulting in the generation of an edge between opposing r and z faces at the termination. This feature is typical of Japan twins and of crystals distorted by twisting around an a-axis. Crystals similarly elongated but with very small prism faces and not tabular are also found. Another common type of distortion is by extension along an edge between two adjoining r and z faces, either at one termination or parallelly at both terminations to give the crystal

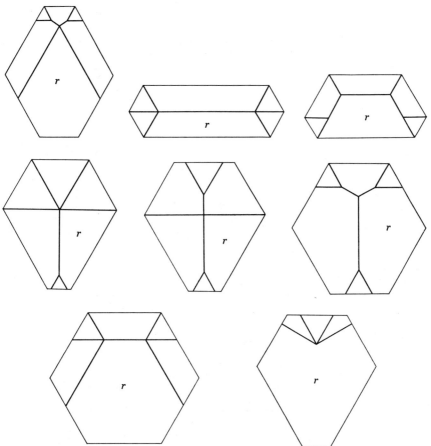

FIG. 34. Some observed types of pseudosymmetrical combinations of *r* and *z*.

a monoclinic appearance. In the absence of *z*, extension also occurs along the edge between two adjoining apical *r* faces, either at one termination or parallelly at both terminations to give the crystal the appearance of a nearly square prism. Some examples of misshapen crystals are shown in Figs. 35–42. The most convenient guide to the orientation of badly misshapen crystals is identification of the prism faces by means of their horizontal striations.

Reference

1. Cross and Shannon: *Proc. U.S. Nat. Mus.* **71**, Art. 18, 1927; Gallitelli: *Per. min. Rome*, **6**, 105 (1935).

Direction of Flow of Solutions. The individual quartz crystals comprising a crust lining the walls of a vein may tend to be misshapen in the same general way and, further, to be oriented spatially in this regard. This feature has been

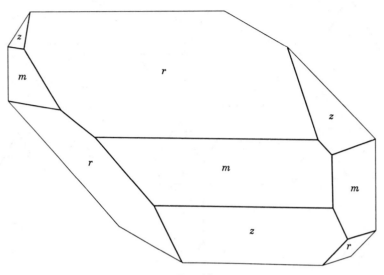

FIG. 35

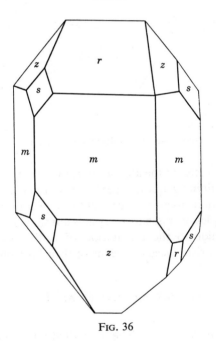

FIG. 36

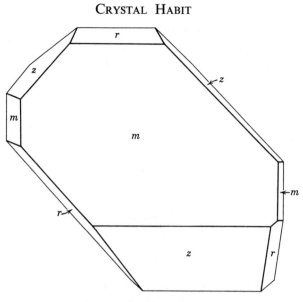

FIG. 37

considered to indicate deposition from a moving solution.[1] A greater amount of material is deposited on the part of the crystal facing the direction of flow of the solution. The interpretation of this in terms of the external morphology can be difficult. In general, when the faces of the terminal rhombohedra of quartz are asymmetrically developed, the larger surfaces face the direction of flow and the smaller faces are on the lee side. The malformation often is most marked when the orientation of the crystal is such as to bring a rhombohedral face perpendicular to the direction of flow. In general it is assumed that the

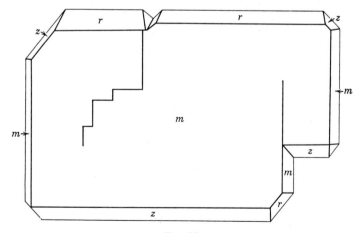

FIG. 38

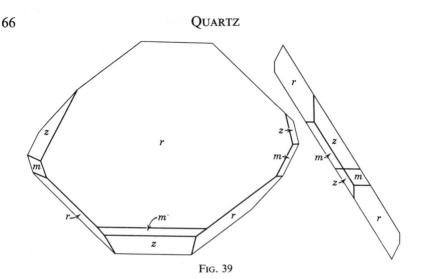

FIG. 39

orientation of the axial system of each crystal, taken during nucleation, is not influenced by the direction of flow. Variations in the thickness of concentric growth zones within the crystal, as revealed by "phantoms" or included films of foreign material, or by etching or irradiation blackening of cut sections, also are relevant but must be used with caution. In the case of crystals wholly bounded by a single closed form, such as a cube of NaCl, or by equivalent faces in a closed zone, such as the prism faces of quartz, the greater thickness of the growth zones is on the side facing the direction of flow. Growth zoning

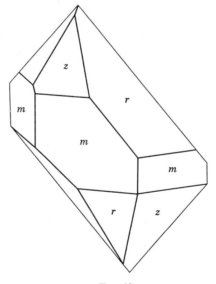

FIG. 40

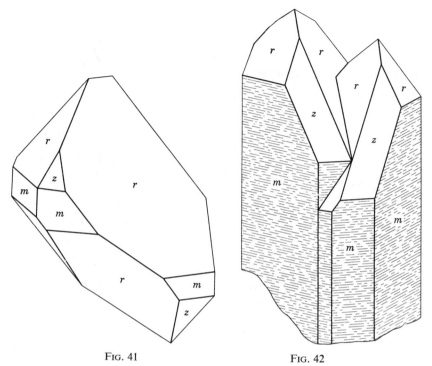

FIG. 41   FIG. 42

concentric to combinations of crystallographically different forms, with inherently different growth rates, is more complicated.

It is often observed that the misshapen crystals have a pseudoplane of symmetry, which is vertically oriented in the vein, or that the larger growth surfaces face directly downwards. This has been interpreted as caused by concentration currents governed by gravity in a body of solution that itself is not moving.[5] The attitude of the axial system of quartz crystals in veins in metamorphic or other rocks in which there is a preferred orientation of quartz grains is influenced by the seeding action of such grains as they are exposed on the walls of the opening.[2] The larger crystals grow from seeds oriented with their c-axes directed into the opening, since these offer the faster growing surfaces to the solution. When a large number of closely spaced nuclei grow simultaneously, geometrical factors result in the suppression of all but a few of the resulting crystals. The crystals with their c-axes more nearly perpendicular to the walls of the fissure survive and grow to the largest size.[4]

Surface incrustations and inclusions of other minerals also may show a selective distribution upon or in quartz crystals, which indicates the direction of gravitational settling or the direction of flow of the depositing solution.[3] Twinning is said to be more abundant on the lee side of the crystal.

68 QUARTZ

References

1. Newhouse: *Econ. Geol.*, **48**, 541 (1953); Stoiber: *Econ. Geol.*, **48**, 541 (1953); Engel: *Econ. Geol.*, **43**, 655 (1948); Laemmlein: *C. R. ac. sci. USSR*, **33**, 415 (1941).
2. Laemmlein: *C. R. ac. sci. USSR*, **22**, 42 (1939).
3. Grigoriev: *C. R. ac. sci. USSR*, **44**, 198 (1944).
4. Laemmlein: *Doklady ac. Nauk. SSSR*, **48**, 177 (1945).
5. Laemmlein: *Doklady ac. Nauk. SSSR*, **33**, 415, 1941; Grigoriev: *Mém. soc. russe min.*, **76**, 275 (1947).

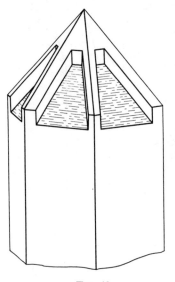

FIG. 43

Skeletal Crystals. Quartz crystals occasionally have depressed faces with the rhombohedral (Fig. 43) or, less often, the prism edges raised as distinct ridges. The interior parts of the faces may then be smooth, but more commonly are rough or are composed of small, more or less merged-together individual crystals forming a coplanar surface. Such crystals also may be cavernous internally, with large irregular openings connecting with the surface or completely grown over and then often partly filled with liquid (see *Inclusions in Quartz*). Highly skeletalized, branching, or dendritic forms of single-crystal growth, such as in NH_4Cl, are not known in quartz.

Quartz phenocrysts in acidic volcanic rocks also may be skeletonized, often with symmetrically arranged inclusions of the ground mass. Crystals of this nature are frequently found in rhyolites, as at Agate Point, Ontario,[1] where quartz phenocrysts contain inclusions of glass as films or plates arranged parallel to the rhombohedral faces. In other occurrences, the phenocrysts may have a deeply embayed outline, as seen in thin section, resembling skeletal growth but produced by partial resorption by the liquid with which they were in contact.[2] These crystals in general have a dipyramidal habit, and probably represent original high-quartz (which see).

References

1. Bain: *Am. Min.*, **10**, 435 (1925); see also Foster: *Am. Min.*, **45**, 892 (1960), and Laemmlein: *Min. Mitt.*, **44**, 470 (1933).
2. McMahon: *Min. Mag.*, **8**, 10 (1888).

Zonal and Sectoral Structure. Quartz crystals, including colorless and transparent types, very commonly are internally zoned. The zones range in thickness from microscopic dimensions up to a millimeter or more. They are arranged concentrically, parallel to the external faces of the crystal, and represent stages of crystal growth. The zones are not apparent to the eye in colorless crystals, but may be revealed by irradiation with x-rays or by etching in solvents such as H_2O at high temperatures in a closed vessel or HF or $(NH_4)HF_2$ (see further under *Etching Phenomena*). The distinctness of the differential etching thus produced may be markedly influenced by the concentration of the solvent, the temperature, and the duration of etching. The best results are obtained on sawn sections that have been lapped flat with coarse abrasive. The zonal etching is very markedly accentuated by etching in an electric field.[1] X-ray irradiation usually differentially pigments the zones in smoky colors (see *Irradiation Effects*). The geometry of the zonal pattern thus produced varies with the crystallographic orientation of the section. The zones as traced by etching or irradiation effects may vary in thickness under particular crystal forms, especially with regard to the terminal r and z rhombohedra, or may terminate or alter in direction or thickness when traced across twin boundaries. The zones also may be confined to particular face-loci, such as those of $r\{10\bar{1}1\}$. Successive sections cut through a quartz crystal when examined in this way may reveal progressive changes in crystal habit.

The zoning doubtless is caused by compositional variation. Direct analyses of individual zones, or measurements of associated changes in the density, indices of refraction, or unit cell dimensions are lacking. In one study,[2] the high-low inversion temperature of quartz taken from successive zones has been shown to vary. The inversion temperature was found to increase in zones taken successively outward from the center of the crystal, indicating a lower content of Al in solid solution attending a decrease in the temperature of crystallization.

Sectoral Structure. A related type of internal heterogeneity in quartz crystals is found in the tendency for Al, Li, and other ions present in solid solution to be selectively distributed in different face-loci. The face-loci appear as sectors in sawn slices or sections. This type of distribution has been established by direct analysis of material removed from different face-loci. It can be demonstrated conveniently by irradiation of the crystal or section. Those face-loci relatively high in their content of Al and other ions in solid solution, provided that these ions act as color centers, develop a relatively dark smoky color by this treatment (see *Irradiation Effects*). A number of studies of this kind have been made on quartz crystals synthesized hydrothermally from solution containing Al and other ions.

In one study,[3] the depth of irradiation color and the content of Al, etc., were found to increase in the order $(0001) < (10\bar{1}0) < (11\bar{2}1) < (51\bar{6}1)$. The overall range in composition of these loci was (in parts per million Si atoms): Al 110–175, Li 37–147, Na 26–212, Mg 9–23, Ge 248. The concentration of Li and Na varied parallelly with Al, but that of Ge did not. The sequence $(0001) < (2\bar{1}\bar{1}0) < (11\bar{2}1) < (51\bar{6}1)$ has been observed,[4] as has[5] the sequence

$(0001) < (01\bar{1}1)$. The non-equivalence of the parallel face-loci of the trigonal prisms, with $(2\bar{1}\bar{1}0)$ or $+X < (\bar{2}110)$ or $-X$, is shown in the following data:[6]

	Atoms per Million Atoms Si		
	Al	Na	Li
(0001)	142	138	4.5
$(2\bar{1}\bar{1}0)$	120	102	2.6
$(\bar{2}110)$	244	157	12

Synthetic crystals[7] colored green or brown by Fe show a deeper coloration and a higher content of Fe on (0001) than on $(10\bar{1}1)$ or $(01\bar{1}1)$. Natural crystals of quartz often show a selective smoky or amethystine color in different face-loci, but analytical data are lacking. The differential irradiation response sometimes observed across the boundaries of Dauphiné twins in natural quartz also is an expression of morphological control, generally between $(10\bar{1}1)$ and $(01\bar{1}1)$.

The selective distribution of material in solid solution in different face-loci of crystals is more familiar and more obvious in cases where it is accompanied by variations in the color of the crystals, as in hour-glass augite and dye-crystal systems. It also has been described[8] in opaque crystals by etching and analytical techniques and in crystals containing U or Th by radiometric methods. A selective fluorescence of face-loci has been observed in colorless crystals.

The very general occurrence of compositional variation between growth zones and face-loci in quartz makes it desirable that crystals to be employed in precision physical devices or as standards be etched or irradiated to ensure that the sample is homogeneous.

References

1. Choong Shin-Piaw: *Am. Min.*, **37**, 791 (1952) and *Nature*, **154**, 464, 516 (1944).
2. Keith and Tuttle: *Am. J. Sci.*, Bowen Vol., 228 (1952).
3. Cohen and Hodge: *J. Phys. Chem. Solids*, **7**, 361 (1958).
4. Cohen: *J. Phys. Chem. Solids*, **13**, 321 (1960).
5. Brown and Thomas: *J. Phys. Chem. Solids*, **13**, 337 (1960).
6. Augustine and Hale: *J. Chem. Phys.*, **29**, 685 (1958).
7. Tsinober and Chentosova: *Kristallografiya*, **4**, 593 (1960).
8. Frondel, Newhouse, and Jarrell: *Am. Min.*, **27**, 726 (1942).

Twisted Crystals. Quartz crystals are occasionally found that are distorted by twisting around an a-axis (Figs. 44 and 45). The crystals in general are composite. They are composed of more or less merged-together separate individuals, ranging in size from a fraction of a millimeter up to a centimeter or more, that are successively offset along the twist axis. When the composite structure is on a very fine scale the crystal faces may appear to be smooth and uniformly curved. Often, more or less distinct crystals with individual rhombohedral terminations are partly merged laterally in offset position. In such specimens the individual crystals or lamellae of the aggregate may be themselves twisted and composite on a finer scale. X-ray study indicates that the crystal structure of the individual elements is not distorted and that the twisting is a property of the aggregate. In polarized light, thin sections show a lamellar or composite structure with slightly divergent extinction in adjoining parts. The irregular lamellae trend vertically in the crystals and are arranged essentially perpendicular to the twisted a-axis.

Twisted crystals of quartz usually are tabular on a pair of prism faces and are elongated along the horizontal axis of twist, giving rise at top and bottom to a serrated or evenly curved edge between opposing faces of the positive and negative rhombohedra; many show a conspicuous development of faces of the trigonal trapezohedron $x\{51\bar{6}1\}$. The twisting seems to be of about equal occurrence in right-handed and left-handed quartz, but the total number of observations is very small. The crystals are attached to the matrix by one end of the twisted a-axis. The free end of the twist axis usually is terminated by two striated but essentially plane prism faces with a reflexed edge between them. The a-axis of quartz is polar, and the distribution of the modifying faces of $x\{51\bar{6}1\}$ on the free end apparently always is such as to identify this end of the a-axis as that which becomes electrically positive when the crystal is compressed along this direction (the X-axis in piezoelectric notation). This rule for the polarity of attachment to the matrix for crystals elongated along the a-axis is true for both right- and left -handed quartz. Some twisted crystals of quartz are elongated along the c-axis, as noted beyond, and these crystals are attached to the matrix by one end of the non-polar c-axis.

The rhombohedral faces parallel to the twist axis may be unequal in size, with the positive rhombohedron predominating, or the two prism faces parallel to the twist axis may be lacking, with the terminal rhombohedral faces then meeting in front in a curved edge. The flattened crystals also may be somewhat elongated along the c-axis rather than along the a-axis of twist. Twisting is relatively rare in crystals of normal habit with elongation along the c-axis and equant cross-section. The twist in this case has been described as around the c-axis; this is identical in morphological expression with a twist of opposite sense around an a-axis. The sense of the twist as taken around the c-axis always is left-handed (counterclockwise) in right quartz and right-handed (clockwise) in left quartz. The terminal edge between $r\{10\bar{1}1\}$ and $m\{10\bar{1}0\}$, parallel to the twisted a-axis, is taken as the direction whose (imaginary) movement defines the twist sense. In a right-handed crystal the twist as described about an a-axis is clockwise, the terminal edge between $(10\bar{1}0)$ and $(01\bar{1}0)$ being taken as the direction whose movement defines the twist sense. The line of sight is taken as from the terminated free end toward the attached end of the a-axis, away from the observer. Some erroneous descriptions of the twist sense have been given in the literature.

The amount of twist varies in different specimens and sometimes in different parts of a single crystal. The usual range is 1–5° per cm. as measured along the twisted a-axis of elongation. An average of 16 Swiss crystals gave

FIG. 44. Twisted crystal of smoky quartz elongated along an *a*-axis. The crystal is oriented with an *a*-axis vertical and the *c*-axis horizontal. The triangular face at top is $x(51\bar{6}1)$; the vertical faces to left and right are *r* and *z* and the vertical face in front is $(10\bar{1}0)$. Right-handed. Crystal about 2 in. long. Switzerland. Amer. Mus. Nat. Hist. coll.

FIG. 45. Twisted crystal of smoky quartz elongated along the c-axis. Crystal approximately 8¼ in. (21 cm) in length. Pikes Peak, Colorado. Amer. Mus. Nat. Hist. coll.

$3.7°$ per cm.[1]; another set of measurements on Swiss crystals gave an average of $1.2°$ per cm.[6] A crystal from Pikes Peak, Colorado, of normal habit elongated along the c-axis showed a total twist in this direction of $45°$ over a total length of 21 cm.[2] An effort has been made[3] to find a uniform description of the twisting in terms of the thickness of the individual members of the aggregate, but the amount of twist seems to be too variable to permit such a description. The amount of twist around the a-axis also has been said to be roughly proportional to the thickness of the crystal.[5]

Twisting typically occurs in dark smoky quartz and is rare in pale smoky, yellowish, or colorless quartz. Both Brazil and Dauphiné twinning appear to be relatively rare in twisted crystals. Twisted quartz crystals are restricted in occurrence. They are known chiefly from the Alpine veins of Switzerland,[4] where they are called Gwendeln by the professional crystal collectors or *Strahlers* of the region. Here they occur as free crystals in cavities, associated with chlorite, adularia, apatite, calcite, sphene, etc. They often occur together with undistorted smoky quartz crystals of ordinary habit; the twisted crystals generally are attached to the matrix by one end of the twisted a-axis, while the normal crystals are attached by an end of the c-axis. Both right- and left-twisted crystals may occur in the same cavity. The crystals usually range in size from an inch or so along the a-axis up to about 10 by 8 in., as measured along the a-axis and c-axis, and 1.5 in. thick. The larger crystals usually are relatively composite. Twisted crystals also are known from the northern Urals,[5] Carrara, Italy,[7] and a few minor occurrences elsewhere. They are quite unknown at many important localities for smoky quartz, including Brazil and the pegmatites of the Appalachian and New England regions in the United States.

The cause of the twisting is not known. It clearly takes place during the growth of the crystal and is not a secondary effect. Precise x-ray, optical, and analytical data that might reveal differences from other quartz are lacking. The twisting has been ascribed to twinning on vicinal faces.[6] It may be related to spiral dislocations in the crystal structure.

Unusual as these twisted crystals may seem, the identical twisted fibers of quartz found in some chalcedony are even more remarkable. These fibers, often only a few microns in cross-section and up to a millimeter or more in length, are elongated along $[11\bar{2}0]$, less commonly along other directions, and are regularly twisted around their length with a pitch of microscopic dimensions (see *Iris Agate*, under *Chalcedony*).

References

1. Frondel: priv. comm., 1959.
2. Frondel: *Am. Mus. Nat. Hist. Novit.* **829**, 1936.
3. Billows: *Rivista min.*, **37**, 3 (1909).
4. Friedlaender: *Beitr. Geol. Schweiz.*, Geotechn. Ser., Lfg. **29**, 1951; Niggli, Koenigs-berger, and Parker: *Die Mineralien der Schweizeralpen*, Basel, 1940; Nowacki:

Schweiz. min. pet. Mitt., **16**, 408 (1936); Rosický: *Publ. fac. sci. univ. Masaryk* **187**, 1933; Spencer: *Min. Mag.*, **19**, 269 (1921).
5. Laemmlein: *C. R. ac. sc. USSR*, 279, 1936; *Bull. ac. sci. USSR, Math.-Nat. Cl.*, 962, 1937.
6. Tschermak: *Denkschr. Ak. Wiss. Wien, Math. Nat. Kl.*, **61**, 35 (1894).
7. Aloisi: *Atti soc. toscana sci. nat., Mem.*, **25**, 3 (1909).

TWINNING

A twin is a geometrical position of intergrowth of two or more crystals of the same species (here quartz) with which is associated a frequency of occurrence greater than that of chance. The description of twinning in quartz in general requires determination of the angular relation between the axial systems of the twinned parts, the relation between the polarity of the a-axes in these systems, and the hand of the twinned crystals. The twins of quartz can be placed into two categories as a matter of convenience: (a) the parallel-axis twins, usually penetration twins, comprising the Dauphiné, Brazil, and Combined Laws (Fig. 46); and (b) the inclined-axis twins, usually contact twins, including the group of twin laws collectively called the Japan Law, together with a number of rare or doubtful laws.

Twinning is the rule rather than the exception in quartz. Des Cloizeaux said that an untwinned quartz crystal was one of the greatest rarities of the mineral kingdom. Almost every crystal contains penetration twinning on one of the parallel-axis laws, and usually several of these laws are present simultaneously. Dauphiné twinning is by far the most common. Brazil twinning infrequently occurs as the only type of twinning in an individual crystal, except in amethyst, and usually is present together with Dauphiné twinning. Twinning on the Combined Law is rare. The relative frequency of these twin laws varies at different localities, as does the ratio of crystals containing large amounts of twinning to untwinned or slightly twinned crystals.[1,2] Twins on the inclined-axis laws are very rare. Of these, twins on the Japan Law are by far the most abundant.

The studies on the relative frequency of the twin laws of quartz and of right- and left-handed crystals (Table 4) are not wholly satisfactory. In most of this work the identification of the twinning and of hand was based largely or entirely on surface features, such as the morphology or natural or artificial etch-figures, with supplementary optical tests in some instances, and the presence of twinning in the interior of the crystals may have gone unnoticed. For this reason the reported proportions of untwinned to twinned crystals probably are considerably in error. The lamellar and satellitic types of Brazil twinning in particular are difficult to recognize by surface features only. In one surface study,[3] only 41 out of 4483 crystals from Brazil were identified as Brazil twins. On the other hand, it is the experience of people who have

FIG. 46. Diagrams illustrating the relation between the axial systems and hand of crystals twinned on the three parallel-axes laws. Arrows indicate hand.

graded very large numbers of Brazilian crystals and sawn sections by optical and etching tests in connection with the manufacture of quartz oscillator-plates that Brazil twinning is extremely common.[4] In other studies Combined twinning either was not sought or was not adequately characterized.

Very little is known of the relation between the incidence of twinning and the crystal habit, the environment of formation, and variation in the chemical composition of the quartz. Brazil twinning is ubiquitous in amethyst, but apparently is relatively rare in smoky quartz, at least from Swiss localities,[2,5] and seems to be absent in rose quartz. It apparently is more common in quartz formed at relatively low temperature.[5] Most Brazil twinning and probably at least some Dauphiné twinning are not imposed on the crystal during nucleation but are developed during the later stages of growth; Dauphiné twinning may also be introduced after the growth of the crystal by mechanical or thermal strain. The tapering-prismatic habits of quartz from Brazilian localities seemingly tend to contain a relatively small proportion of twinning.[1] Composition is an important factor in artificially producing or removing Dauphiné twinning in quartz. Twinning has been said[43] to be less common in the quartz grains of igneous rocks than in vein quartz.

A. QUARTZ TWINS WITH PARALLEL AXES

1. Dauphiné Law. (Syn: orientational twinning; electrical twinning; Swiss Law). The twinned parts are related geometrically as by a rotation of 180° around the *c*-axis and are of the same hand. The crystal axes are parallel, but the electrical polarity of the piezoelectric *a*-axes is reversed in the twinned parts—hence sometimes the name "electrical twinning." Right-right and left-left Dauphiné twins appear to be equally frequent. Dauphiné twins almost invariably are of the interpenetration type. The composition surfaces generally are irregular and trend vertically in the crystal (Fig. 47). Commonly the twinned parts are complexly intergrown, sometimes in sponge-like fashion (Fig. 48) or with a crude three-fold sectoral distribution as seen in sawn and etched basal sections[6] (Fig. 49). The distribution of the sectors seems to be

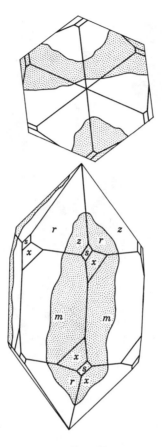

FIG. 47

FIG. 48. Etched basal sections of quartz, showing mosaic type of Dauphiné twinning (right) and Brazil twinning (left). Leydolt, 1855.

Fɪɢ. 49. Three-fold distribution of natural Dauphiné twinning in basal sections of Brazilian quartz, as revealed by etching. About one-sixth natural size.

related to the terminal faces or edges in prismatic crystals. The twin boundaries as seen on the terminal r and z faces usually trend irregularly, although locally they may be straight; they may run along or near an r z edge and then frequently pass downward to the prism faces very near to an m r z corner and carry on near and roughly parallel to a prism edge. If a face of a trigonal trapezohedron is present, such as x, a re-entrant angle bounded by x and a (twinned) face of m may form. Naturally etched crystals may show scalloped grooves along the twin boundary.

Rarely, two individuals of about equal size may be united with an irregular, nearly vertical composition surface. Euhedral crystals of this nature may appear as two individuals joined as contact twins with large re-entrant angles at the terminations and laterally: the separate r and z faces in the two parts are parallel. Morphological specialization of such twins is possible according to the nature of the re-entrant angle at the termination: this may be constituted by facing pairs of r faces, z faces, or r z edges. Euhedral, wholly merged penetration twins of two individuals each terminated by r only have been observed[41] in which each face of r of one individual contains the edge of the opposing rhombohedron of the other individual as a projection or gable (see Figs. 50 and 51).

Dauphiné twinning cannot be recognized by optical means but may be detected by etching either the crystal or sawn sections thereof;[7] also it may be evidenced by the insertion of areas of z (usually dull) into r and vice versa, by an interruption of the horizontal striations on $\{10\overline{1}0\}$ usually as an irregular vertical strip toward an edge of the prism face, and by the morphological distribution of the faces of s, x, or other general forms. X-ray Laue photographs can be applied to the study of this and of other twin laws in quartz.[8] The electrical charges induced by pressure or by change of temperature at the ends of the a-axes also are opposite in sign in the twinned parts, but these tests cannot be made conveniently other than in properly oriented sawn sections. Surface boundaries due to lineage structure or to subparallelly intergrown crystals can easily be mistaken for twin boundaries. In an idealized Dauphiné twin with vertical composition surfaces the trigonal trapezohedra are repeated on each of the six prism faces above and below at opposite corners, and the trigonal pyramid s similarly is repeated on each of the twelve $m\ r\ z$ corners (Figs. 52 and 53). The full complement of twinned faces of general forms above and below has been rarely if ever seen; usually only a few faces of unequal size are present. The lower faces are generally lacking because the prismatic crystals tend to be attached to the matrix by one end of the c-axis. The regular repetition of the faces follows from the threefold sectoral distribution of the twinned parts earlier described. Dauphiné twins also are illustrated in Figs. 18, 54, and 55.

Dauphiné twinning is of greater consequence than Brazil twinning on the morphology and on the practical utilization of quartz, because the twinned parts generally comprise more or less equal volumes of the crystal. Brazil twins ordinarily involve relatively small inserts in the peripheral parts of a crystal of opposite hand. Precise x-ray goniometric measurements usually show a divergence of a few minutes of arc or more across the boundaries of natural Dauphiné twins as they appear on sawn and lapped surfaces.

Dauphiné twinning in quartz was recognized in 1816 by Weiss,[9] who found penetration twins with re-entrant angles such as shown in Fig. 51. Earlier, Dauphiné twins of the usual type without re-entrants were unknowingly described by Haüy.[10] A Dauphiné twin also appears in a crystal figured by Cappeler (1723).[44] Haüy's fig. 15 on Plate 57 of *quartz plagièdre* in his *Traité de minéralogie* is a left-handed twin with faces of $x\{51\overline{6}1\}$ repeated equally on all corners above and below (doubtless idealized). A Dauphiné twin of this habit also appears to have been unknowingly employed by Herschel,[11] who reprints Haüy's figure, in his study of optical rotation in quartz and its relation to the morphology. The twinning has no effect on the optical rotation because the hand and the orientation of the optical indicatrix are identical in the twinned parts. The same figure is reproduced without interpretation in

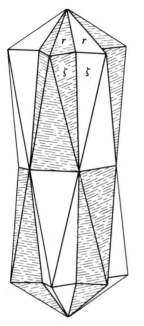

FIG. 50

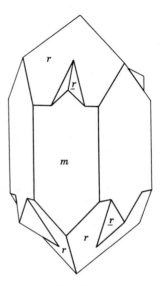

FIG. 51

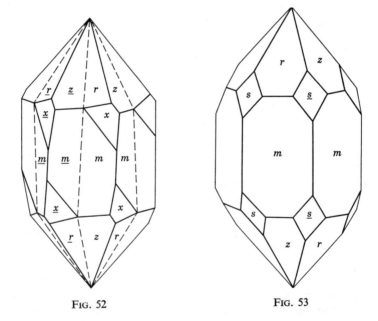

FIG. 52

FIG. 53

FIGS. 50–53. Dauphiné twins. Figures 52 and 53 show the idealized repetition of the $s\{11\bar{2}1\}$ and the $x\{51\bar{6}1\}$ faces.

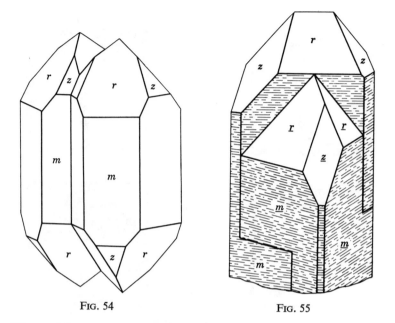

FIG. 54   FIG. 55

all editions of Dana's *System of Mineralogy* up to the sixth (1892), when it was replaced by a similar figure of opposite hand showing the composition surfaces. The recognition of Dauphiné twinning on purely morphological grounds in the absence of re-entrant angles is often difficult, and the extreme commonness of these twins was not realized until Leydolt in 1855 showed how they could be recognized by etching the crystals in a solvent.

Secondary Dauphiné Twinning. Dauphiné twinning may be produced artificially in various ways: by inversion[12] from high-quartz when cooled down through the inversion point at *ca.* 573°; by thermal shock[13] produced by quenching from temperatures as low as 200°; by the application of an intense electrical field;[14] by local pressure at room temperature or higher;[15] by putting a thermal gradient in a cut plate as it cools through the inversion point;[16] and by pure bending or torsion[16] applied to cut plates or bars at temperatures from 573° down to at least 300°. Dauphiné twinning disappears when quartz is heated over the 573° inversion temperature because the twin operation around the three-fold axis of low-quartz becomes an operation to identity in the six-fold axis of high-quartz. On cooling, Dauphiné twinning may or may not reappear. The amount and distribution of inversion twinning is influenced by the rate of cooling, temperature gradients in the sample, the size and shape of the sample, the presence or absence of Dauphiné twinning in the initial unheated sample, and the composition of the quartz. Irregularities along the edges of cut plates and inclusions also tend to localize the twinning. The inversion twinning is primarily a consequence of strain produced in

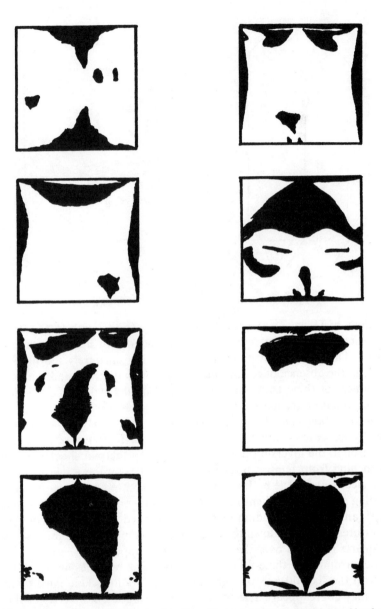

FIG. 56. Symmetrical patterns of secondary Dauphiné twinning (black) produced by thermal strain in thin plates of quartz, as revealed by etching. Oriented with X horizontal, Z' ($-49°$) vertical. About half natural size.

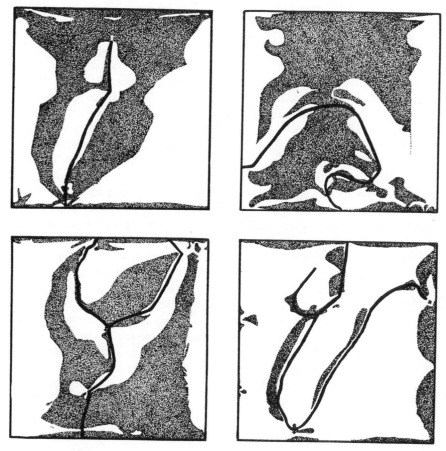

FIG. 57. Relation of secondary Dauphiné twinning (stippled) to cracks (dark lines), as produced by thermal strain in thin quartz plates.

the quartz and is related to thermal cracking (Figs. 56 and 57). With relatively thin plates, decrease in the rate of cooling tends to increase the amount of twinning and to decrease the amount of cracking. The incidence of twinning and of cracking increases if whole crystals or angular fragments are employed, becoming more marked with increase in size. The Dauphiné twinning produced in dimensioned plates generally has a pattern of distribution, revealed by etching, that reflects gradients of elastic strain produced during the heating and cooling cycle. Thermal cracks produced in the plates may be bordered by Dauphiné twinning or may be distributed relative to the overall pattern of twinning.[13] Twin patterns also can be produced by mechanically bending or twisting the plates when they are at temperatures below 573° or as they cool through the inversion point.

Dauphiné twinning also can be produced by rapidly cooling thin plates that have been heated to temperatures of about 200° or higher but below the inversion point. The incidence of twinning and of cracking increases with increasing quenching range. The twinning develops in the uncracked areas, sometimes in definite patterns. Very closely cracked plates do not develop twinning. Thin films of Dauphiné twinning may form on the sawn surface of quartz crystals if the sawing operation is carried on in such a way that much frictional heat is developed.[17] The twinning produced in this way may be confined to particular growth zones in the crystal, and the tendency to twin varies markedly in different crystals; it is more readily produced in quartz that apparently is relatively pure.

Small geometrically shaped areas of Dauphiné twinning can be produced at room temperature by pressing steel spheres or rounded metal points against plane surfaces of quartz.[15] The pressures needed are low, of the order of 10 kg. per mm.[2], and decrease with increasing temperature. The shape of the twinned areas varies with the orientation of the surface (Figs. 58 and 59). The surface must be lapped down slightly and then etched to reveal the twinning clearly. The boundaries of the twinned areas lie along {10$\bar{1}$0} and {0001} planes. The extent of the twinning increases with the magnitude and the duration of the pressure, smaller loads requiring longer time to produce an equal effect. Each crystallographic surface has a characteristic minimal load and duration necessary to produce twinning. The planes {10$\bar{1}$0} and {11$\bar{2}$0} require less pressure and shorter duration than does {0001}. The shape of the areas of secondary Dauphiné twinning produced

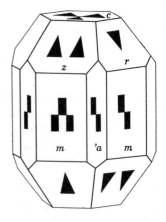

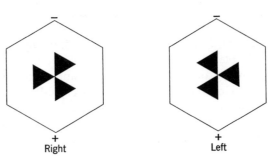

FIG. 58. Form of the secondary Dauphiné twinning produced on different plates of quartz by long-applied local pressure.

FIG. 59. Relation of secondary Dauphiné twinning produced by local pressure on (0001) to the hand of the quartz.

within the crystal by local pressure tends to be geometric. An idealized representation of the shape of the twinned areas as produced by a force acting radially from the center of a crystal is shown in Fig. 60. The twinned areas consist of twelve trigonal prisms arranged alternately above and below the point of pressure. The arrangement recalls the three-fold distribution of natural Dauphiné twinning often seen in etched basal sections (Fig. 49). The experimental observations suggest that secondary Dauphiné twinning should be common in stressed quartzose rocks. An untwinned single-crystal plate can be con-verted to an untwinned single-crystal plate that nevertheless is a twin in the sense that the crystal structure has been converted to a new orientation at 180° to the first.[16,18,19] Thus a plate cut parallel to $\{01\bar{1}1\}$ can be converted by heating over the inversion temperature, or by mechanical stress at lower temperatures, to a plate parallel to $\{10\bar{1}1\}$. In this case the "twinning" can be identified by crystallographic tests. In some other orientations such twinning could be recognized only by reference to an arbi-trary datum.

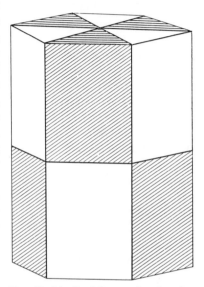

FIG. 60. Idealized figure, showing the form and distribution of the second-ary Dauphiné twinning produced by a force acting uniformly in all direc-tions outward from the center of the crystal.

Much research has been done on the practical problem of eliminating Dau-phiné (electrical) twinning from indus-trial quartz crystals and from fabricated quartz plates and bars. Considerable success has been attained, but no method has been found that is 100 per cent successful. The principal factors involved are the size and shape of the quartz plate, the presence of thermal or stress gradients in the plate and their crystallographic orientation to the quartz, the temperature, and both the composition and the growth history of the quartz crystal. Purely thermal methods are least successful; in general a stress gradient such as is produced by bending or torsion must be applied. It is not possible to obtain strictly reproducible results in identical plates that have been cut from different quartz crystals, presumably because of a variation in the quartz itself. The untwinned or twinned areas produced tend to have a rather definite pattern of distribution in the plate that is deter-mined jointly by the crystallographic orientation of the plate and the stress distribution therein. Longitudinal compression or extension apparently has no control over the movement of twin boundaries. The relation between the crystallographic orientation of plates and the effectiveness of an applied torque in detwinning quartz has been partly determined.[16] The detwinning vector is zero for discs or square plates cut perpendicular to $\{0001\}$ (Z-cut), $\{11\bar{2}0\}$ (Y-cut), or $\{10\bar{1}0\}$ (X-cut), and is small for X-cut rectangular plates with the elongation tilted in the $\{11\bar{2}0\}$ plane at angles approaching $0°$ or $90°$ to the c-axis. In the so-called rotated Y-cuts, which include most of the orientations of interest for frequency-control applications, the detwinning vector is relatively large for angles of tilt to c of about 20–70°.

Secondary Dauphiné twinning in quartz appears to be an example of the effect called piezocrescence, i.e., the change of atomic position as induced by the application of stress.[16] The atomic movements in the Dauphiné twin operation are small and do not involve the breaking of Si-O bonds. There is only a slight difference in bond angles across the composition surface of the twin. The effect of an applied stress of sufficient magnitude, generated by an outside force or by an internal thermal gradient, is to cause movement of the transitional zone at the twin boundary in such direction as to bring the crystal structure into equilibrium with the new stress system. Dauphiné inversion twinning also has been viewed as a chance effect produced at the inversion point by nucleation and growth at separate centers differing by a 180° rotation. The element of chance arises in the nature of the structural relation of high- to low-quartz. This can be visualized[18] as involving a change from a six-fold linear arrangement of the Si atoms in $\{0001\}$ of high-quartz to a three-fold zigzag arrangement in low-quartz. The change in bond angle during inversion from high-to low-quartz can be effected in two equivalent ways, differing by a 180° rotation. If both ways happen simultaneously in different parts of the crystal, a Dauphiné twin results. The associated strain phenomena, the more or less regular stress pattern in the twinning produced, and the observed influence of compositional variation in the quartz itself suggest that the actual controlling mechanism in inversion twinning is piezo-crescence. In any case, the growth of the twin boundaries continues as the quartz cools below the inversion point.

The tendency of quartz to acquire secondary Dauphiné twinning, the depth of smoky coloration produced by x-ray irradiation, and the composition of the quartz in terms of minor elements are found to be related.[16,38] Different crystals vary markedly in their twinning response. Quartz that contains relatively large amounts of other elements in solid solution has a smaller tendency to twin, and also resists cracking by thermal shock, presumably because of an increase in the elastic limit produced by the minor constituents. Such quartz also becomes relatively deeply colored by x-ray irradiation. Furthermore, when Dauphiné twinning is initially present the original twin boundaries tend to be retaken when the quartz is heated over the inversion point and cooled.[20] On the other hand, pure quartz twins readily and is less affected by x-ray irradiation; also, memory effects are less marked. Quartz crystals that possess a zonal or segmental structure, with compositional variation therein, may show a complicated response. In experiments involving heat treatment under torsional stress a comb type of secondary twinning was observed to develop along growth zones that were deeply colored by x-ray irradiation.[16]

2. Brazil Law. (Syn.: Chiral twinning; optical twinning). The twinned parts are related as by reflection over $\{11\bar{2}0\}$ and are of opposite hand. The crystal axes are parallel, but the electrical polarity of the a-axes is reversed in the twinned parts (Fig. 46). Brazil twins commonly appear as thin plates or bodies of restricted size (satellites), often with a geometric outline, inserted in a large crystal of opposite hand (Fig. 61). They tend to be more common toward the peripheral parts of the host crystal. The plates have the appearance, when viewed along the c-axis in polarized light, of either right triangles (with the base of the triangle parallel to a prism face, the hypotenuse parallel to a prism face at 120° to the first, and the third side parallel to $\{11\bar{2}0\}$), or of equilateral triangles with all three sides parallel to traces of prism faces (Fig. 48). In three dimensions, the twinned inserts may appear as irregularly developed polyhedra bounded by one or two faces each of forms such as $\{10\bar{1}0\}$, $\{10\bar{1}1\}$, $\{01\bar{1}1\}$, $\{0001\}$, and $\{11\bar{2}1\}$; see Fig. 61. The twinned inserts frequently are very thin; the twinning also develops as extended laminae, or is polysynthetic (Figs. 62 and 63). The plane of flattening or the composition plane of the lamellar twins generally is $r\{10\bar{1}1\}$. In polysynthetic twins one or two of the three equivalent planes of this form may predominate.

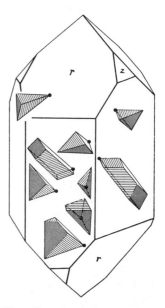

FIG. 61. Morphology of isolated Brazil twins within a crystal of quartz.

The composition plane also may be z, and then in association with lamellae with composition plane r. Crystals have been found with one or two even layers of Brazil twinning a centimeter or more thick transecting the crystal parallel to an r face. Apparently amethyst invariably is composed of very regular polysynthetic Brazil twins as thin alternating lamellae arranged parallel to the terminal r or r and z faces (see further under *Amethyst*). Brazil twinning almost invariably is of the penetration type, but occasionally two complete right and left individuals may be joined as contact twins with large re-entrant angles.[46] On the average, right and left quartz appears to be equally frequent in Brazil twins, although in individual crystals either the right or left component usually greatly predominates. Both Brazil and Dauphiné twinning are present simultaneously in most quartz crystals.

FIG. 62

FIG. 63

FIGS. 62–63. Polysynthetic Brazil twinning in etched −49° (BT) sections of colorless Brazilian quartz. Viewed in reflected light. About one-third natural size.

Brazil twinning can be recognized by examination in polarized light (hence the name "optical twinning") because the plane of polarization is rotated in an opposite sense in the twinned parts. Groth (1869)[45] drew attention to the utility of optical tests in examining Brazil twins. The twinning is not noticeable, however, in the quartz grains of rocks as examined in thin sections under the microscope. The amount of optical rotation in a basal section 0.03 mm. thick is 0.65° for Na. The twinning is best examined by immersing large crystals in a transparent container which is filled with liquid with an index of refraction near that of quartz and which has been placed between crossed polarizers. The crystal should be viewed at a slight angle to the optic axis. The Brazil twins generally are wedge-like and have such thickness as to show several orders of interference colors, while the host crystal is so thick as to provide only a uniform background of extremely high order: the Brazil twins, of opposite hand, then stand out as brightly colored triangular areas. The interference colors of the host crystal also may be seen as bands fringing the base and apex of the crystal and expressing the irregular contour of these surfaces. Very thick and uniform Brazil laminae are less readily recognized optically, although optical effects may be observed along the surfaces of juncture. Combined twins also affect polarized light, and could be confused with Brazil twinning on the basis of optical examination alone. The difference in size and particularly in shape of the twinned areas in these two types of twins probably is a practical if not rigorous guide to their distinction by optical means.

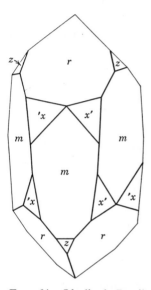

FIG. 64. Idealized Brazil twin, showing repetition of the $x\{51\bar{6}1\}$ faces.

Brazil twins also can be identified by their effect on the morphology of the crystal and by etch tests. However, the small size of the twinned inserts, especially when polysynthetic, generally precludes such tests. In an idealized Brazil twin the faces of a trigonal trapezohedron are repeated in pairs above and below on alternate prism faces. The full complement of twinned faces as shown in the classical drawing of Rose (1846)[21] (see Fig. 64) probably never has been observed. Usually only a few twinned faces of unequal size are present, and these rarely as a facing pair. A Brazil twin with a ditrigonal prism repeated in pairs on each consecutive prism edge has been noted.[47] Brazil twins usually appear on the prism faces of a crystal as small horizontal strips or polygonal areas that differ in surface appearance from the adjoining coplanar m face. They are more apparent on crystals that have been deeply etched. When the twinning is polysynthetic, horizontal striations may appear on the prism faces, and striations parallel to the $r\,z$ or $r\,r$ edges may appear on the terminal rhombohedron faces; a rippled appearance may be seen on cross-fractures.

Brazil twinning has not been obtained artificially and cannot be removed by heating over the 573° inversion point to high-quartz. Brazil twins were first described and explained by G. Rose in 1846. He did not employ optical tests. Earlier, both Herschel (1821) and Weiss (1816) noticed crystals that probably were twins on this law, and Brewster (1823)[22] described the intergrowths of right and left quartz in amethyst that we now know to be Brazil twins. The name alludes to the occurrence of this twin law in quartz from Brazil, but Brazil twinning is equally common over the world.

3. Combined Law. (Syn.: Leydolt twinning; Liebisch twinning; Compound twinning; Dauphiné-Brazil twinning). The twinned parts are related geometrically as by a combination of rotation of 180° around the c-axis and by reflection over $\{11\bar{2}0\}$, or simply as a reflection over $\{0001\}$. The crystal axes are parallel, but, unlike Dauphiné and Brazil twins, the electrical polarity of the a-axes is not reversed in the twinned parts (Fig. 46). This rare twin law has long been known[23] but was not clearly described in quartz until World War II, when a small number of examples were found,[24] mostly in etching studies connected with the use of quartz in the manufacture of oscillator-plates. Twinned euhedral crystals showing a repetition of general forms apparently have not been observed. The idealized habit of penetration twins on the Combined Law varies with assumptions made as to the position of the composition surfaces. If the composition surfaces are taken as cyclic on $\{10\bar{1}0\}$ planes, a twinned pair of left and right trigonal trapezohedral faces would be situated above and below across alternate prism edges; the termination would be six-sided and composed of three right r and three left r faces alternating. If the composition surfaces are taken as cyclic on $\{11\bar{2}0\}$ planes, the trigonal trapezohedral faces would be developed as on a single untwinned crystal, either right or left; the termination would be six-sided, with each face composed of coplanar right r and left z surfaces. Some hypothetical habits have been figured together with representations of the etch-figures on the faces.[25] In sawn and etched sections Combined twins appear as relatively large areas bounded irregularly or by planar composition surfaces parallel to a terminal r face. Dauphiné twinning usually is also present. The twinning is conveniently identified in sawn sections by examination of the light-figures produced when a point source of light is viewed through the etched surface;[26] the light-figures obtained on opposite sides of the twin boundary differ only in hand (Fig. 89).

Twins on the Combined Law, like those on the other parallel-axis laws, involve two adjoining crystals. A Brazil twin wholly situated in one part of a Dauphiné twin is related to the other part of that twin in the manner of the Combined Law, but the aggregate does not have genetic significance as an example of that law. Associations of this type are common, but the true dual twin is rare. Twins on the Combined Law have been produced artificially[27] by introducing Dauphiné twinning, e.g., by cooling through the 573° inversion point, into quartz already twinned on the Brazil Law. The Combined Law results where the secondary Dauphiné twinning cuts across the initial Brazil twin.

B. QUARTZ TWINS WITH INCLINED AXES

1. Japan Law. Four related types of twins are grouped under this name. All are contact twins of two individuals, with the c-axes inclined at 84°33′. A pair of prism faces in the two individuals are coplanar in the twin. The composition plane in all instances is (11$\bar{2}$2). In one of the four twin types this plane is a twin plane, while in the others it is a pseudotwin plane in that it describes only the angular relation between the axial systems of the twinned parts. The further description of these twins rests on the hand, which may be the same or different in the twinned individuals, and on the relation between the polarities of the a-axes in the twinned axial systems (Fig. 65). Alternate descriptions can be based on a combination of hand and the identity of the particular faces of the form {11$\bar{2}$2} that are parallel to the idealized composition plane; or on a combination of hand and the identity of the unit rhombohedron faces that separately terminate the coplanar prism faces of the twinned individuals. The four types of Japan twin further can be considered as comprising a basic operation, that of a 180° rotation around the normal to (11$\bar{2}$2), which appears either alone (Type I), or in combination with the Dauphiné Law twin operation (Type II), the Brazil Law twin operation (Type III), or the Combined Law twin operation (Type IV).

Three of the four types can be specialized into subtypes on morphological grounds. Japan twins generally consist of two individuals joined by ends of the c-axes to give a V-shaped appearance (often much misshapen, as noted beyond). The twins generally are attached to the matrix at the apex of the V. The V-shaped twins can be considered as opposing halves of an X-shaped twin of two individuals. The two halves are identical in terms of operation but are separately identifiable in all but Type I twins. If analogous right-right and left-left combinations or twin enantiomorphs are not counted separately, the total number of Japan twin types and subtypes that can be recognized is seven. If twin enantiomorphs are counted separately, the total is ten; all ten are illustrated in Fig. 65. A description of the four main types of Japan twins follows:

(I) *Japan*-I. *Subtypes* I(R) *and* I(L). The twinned individuals are of the same hand, right or left, with identical rhombohedral faces r and r, or z and z, terminating the coplanar prism faces. The twin axis is perpendicular to (11$\bar{2}$2). Two unlike planes of {11$\bar{2}$2} are parallel to the composition plane.

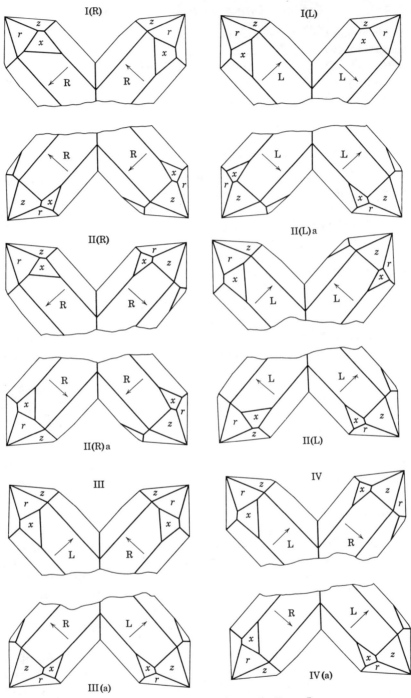

FIG. 65. Idealized types of twins on the Japan Law.

92

Examples apparently of this type are known from La Gardette, France, and also from Japan.

(II). *Japan*-II. *Subtypes* II(R), II(R)a, II(L), II(L)a. The twinned individuals are of the same hand, right or left, with unlike rhombohedral faces, r and z, terminating the coplanar prism faces. Two identical planes of $\{11\bar{2}2\}$ are parallel to the idealized composition surface. Also described as having the twin axis $[1\bar{2}13]$ (the edge between $(01\bar{1}1)$ and $(\bar{1}101)$). Twins of this type are known from Japan, Serra do Cristaes in Brazil, La Gardette, and other localities.

(III). *Japan*-III. *Subtypes* III *and* IIIa. The twinned individuals are of different hand, right and left, with identical rhombohedral faces, r or z, terminating the coplanar prism faces. The x and s modifying faces are symmetrically placed with regard to the composition plane. Twin plane $(11\bar{2}2)$. Two unlike planes of $\{11\bar{2}2\}$ are parallel to the composition face. Examples are known from Japan, La Gardette, and probably other localities.

(IV). *Japan*-IV. *Subtypes* IV *and* IVa. The twinned individuals are of different hand, right and left, with unlike rhombohedral faces r and z terminating the coplanar prism faces. Two identical faces of $\{11\bar{2}2\}$ are parallel to the idealized composition surface. Also described as having a twin plane (irrational) perpendicular to $(12\bar{1}3)$. Apparently known only from La Gardette.

The relative frequencies of the main types or subtypes of Japan twins are not known. The identification of a particular type or subtype of Japan twin often is difficult, and not all the Japan twins described in the literature can be identified as of a particular type for lack of critical data on the morphology and hand. In general the twins are best identified by study of the symmetry of artificial etch pits.[28,29] The separate individuals of Japan twins may themselves be further twinned on the Dauphiné, Brazil, or Combined Law, giving rise to compound twins often of great complexity.[29] In this connection, it is not certain whether all the reports of Japan twin types refer to a combination of only two individuals. In some instances it appears as if a third individual of small size occurs sandwiched between the two main parts and contributes to the multiplicity of the twin operations.

Japan twins very commonly are flattened parallel to the coplanar prism faces to give a tabular appearance. Such crystals often are misshapen by development of an extended edge between the opposing r and z faces that terminate the coplanar prism faces. Flattened crystals also occur misshapen by elongation parallel to the composition surface, through extension of a pair of rhombohedral edges approximately parallel thereto, and then usually with a relatively small re-entrant angle. Some examples of Japan twins are shown in Figs. 66–69. When misshapen Japan twins occur with untwinned quartz crystals, the latter have their normal equant habit. The distortion is associated with the twinning, as is observed in many other twinned species. Japan

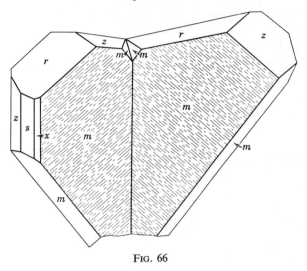

FIG. 66

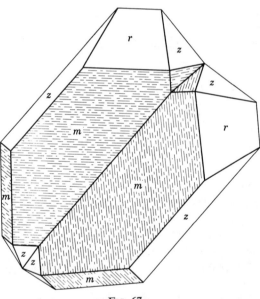

FIG. 67

twins of crystals with equant hexagonal cross-section are unusual, as are twins in which a prismatic individual contains a smaller crystal projecting in twin position from its side. Chance intergrowths of quartz crystals that simulate Japan twins are common. They usually involve individuals of equant cross-section that are joined together by prism surfaces.

The composition surface of Japan twins generally is plane, but may be

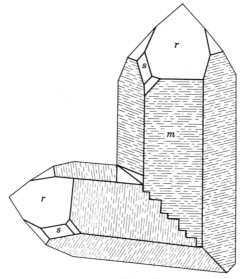

FIG. 68

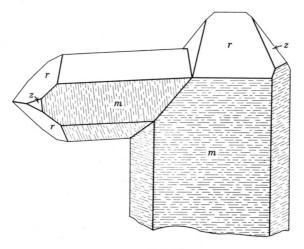

FIG. 69

FIGS. 66–69. Twins on the Japan Law.

irregular locally or have a zigzag development. It is best examined in sawn cross-sections or by lapping down the coplanar prism surface. The two individuals of Japan twins generally are of equal thickness (and of more or less equal size), but when one is thicker it may envelop the other individual over the initial composition surface.

The name Japan Law alludes to the remarkable development of these twins at certain Japanese localities. The name first came into general use in the literature after 1900. The general law was first recognized by Weiss[30] in 1829 in crystals from La Gardette, and had been variously termed the La Gardette Law, Weiss' Law, and the $\{11\bar{2}2\}$-Law. The most famous locality for these twins is at Otomezaka in Yamanashi Prefecture, Japan, where beautiful twins up to half a meter across occur in granite pegmatite and quartz pegmatite. Another notable Japanese locality is at Narushima, Nagasaki Prefecture, where the twins occur in the central cavities of quartz veins cutting sandstone, Japan twins also occur at the epidote locality on Prince of Wales Island, Alaska; in cavities in quartz porphyry at Saubach, Saxony;[29] at Llallagua, Bolivia, in sulfide veins; and at Hamburg, Cochise County, Arizona. Small twins up to 1.5 mm. in size occur in the sulfide veins at Felsöbanya, Hungary[32]. Japan twins (and other inclined-law twins) have been identified by optical study of quartz grains in sandstone.[33] A Japanese twin 26 cm. high and 6 kg. in weight was found at Gudshiwass in the Pamirs, USSR,[34] and another large twin has been described[35] from a cavity in itacolumite sandstone at Cristalina, Goiaz, Brazil. Japanese twins have also been reported from Crummendorf, Silesia,[39] and Désakna, Romania.[40] Numerous other localities are known,[31] but the Japan Law is the least common of the four well-established twin laws in quartz.

Additional Laws. A number of additional inclined-axis twin laws have been reported[36] in quartz. Many of these are of doubtful validity, although the Zinnwald and Zwickau Laws are relatively well established. In some instances there has been confusion between low-quartz and high-quartz. Some of the supposed laws rest on but a single observation; it should be emphasized that the recognition with generality of a definite geometrical position of intergrowth between crystals of the same species (twin) requires statistical proof that the position in question has a frequency greater than chance. The laws listed below fall into three main types: (1) contact twins with a pair of coplanar prism faces, analogous to the Japan Law, with the c-axes inclined at particular angles; (2) contact twins with a pair of coplanar rhombohedral faces, the twin being defined by statement of the angle of rotation about the normal to the joined rhombohedral face; and (3) contact twins with a prism face and rhombohedral face coplanar, the twin being defined by zonal relations or the angle of rotation about the normal to m (or r); the zero position is taken with zones mrz and $\underline{mrz}$ parallel. Most of the described examples of these twins consist of a small, well-individualized crystal resting upon a face of a much larger crystal. All the laws can be specialized with regard to hand and the relation between the polarity of the a-axes in the axial systems of the twinned parts. The face symbols used for the description of the zonal relation in these twins follow the convention of Zyndel[36] and other writers on inclined twins: $b_1(01\bar{1}0)$, $b_2(10\bar{1}0)$, $b_3(1\bar{1}00)$, $r_1(01\bar{1}1)$, $r_2(10\bar{1}1)$, $r_3(1\bar{1}0\bar{1})$, $r_4(0\bar{1}11)$, $\rho_1(01\bar{1}1)$, $\rho_2(10\bar{1}\bar{1})$. $\rho_3(1\bar{1}01)$, $o(0001)$.

2. Zwickau Law. The twinned individuals have two coplanar prism faces with the c-axes inclined at $42°17'$. The zones of the faces $b_1b_2b_3$ and $\rho_1b_2r_3$, and of $\rho_1b_2\rho_3$ and $b_1b_2b_3$, are parallel in the twin. Originally from Zwickau, Saxony; also from Seedorf, Disentis, and Finsteraarhorn in Switzerland.

3. Goldschmidt's Law. Two coplanar prism faces with the c-axes inclined at $47°43'$. The zones of the faces $r_3b_2\rho_1$ and b_2r_2o, and of b_2r_2o and $r_1b_2\rho_3$, are parallel in the twin. Reported from Brusson, Piedmont, and Dauphiné, France (?).

4. Breithaupt's Law. Twin plane $(11\bar{2}1)$. The twinned individuals have two coplanar prism faces with the c-axes inclined at $48°54'$. The zones of the faces $b_1r_2\rho_3$ and $b_3r_2\rho_1$, and of $b_3\rho_2r_1$ and $b_1\rho_2r_3$, are parallel. Reported from Dauphiné and an unknown locality.

5. Friedel's Rectangular Law. Two coplanar prism faces with the c-axes inclined at $90°$. Also interpreted as a twin on $(44\bar{8}1)$, with ρ $45°39'$. Observed on artificial crystals and at an unstated natural occurrence. Also from Switzerland.[42]

6. Reichenstein-Grieserntal Law. Twin plane $(10\bar{1}1)$. The c-axes of the twinned individuals are inclined at $76°26'$. The rotation angle is $180°$, and faces of the zones b_2r_2o and b_2r_2o, $b_1r_2\rho_3$ and $b_1r_2\rho_3$, and $b_3r_2\rho_1$ and $b_3r_2\rho_1$, are parallel. Early described examples of this law from Reichenstein, Saxony, and other localities were later shown to be oriented growths of quartz upon $\{01\bar{1}2\}$ of calcite joined together laterally (pseudo twins). Later observed from Grieserntal and the Göscheneralp, Switzerland. This law is identical with the Esterel Law of high-quartz, and examples of the latter have been erroneously placed under low-quartz by some authors.

7. Sella's Law. Twin plane $(10\bar{1}2)$. The rotation angle is $86°4'$, and the c-axes are inclined at $64°50'$. The zones of the faces $b_3\rho_3o$ and $b_1\rho_4o$ are parallel. First described from Sardinia (?); it is not certain whether the original material was high- or low-quartz. Also from Marmarosch, Hungary. This law is equivalent to the Sardinian Law of high-quartz.

8. Zyndel-A Law. Two coplanar $\{10\bar{1}1\}$ rhombohedral faces, with a rotation angle of $70°21'$. The zones of the faces $b_1r_2\rho_3$ and b_2r_2o and of b_2r_2o and $b_3r_2\rho_1$, are parallel in the twin. Two examples, from Hesselkulla, Sweden, and Hungary.

9. Tiflis Law. (Zyndel-"A"). Two coplanar $(10\bar{1}1)$ rhombohedral faces, with a rotation angle of $109°39'$. The zones of the faces $b_3r_2\rho_1$ and b_2r_2o, and of b_2r_2o and $b_1r_2\rho_3$, are parallel in the twin. From Tiflis (Tbilisi), Trans-caucasia.

10. Zinnwald Law. A $(10\bar{1}1)$ rhombohedral face of each individual parallel to a prism face of the other, one set being in contact, with the c-axes falling in the plane $(11\bar{2}0)$ at an angle of $38°13'$. Rotation angle $0°$. Reported from Zinnwald, Saxony; also from Disentis, Viamala, and Schams in Switzerland

and from Offerdal, Sweden; Dauphiné, France; Trenton Falls, U.S.; Snarum, Norway; and Hungary.

11. Lötschental Law. A $(10\bar{1}1)$ rhombohedral face and a prism face in contact, with a rotation angle of $47°43'$. The zones of the faces $r_3b_2\rho_1$ and $\underline{o}\rho_2b_2$ are parallel in the twin. Known from Lötschental, Disentis, and Seedorf, Switzerland.

12. Zyndel-L Law. A $(10\bar{1}1)$ rhombohedral face and a prism face in contact, with a rotation angle of $70°21'$. The zones of the faces b_2r_2o and $\underline{r}_1\rho_2b_3$ are parallel in the twin. One example, from Seedorf, Switzerland.

13. Seedorf-I Law. A $(10\bar{1}1)$ rhombohedral face and a prism face in contact, with a rotation angle of $19°39'$. The zones of the faces $b_1b_2b_3$ and $b_1\underline{\rho}_2r_3$ are parallel in the twin. One example, from Seedorf.

14. Seedorf-II Law. A $(10\bar{1}1)$ rhombohedral face and a prism face in contact, with a rotation angle of $90°$. The zones of the faces $b_1b_2b_3$ and $\underline{o}\rho_2b_2$ are parallel in the twin. One example, from Seedorf.

15. Disentis Law. A $(10\bar{1}1)$ rhombohedral face of each individual parallel to a prism face of the other, one set being in contact. Rotation angle $22°38'$. The zones of the faces $\underline{r}_1b_2\rho_3$ and $\underline{r}_1\underline{\rho}_2b_3$ are parallel. One example, from Disentis, Switzerland.

Additional, very dubious twins have been reported:[37] $\{05\bar{5}8\}$ from Virginia; $\{0.9.\bar{9}.12\}$, $\{0.2.\bar{2}.11\}$, and $\{0.7.\bar{7}.20\}$ from Virginia; $\{31\bar{4}1\}$ from Arkansas and the Caucasus; $\{03\bar{3}4\}$ or $\{9.0.\bar{9}.11\}$ from Germany; $\{17.0.\bar{17}.1\}$ (?) from Russia.

References

1. Gault: *Am. Min.*, **34**, 142 (1949).
2. Brandenstein and Heritsch: *Min. Mitt.*, ser. 3, **2**, 425 (1951).
3. Trommsdorf: *Jb. Min.*, Beil. Bd. **72**, 464 (1937).
4. Hurlbut: *Am. Min.*, **31**, 443 (1946); Gault (1949).
5. Friedlaender: *Beitr. Geol. Schweiz, Geotechn. Ser.*, Lfg. **29**, 1951.
6. The three-fold arrangement has been figured by Leydolt, *Ak. Wiss. Wien. Sitzber.* **15**, 59 (1855); Friedlaender (1951); Frondel, *Am. Min.*, **30**, 447 (1945); Johnston and Butler, *Bull. Geol. Soc. Amer.*, **57**, 602 (1946); and other authors.
7. Leydolt (1855). Figures of etched Dauphiné twins and descriptions of the etching criteria are found in many technological and mineralogical publications; see Parrish and Gordon, *Am. Min.*, **30**, 205 (1945), for summary and partial literature.
8. Heritsch: *Min. Mitt.*, **2**, 432 (1951).
9. Weiss: *Mitt. Ges. Naturforsch. Freunde Mag. Berlin*, **7**, 163 (1816).
10. Haüy: *Traité de minéralogie*, Paris, 1801.
11. Herschel: *Trans. Cambridge Phil. Soc.*, **1**, 43 (1821).
12. Zinserling: *C. R. ac. sci. URSS*, **33**, no. 5, 365 (1941) and *Trudy Inst. Krist. Ak. Nauk SSSR*, no. 12, 141, 144 (1956); Frondel (1945); Wright and Larsen: *Am. J. Sci.*, **27**, 421 (1909); Mügge: *Jb. Min.*, Festband, 181, 1907.
13. Zinserling (1941) and *J. Tech. Phys. (USSR)*, **12**, 404 (1942); Frondel (1945).
14. Zinserling: *C. R. ac. sci. URSS*, **33**, no. 6, 421 (1941) and *Dokl. Ak. Nauk SSSR*, **95**, 80 (1954).

15. Zinserling and Schubnikov: *Trav. inst. Lomonossoff géochem.*, *crist.*, *min.*, 57, 1933, and *Zs. Kr.*, 85, 454 (1933); Schubnikov and Zinserling: *Trav. inst. Lomonossoff géochem.*, *crist.*, *min.*, no. 3, 5, 1933; Zinserling: *Trav. lab. crist. ac. sci. URSS*, 2, 149 (1940).

16. Wooster, Wooster, Rycroft, and Thomas: *J. Inst. Elect. Eng.*, 94, Pt. IIIA, 927 (1947); Thomas and Wooster: *Proc. Roy. Soc. London*, 208A, 43 (1951). See also Stepanov: *Zhur. Exp. Theor. Phys.*, 20, 438 (1950); Zinserling: *Trudy Inst. Krist. Ak. Nauk. SSSR*, no. 11, 165, 1955.

17. Frondel: *Am. Min.*, 31, 58 (1946).

18. Frondel (1945).

19. Zinserling and Laemmlein: *C. R. ac. sci. URSS*, 33, 419 (1941).

20. Armstrong: *Am. Min.*, 31, 456 (1946).

21. Rose: *Abh. Ak. Wiss. Berlin*, 217, 1846.

22. Brewster: *Trans. Roy. Soc. Edinburgh*, 9, 139 (1823).

23. Leydolt (1855); Liebisch: *Phys. Krist.*, Leipzig, 1896; Heide: *Zs. Kr.*, 66, 239 (1927); Lewis: *Treatise on Cryst.*, Cambridge, 1899; Ivanov and Shafranovsky: *Mem. soc. russe min.*, 67, 435 (1938).

24. Booth and Sayers: *Post Office Elect. Eng. J.*, 31, 245 (1939); Thomas: *Nature*, 155, 424 (1945); Willard: *Bell Syst. Tech. J.*, 23, 11 (1944); Gordon: *Am. Min.*, 30, 269 (1945); Gault (1949); Heritsch (1951).

25. Ivanov and Shafranovsky (1938).

26. Gordon (1945).

27. Frondel: (1945).

28. Kozu: *Am. J. Sci.*, Bowen Vol., Pt. 1, 281, 1952.

29. Heide (1927); Brauns: *Jb. Min.*, I, 29 (1919).

30. Weiss: *Ak. Berlin*, Abh. 77, 1829.

31. Brauns (1919); Heide (1927); Tokody: *Zs. Kr.*, 99, 56 (1938).

32. Tokody (1938).

33. Borg: *Am. Min.*, 41, 792 (1956).

34. Zakharchenko, Prozorov, and Shafranovsky: *Mem. soc. russe min.*, 75, 167 (1946).

35. Amaral: *Min. e metalurgia* (*Rio de Janeiro*), 12, 215 (1948).

36. Zyndel: *Zs. Kr.*, 53, 15 (1913); Friedel: *Bull. Soc. Min.*, 46, 79 (1923); Goldschmidt: *Min. Mitt.*, 24, 167 (1905) and *Zs. Kr.*, 44, 407 (1908); Drugman: *Min. Mag.*, 16, 112 (1911); Aminoff: *Ark. Kemi, Min., Geol.*, 6, no. 22 (1917); Laemmlein: *Zs. Kr.*, 63, 291 (1926); Parker: *Schweiz. min. pet. Mitt.*, 27, 35 (1947).

37. Brown: *Am. J. Sci.*, 30, 191 (1885); Huntington: *Proc. Am. Ac. Sci.*, 225, 1885; Prendl: *Mem. Neuruss. Wiss. Ges. Odessa*, 26, 3 (1904); Kaiser: *Cbl. Min.*, 94, 1900; Gerasimov: *Bull. com. géol. Leningrad*, 38, 561 (1935).

38. Zinserling: *C. R. ac. sci. URSS*, 33, 368, 419 (1941); Armstrong (1946).

39. Wagner: *Zs. Kr.*, 93, 409 (1936).

40. Tokody: *Mat. Termés. Ért.*, 48, 776 (1932).

41. Weiss (1816); Böggild: *Medd. om Grønland*, 149, no. 3, 115 (1953).

42. Friedlaender: *Schweiz. min. pet. Mitt.*, 25, 25 (1944).

43. Mügge: *Jb. Min.*, I, 1 (1892).

44. See Weber: *Schweiz. min. pet. Mitt.*, 3, 113 (1923).

45. Groth: *Ann. Phys.*, 137, 435 (1869).

46. See Groth: *Ann. Phys.*, 158, 220 (1876).

47. Hidden and Washington: *Am. J. Sci.*, 33, 501 (1887).

ORIENTED GROWTHS

The known oriented growths involving quartz are summarized[1] in Table 13. Oriented growths of two minerals represent positions of geometrical coincidence with which is associated a frequency of occurrence greater than that of chance. The position of orientation is described by the identity of the planes in the contact surface and the mutual attitude of a crystallographic direction in each of these planes. A definite rather than a random orientation is taken because the energy of nucleation of the crystallizing phase is then in part provided by structural coincidences between the two phases across the interface. The orientation is effected through ordering of units of structure in the adsorption layer at the interface preceding crystallization. In many of the known instances of oriented growth involving quartz the position of orientation is known incompletely.

Almost all the oriented growths with quartz have formed by the crystallization of another substance upon the surface of quartz or vice versa. One instance is known of precipitation from solid solution in quartz, involving rutile. Precipitation from solid solution in high-quartz might be expected to accompany the inversion to low-quartz but has not been observed in natural material. A precipitate of a LiAl silicate (?) has been reported[19] in low-quartz containing much Al and Li that had been heated over 573° and cooled. The precipitation of quartz from the breakdown of a solid solution in another substance also has not been observed.

Among the best-known examples of oriented growths with quartz are those with calcite, rutile, and feldspar. Quartz occurs as oriented overgrowths[2] upon calcite, with $\{10\bar{1}1\}$ or $\{01\bar{1}1\}$ parallel to calcite $\{01\bar{1}2\}$ and the edges between the rhombohedra and the $(10\bar{1}0)$ prism faces of the two minerals parallel. The quartz crystals may occur as isolated crystals in parallel position upon calcite $(01\bar{1}2)$, but commonly they are united laterally to form a continuous layer with a smooth surface. Alternatively, one single quartz crystal covers each rhombohedral face of the calcite, and the whole calcite crystal may be so enveloped. In the latter case, the individual quartz crystals simulate a twin relation, as in a trilling, but with irrational twin elements. Such growths early were erroneously described as twins under the name Reichenstein Law (from the locality in Saxony). Rutile frequently occurs as oriented microscopic needles in quartz, especially in the rose quartz of pegmatites and in the blue quartz of certain igneous and metamorphic rocks (see under *Varieties of Quartz*). A large number of different orientations have been observed (Table 13). The oriented rutile inclusions may give rise to asterism in spheres or properly oriented polished sections. The rutile has exsolved from a substitutional solid solution of Ti for Si in the quartz structure formed at relatively high temperatures. The structural control of the

TABLE 13. ORIENTED GROWTH WITH QUARTZ

1. Rutile[3] in quartz with rutile [001] parallel the following directions (edges) of quartz: [0001] = (10$\bar{1}$0) $\wedge$ (11$\bar{2}$0); [1$\bar{2}$10] = (10$\bar{1}$0) $\wedge$ (0001); [1$\bar{1}$00] = (11$\bar{2}$0) $\wedge$ (0001); [1$\bar{1}$03] = (11$\bar{2}$0) $\wedge$ ($\bar{1}$2$\bar{1}$1); [10$\bar{1}$3] = ($\bar{1}$2$\bar{1}$0) $\wedge$ ($\bar{2}$111); [1$\bar{2}$16] = (10$\bar{1}$0) $\wedge$ ($\bar{1}$2$\bar{1}$1); [1$\bar{2}$13] = (10$\bar{1}$0) $\wedge$ (01$\bar{1}$1); [1$\bar{2}$12] = (10$\bar{1}$0) $\wedge$ ($\bar{1}$2$\bar{1}$3); [2$\bar{4}$23] = (10$\bar{1}$0) $\wedge$ (01$\bar{1}$2); [2$\bar{7}$56] = (01$\bar{1}$2) $\wedge$ ($\bar{2}$111);

 [1$\bar{5}$43] = (10$\bar{1}$1) $\wedge$ ($\bar{1}$102) $\wedge$ ($\bar{2}$111); [4.$\bar{11}$.7.3] = (10$\bar{1}$1) $\wedge$ (01$\bar{1}$6); [1$\bar{1}$02] = (11$\bar{2}$0) $\wedge$ ($\bar{1}$101); [1$\bar{1}$01] = (11$\bar{2}$0) $\wedge$ ($\bar{1}$102).
2. Quartz upon calcite,[2] with quartz (10$\bar{1}$1)[1$\bar{2}$10] parallel calcite (01$\bar{1}$2)[$\bar{2}$110].
3. Quartz upon fluorite,[8] with quartz (10$\bar{1}$1), (10$\bar{1}$0), (11$\bar{2}$0), (0001), (10$\bar{1}$2), or (11$\bar{2}$1) parallel fluorite (001), but the position of orientation not further specified.
4. Quartz upon hematite,[9] with quartz (10$\bar{1}$0)[0001] parallel hematite (0001)[11$\bar{2}$0]; also (11$\bar{2}$0)[0001] parallel (0001)[10$\bar{1}$0]; (10$\bar{1}$0)[(10$\bar{1}$0) $\wedge$ (11$\bar{2}$1)] parallel (0001)[10$\bar{1}$0]; (10$\bar{1}$1)[(10$\bar{1}$1) $\wedge$ (11$\bar{2}$0)] parallel (0001)[10$\bar{1}$0].
5. Quartz inclusions in muscovite,[10] with quartz (10$\bar{1}$0)[11$\bar{2}$0] parallel muscovite (001)[100]; (10$\bar{1}$1)[11$\bar{2}$0] parallel (001)[100]; (10$\bar{1}$2) parallel (001); (0001)[11$\bar{2}$0] (?) parallel (001)[100]; (10$\bar{1}$1)[1$\bar{2}$13] parallel (001)[100]; (10$\bar{1}$1)[11$\bar{2}$0] parallel (100)[110].
6. Biotite upon (or replacing) quartz,[15] with biotite (001)[100] parallel quartz (0001)[11$\bar{2}$0].
7. Quartz and sphalerite, of unknown orientation.[11]
8. Quartz and cordierite,[12] with the c-axes of the two minerals inclined at 55–58°.
9. Quartz overgrown by AlPO$_4$ (synthetic)[13] in parallel position.
10. Quartz and feldspar (see text).
11. Cassiterite upon quartz,[17] with (111)[$\bar{1}$12] parallel (10$\bar{1}$0)[0001].

orientation has been discussed.[3] In addition, rutile very frequently occurs as randomly oriented hairs and needles, easily visible to the unaided eye, that have been mechanically included in quartz crystals during their growth (see *Inclusions in Quartz*). What seems to have been an oriented growth of rutile with quartz was described[7] as early as 1827.

In the so-called graphic granite of pegmatites, tapering rod-like crystals of quartz occur intergrown in microcline or orthoclase. The rods often range up to a half-inch in thickness and to a foot or more in length; they also occur on a microscopic scale. The rods have an open or skeletal polygonal cross-section, with angles of approximately 60°, and on cross-fractures somewhat resemble hieroglyphics in appearance (Fig. 70). The lateral surfaces of the rods are irregular and tapering, with cross-striations, and are not true crystal faces but have formed by mutual interference with the feldspar. The quartz rods tend to lie perpendicular to the bounding face of the feldspar; they sometimes protrude from the surface of euhedral feldspar crystals into cavities and may then be enlarged and terminated by crystal faces (Fig. 71). In most instances the rods are elongated parallel to the direction of a rhombohedral

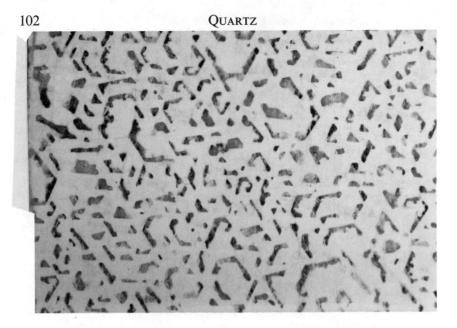

FIG. 70. Graphic granite. Freeport, Maine. About half natural size.

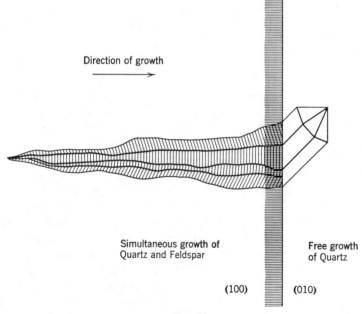

Direction of growth

Simultaneous growth of
Quartz and Feldspar

Free growth
of Quartz

(100)   (010)

FIG. 71

edge rz and not to the c-axis. The quartz usually is more or less smoky in color. The mutual crystallographic relations of the quartz and the feldspar have been the subject of much investigation.[4] Adjacent but isolated quartz rods often extinguish simultaneously in polarized light. Frequently the quartz is oriented so that an rz edge is parallel to the c-axis of the feldspar, the c-axes of the two minerals inclined at $42°16'$ (known as Fersman's Law), with the c-axis of the quartz falling in (010) or (100) of the feldspar or otherwise; also at other angles of inclination of the c-axes, with the c-axis of the quartz falling in (010), (100), or (110) of the feldspar.[14] In other instances the quartz and feldspar apparently are randomly oriented.

The formation of the intergrowths has been attributed to simultaneous crystallization of the quartz and feldspar. A eutectic relation has been postulated because the relative proportions of the quartz and feldspar are approximately constant in some occurrences, with about 26 weight per cent quartz and 74 per cent feldspar, but widely different proportions have been observed. In some instances the quartz intergrowths are confined to particular parts of feldspar crystals. An origin by replacement of the feldspar with introduced quartz also has been suggested. Many graphic granites contain abundant albite or oligoclase which has partly or completely replaced the original microcline: intergrowths of feldspar and quartz that simulate eutectic and graphic texture have been obtained experimentally by the hydrothermal recrystallization of glass.[16] Quartz also occurs as "graphic" intergrowths with tourmaline,[5] spodumene,[18] cordierite,[12] and muscovite. The mutual orientation of the quartz with the tourmaline has not been investigated.

Oriented overgrowths of organic substances, including anthraquinone, amino acids, and α-glycine, have been obtained[6] by crystallization from solution or by sublimation upon polished surfaces of quartz that have been activated or cleaned by treatment with HF. With α-glycine, no effects of enantiomorphism were observed on the position of orientation, indicating the predominance of one-dimensional structural analogies on the superdeposition.

References

1. From von Vultée: Jb. Min., Abh. 87, 389 (1955). Reviews of the mutual orientation phenomena of natural and synthetic crystals are given by Seifert: Structure and Properties of Solid Surfaces, ed. by Gomer and Smith, Chicago, 1953, Ch. 9, p. 318, and Fortschr. Min., 19, 103 (1935), 20, 324 (1936), and 22, 185 (1937); also Mügge: Jb. Min., Beil. Bd. 16, 335 (1903), and Wallerant: Bull. soc. min., 25, 180 (1902).
2. Breithaupt: Jb. Min., 576, 1861; Eck: Zs. deutsch. geol. Ges., 18, 428 (1866); Frenzel and vom Rath: Ann. Phys., 155, 17 (1875); Meixner: Jb. Min., 1, 56 (1937) (abstr.).
3. von Vultée: Zs. Kr., 107, 1 (1956).
4. Fersman: Zs. Kr., 69, 77 (1929); Woitschach: Zs. Kr., 7, 82 (1881) (abstr.); Bobkova: Publ. Fac. Sci. Univ. Masaryck, Brno. no. 204, 1935 (Jb. Min., 1, 173, 1937); Kossoy: C. R. ac. sci. USSR, 19, 273 (1938); Drescher-Kaden: Chem. der Erde, 14, 157 (1942); Wahlstrom: Am. Min., 24, 681 (1939); also von Vultée (1955).

5. Newhouse and Holden: *Am. Min.*, **10**, 42 (1925); Wahlstrom (1939); Drescher-Kaden (1942); Hogbom: *Bull. Beol. Inst. Upsala*, **3**, 436 (1897); Tilley: *Trans. Roy. Soc. South Australia*, **43**, 156 (1919).
6. K. F. Seifert: *Zs. Kr.*, **114**, 287, 361 (1960); H. Seifert: *Int. Congr. Reactivity of Solids, Madrid*, **3**, 93 (1959), and *Naturwiss.*, **46**, 261 (1959); Willems: *Naturwiss.*, **41**, 302 (1954).
7. Marx: *Kastner's Archiv Naturlehre*, **12**, 220 (1827).
8. Kalb: *Cbl. Min.*, 201, 1916; Mügge (1903).
9. Gliszczynski and Stoicovici: *Zbl. Min.*, 82, 1938A.
10. Kalb (1916); Kurbatov and Kurbatov: *Zbl. Min.*, **1**, 50 (1943) (abstr.); Frank-Kamenetski: *Chem. Abs.*, **49**, 7456 (1955).
11. Kreutzwald: *Jb. Min.*, Beil. Bd. 70, 234 (1935).
12. Shibata: *Japan. J. Geol. Geogr.*, **13**, 205 (1936).
13. Gruner: *Am. Min.*, **31**, 196 (1946).
14. See summary in Drescher-Kaden (1942).
15. Kostov: *Min. Mag.*, **31**, 333 (1956).
16. Schloemer: *Schweiz. min. pet. Mitt.*, **33**, 510 (1953).
17. Ramdohr: *Cbl. Min.*, 200 1923.
18. Quensel: *Geol. För. Förh.*, **68**, 47 (1946).
19. Cohen and Hodge: *J. Phys. Chem. Solids*, **7**, 361 (1958); Cohen: *J. Phys. Chem. Solids*, **13**, 321 (1960).

PHYSICAL PROPERTIES

CLEAVAGE AND FRACTURE

Cleavage. Cleavage is not ordinarily observed in quartz and is not easily obtained by fracturing crystals by impact, as with a hammer. It has been observed on the following seven forms: $r\{10\bar{1}1\}$, $z\{01\bar{1}1\}$, $m\{10\bar{1}0\}$, $c\{0001\}$, $a\{11\bar{2}0\}$, $s\{11\bar{2}1\}$, and $x\{5\bar{1}\bar{6}1\}$. The best cleavages are those on r, z, and m. Of these three forms, the rhombohedral cleavages are higher in quality and easier to produce than that on the prism. The cleavage on r is of higher quality and apparently more readily produced than that on z. The two cleavages, however, have been sometimes described as equal. Both the r and z cleavage surfaces are rather rough and interrupted. Many of the descriptions of natural and of experimentally produced rhombohedral cleavage in quartz are incomplete in that there is not a rigorous identification of r and of z. A single rhombohedral cleavage usually is stated to be r. It appears from the experimental evidence that cleavage may be developed on either r or z or on both simultaneously, depending on the method employed.

A distinct and easy rhombohedral cleavage is sometimes observed in anhedral quartz from pegmatites.[1] Rhombohedral cleavage surfaces 2–9 ft. across have been observed in anhedral crystals in the cores of pegmatites in western Arizona and southern California. This cleavage often is relatively prominent on one of the three symmetrically equivalent planes, giving rise to a lamellar or sheeted appearance. It is sometimes described as a parting, and this description is more apt when, as is often the case, the crystal tends to separate into relatively thick plates which are themselves further cleavable only with difficulty. The effect probably is due to mechanical deformation which produced a degree of separation along widely spaced planes. A rhombohedral parting in quartz also has been ascribed to lamellar Brazil twinning with composition planes parallel to r and z, but amethyst, which possesses such a structure to a marked degree, does not show this parting. Natural cleavage on the r and z rhombohedra and on other planes also has been observed in deformed vein quartz.[2] Single-crystal quartz pebbles with good rhombohedral cleavage occur in the Lafayette formation of South Carolina.[14] Oscillator-plates have been observed to fracture on a rhombohedral cleavage when strongly driven.

The anisotropy of fracture in quartz has been studied[15] by examining crushed fragments in the 100–200 mesh size range by universal stage techniques to determine the angle between the optic axis and the normal to the plane of rest of the grains. Analysis of several thousand such measurements indicated the best cleavage to be r or z, and the next best to be m, with some evidence of cleavage on c, s, and x. Cleavages on r, z, and m have been observed[16] on the grains of quartz sands deformed at 270–320° and 26,000–32,000 p.s.i. in slightly alkaline solutions.

Pits bounded by rhombohedral cleavage faces on both r and z can be obtained by rocking and crushing sawn basal sections upon a lap coated with very coarse abrasive. A hexagonal light-figure with the edges parallel to the traces of the prism faces is seen when a pinhole source of light is viewed through such a surface.[3] All sides of the hexagon are of equal intensity, indicating that the r and z cleavages are equally developed. These light-figures may show faint details with a three-fold rotatory symmetry that reveals the hand of the quartz, presumably through the development of a minor cleavage on a general form that may be $x\{51\bar{6}1\}$. A rough ground sphere of quartz[11] also shows surface reflections in a beam of light that correspond to the equal development of cleavages on r and z.

The rhombohedral and prismatic cleavages are best produced by thermal shock,[4] especially in thin slices, or by high local pressure or impact on natural faces or sawn sections. The rhombohedron $\{10\bar{1}1\}$ was taken by Haüy as the *integral molecule* of quartz on the basis of the cleavage developed by quenching a heated fragment in water or by percussion. Both the r and z cleavages are said to be produced by the pressure of a rounded metal point on a thin slice cut parallel to $\{11\bar{2}0\}$. In a $(11\bar{2}0)$ or Y section parallel pairs of faces, such as $r(10\bar{1}1)\,(\bar{1}01\bar{1})$ and $z(10\bar{1}\bar{1})\,(\bar{1}011)$, are perpendicular to the plane of the section, and the other parallel pairs of faces of these forms are inclined to the section. All these parallel pairs of faces or cleavage planes can be separately identified if the hand of the quartz and the polarity of the X-axis perpendicular to the section is known.

Cleavage on $m\{10\bar{1}0\}$ is sometimes observed[5] on natural fracture surfaces, occasionally to the exclusion of the rhombohedral cleavages. The cleavage surfaces of $\{10\bar{1}0\}$ tend to be undulatory and to pass into conchoidal fractures. The $\{10\bar{1}0\}$ cleavage can be readily produced by thermal shock, i.e., by the slow edgewise immersion of a thin (0001) section heated to 300–400° into cold water (Fig. 72). A particular prism cleavage direction can be produced or altered in course by varying the direction of entrance of the section into the water. The method has been used in connection with optical tests to orient anhedral quartz.

Cleavage on $\{0001\}$ is difficult to produce.[6] It may be observed, together with cleavage on $\{10\bar{1}0\}$ and the well-marked cleavages on r and z, by local pressure on $(11\bar{2}0)$ sections. Both the $\{0001\}$ and $\{10\bar{1}0\}$ cleavages may in part or entirety represent coplanar surfaces composed of cleavages on r and z. A perfect $\{0001\}$ cleavage has been produced in quartz, however, by the action of an electric field at high temperatures.[13] Ill-defined cleavages also have been reported[7] on $s\{11\bar{2}1\}$, $m\{11\bar{2}0\}$, and $x\{51\bar{6}1\}$. These and the other cleavages of quartz are best demonstrated experimentally by special techniques, such as high local pressure by a rounded steel point or properly oriented chisel edge on a thin slice cut perpendicular to the cleavage to be tested, or striking a chisel edge applied against a notch or groove cut into the

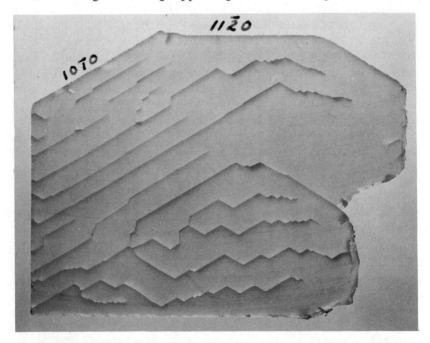

FIG. 72. (0001) section of quartz containing cleavage cracks on $(10\bar{1}0)$ produced by thermal shock.

surface of a sawn, oriented block or crystal. Large crystals or sawn blocks of quartz that have been passed through the high-low inversion at about 573° generally show only rhombohedral cleavage cracks or irregular fractures. The fracture of quartz under high confining pressures also produces cleavages (see *Plastic Deformation*). So-called cap-quartz is separable into pyramidal layers parallel to *rz* because of the deposition of a little foreign material at more or less regular[12] intervals in the growth of the crystal. This is a mechanical separation and not a true cleavage.

Cleavage in quartz regardless of orientation ruptures Si-O bonds, since the crystal structure is a three-dimensional linkage of (SiO_4) tetrahedra. Calculation of the number of bonds ruptured per unit area[8] indicates that there is a minimum cohesion across *r* and *z* planes, with increasingly greater cohesion across *m*, *c*, and *a*. The bonding across *r* and *z* is identical, and the superiority of the *r* over the *z* cleavage is probably due to the greater "smoothness" of the *r* plane as seen on an atomic scale in the structure. In any given experiment with directed pressure, the observed relative ease of rupture in different structural planes is determined primarily by the crystallographic resolution of the forces applied.

Fracture. The fracture of single-crystals of quartz is conchoidal to subconchoidal. The fracture surfaces tend to be broader and more even when cutting across the *c*-axis at large angles. Conchoidal fractures a millimeter or so in size have been observed, radiating out from small crystals of other minerals wholly enclosed in quartz crystals. The cracking results from volume change attending change of temperature or alteration. Thin plates that have been subjected to severe thermal shock or passed through the high-low inversion may develop cracks resembling sine-curves or arranged in intricate, closely spaced pattern systems that suggest fingerprints.[9] Such cracking may be associated with or bordered by secondary Dauphiné twinning (which see).

The fracture of the massive, microgranular types of quartz is flat conchoidal to uneven or splintery; that of the fibrous varieties is uneven, hackly, or splintery.

Percussion and Pressure Figures. The small fracture patterns or figures obtained by dropping or pressing a steel ball on natural faces or sawn sections of quartz crystals are characterized[10] by the development of cleavage cracks on the rhombohedron $z\{01\bar{1}1\}$. No difference is observed between the figures obtained by percussion and those obtained by pressure except that the latter tend to be somewhat more regular and are accompanied by the development of secondary Dauphiné twinning. Cleavage on $r\{10\bar{1}1\}$ or on other forms has not been observed, except surficially on (0001) sections where traces of cleavage on $r\{10\bar{1}1\}$ may be observed.

The figures vary in their shape on surfaces of different crystallographic orientation (Figs. 73 and 74). The shape in all instances is roughly described by the surface of intersection of the tested plane with the faces of an idealized

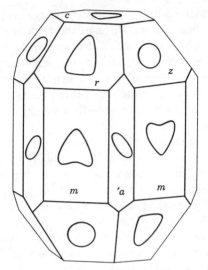

FIG. 73. Idealized figure, showing the form of the percussion figures produced on different planes of quartz.

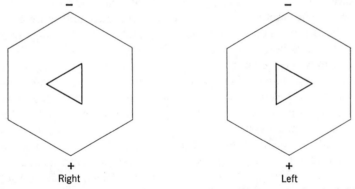

FIG. 74. Relation of the percussion figure produced on (0001) to the hand of the quartz.

$z\{01\bar{1}1\}$ rhombohedron. On $\{0001\}$ the percussion figure consists of two parts: an inner essentially circular area of concentric cracks, and an outer more or less rounded hexagonal rift. When the surface is progressively lapped down, the figure appears as a triangle, formed by a three-sided pyramid bounded by z, with rounded edges and slightly concave sides, extending down into the quartz with its truncated apex at the point of impact. A similar figure is obtained on glass except that the internal figure is conical, not pyramidal, with an annular ring extending out along the surface of the glass around the apex of the cone and corresponding to the central circular part of the quartz figure. The hexagonal shape of the surface figure is caused by the presence of shallow, inwardly dipping cracks on both r and z. On $\{01\bar{1}1\}$

itself, the percussion figure is essentially circular, identical with the figure obtained on glass, while the figure on $\{10\bar{1}1\}$ is geometric.

The shape of the figures is independent of size (or applied pressure). The size of the pressure figures depends both on the amount and on the duration of the pressure. The percussion figures are readily produced by dropping steel balls of a weight in the range from 10 to 30 grams from a height of a few meters upon a thick, firmly held section of quartz.

References

1. Walker: *Univ. Toronto Stud., Geol. Ser.*, no. 38, 31, 1935; Rogers: *Proc. Geol. Soc. Amer.*, 347, 1934; Jahns: *Am. Min.* **38**, 563 (1953); Murdoch and Webb: *Am. Min.*, **23**, 349 (1938); Scheerer: *Ann. Phys.*, **65**, 295 (1845).
2. Anderson: *Bull. Geol. Soc. Amer.*, **56**, 409 (1945); Grigoriev: *Mem. Mineral. Soc. USSR*, **86**, 539 (1958).
3. Rivlin: *Nature*, **146**, 806 (1940).
4. Kenngott: *Uebers. Min. Forsch.*, 170, 1844–49; Booth: *J. Inst. Elect. Eng.*, **88**, Fig. 34 (1941); Zinserling: *C. R. ac. sci. URSS*, **33**, 421 (1941); Savart: *Ann. chim. phys.*, **40**, 5, 113 (1829); Scheerer (1845).
5. Mallard: *Bull. soc. min.*, **13**, 61 (1890); Drugman: *Min. Mag.*, **25**, 259 (1939); Fairbairn: *Am. Min.*, **24**, 351 (1939).
6. Zinserling (1941).
7. Ichikawa: *Am. J. Sci.*, **39**, 472 (1915).
8. Fairbairn (1939).
9. Mügge: *Jb. Min.*, Festband, 181, 1907; Nacken: *Jb. Min.*, I, 71 (1916); Frondel: *Am. Min.*, **30**, 447 (1945).
10. Schubnikov and Zinserling: *Zs. Kr.*, **83**, 243 (1932); Zinserling and Schubnikov: *Trav. inst. Lomonossoff géochem.*, no. 3, 5, 67, 1933.
11. Denning and Conrad: *Am. Min.*, **44**, 423 (1959).
12. Liesegang: *Naturwiss.*, **3**, 500 (1915).
13. Zinserling: *J. Tech. Phys. USSR*, **12**, 552 (1942).
14. Stephen Taber: priv. comm., 1957.
15. Bloss: *Am. J. Sci.*, **255**, 214 (1957).
16. Borg and Maxwell: *Am. J. Sci.*, **254**, 71 (1956).

MECHANICAL DEFORMATION

Plastic Deformation. Experimental studies of the mechanical deformation of quartz have shown[1] that this substance remains brittle and elastic up to the moment of fracture, without evidence of plastic flow by translation gliding or twin gliding, at the maximum pressures, approximately 138,000 atm., applied in the laboratory. The strength of quartz increases with increase in the confining pressure. At 25,000 atm. confining pressure, at room temperature, quartz is found to have a strength of 150,000 kg. per cm.^{2}, comparing to the normal strength at atmospheric pressure of 24,000 kg. per cm.^{2} The strength is found to be materially reduced when the quartz is immersed in an aqueous solution; at 400° under a few hundred atmospheres of confining pressure quartz yields by rupture at a differential stress of 4000 kg. per cm.^{2} again without evidence of plastic flow.

The quartz fragments produced by rupture at high pressures[2] are needle-like with rather smooth lateral surfaces. These surfaces are parallel or sub-parallel to the c-axis, to the r or z rhombohedra or to $\{0001\}$, or are irrational planes, depending on the crystallographic orientation of the needle axis. The direction of the needle axis, which in general is roughly parallel to the direction of pressure regardless of the crystallographic orientation of the pressed quartz cylinder, can be parallel to the c-axis or to any rz, rm, zm, rr, or zz edge.

Petrofabric studies of quartz in metamorphic rocks indicate that it is extremely susceptible to reorientation when subjected to deformation. The quartz grains of deformed rocks may show a high degree of preferred orientation, and efforts have been made to interpret the types of orientation shown by petrofabric diagrams by postulating plastic flow by translation gliding on various planes and directions. The translation gliding elements that have been postulated[3] include glide plane $\{0001\}$, glide direction parallel to an rm edge; glide plane $\{10\bar{1}1\}$ or $\{01\bar{1}1\}$, glide direction an rz edge; glide plane $\{11\bar{2}2\}$, glide direction an rz edge; a glide plane in the prism zone, with glide directions parallel to an rm edge or parallel to the c-axis; glide planes near $\{01\bar{1}3\}$ and $\{01\bar{1}2\}$ with the glide direction parallel to the trace of the c-axis; and glide planes on $\{10\bar{1}2\}$, $\{10\bar{1}4\}$, $\{3\bar{1}\bar{2}2\}$, and $\{\bar{1}2\bar{1}6\}$. It also has been suggested[4] that translation gliding might take place with the glide direction parallel to an rm edge in a plane or planes with this direction as zone axis but not necessarily rational. The glide direction rm produces a minimum disturbance of the bonding between the (SiO_4) tetrahedra composing the structure. The belief has been expressed[5] that the lamellar Brazil twinning in amethyst is secondary and produced by mechanical deformation, but this feature definitely is formed during crystal growth. The striations on the boundary surface between adjoining crystals of quartz, produced by mutual interference during crystal growth, also appear to have been confused[5] with displacement striae such as are formed by translation gliding.

The so-called Boehm lamellae[6] or deformation lamellae often seen in the quartz grains of deformed rocks have been much discussed[7] in this connection. These microscopic, planar laminae, not to be confused with planes of liquid inclusions or fractures, occur in virtually parallel, closely spaced array. They often are gently undulating. The laminae may occupy part or all of a grain, as seen in thin section, and two or more sets of laminae may be present at different angles. Individual laminae commonly are slightly lens-shaped and may pinch out before reaching the edge of the grain. They are best seen when their plane is nearly parallel to the axis of the microscope. The laminae have been variously described as having a very slightly higher or lower index of refraction than the surrounding quartz, indicating that they are in angular disorientation (or have a different chemical composition). The Boehm lamellae are not parallel to any definite crystallographic plane or zone axis of the quartz but do appear to form more readily at low angles to $\{0001\}$. It is not certain whether the laminae are an indication of the ruptural deformation of the quartz or are traces of translation gliding planes. Their orientation apparently is controlled by the stress pattern in the deformed rock. The formation of the lamellae also has been interpreted[9] as being caused by stresses acting after the preferred orientation of the quartz grains had already been effected. The stresses, acting on particular grains oriented to give a high resolved shear along [0001], produced kink-bands at high angles to [0001], resulting from movement parallel to [0001] along irrational planes in this zone.

The curious "blue needles" found in euhedral quartz crystals (see *Inclusions in Quartz*) may be of interest in this connection. They are needle-like voids occurring in V-shaped clusters, with the plane of the V parallel to an *r* or *z* rhombohedral face and the sides parallel to the *rz* edges. Their origin is unknown. A coarse lamellar growth structure also is found in imperfect quartz crystals (see *Optical Anomalies*). Observations on the plastic deformation and cleavages of weakened models of quartz, such as $AlPO_4$, $AlAsO_4$, and BPO_4, are lacking. Secondary Dauphiné twinning is readily produced in quartz by pressure (see *Twinning*). Both Dauphiné and Brazil twinning are extremely common in quartz, but neither type can be recognized by optical examination in thin section. Their bearing on the plastic deformation of quartz has not been investigated.

An optical effect in quartz that has been attributed to strain[8] appears as a mosaic of rhomboidal areas that do not extinguish uniformly. It may be related to sets of cross-fractures on rhombohedral planes.

References

1. Griggs and Bell: *Bull. Geol. Soc. Amer.*, **49**, 1723 (1938); Griggs: *Bull. Geol. Soc. Amer.*, **51**, 1001 (1940); Bridgman: *Phys. Rev.*, **57**, 342 (1940).
2. Griggs and Bell (1938).
3. Schmidt: *Fortschr. Min.*, **11**, 334 (1927); Holmquist: *Geol. För. Förh.*, **48**, 411 (1926); Sander: *Gefügekunde der Gesteine*, Vienna, 1930; Savul: *Bull. inst. polytech. Jassy*, **3**, 375 (1948); Fischer: *Zbl. Min.*, **1925A**, 210.
4. Fairbairn: *Bull. Geol. Soc. Amer.*, **52**, 1265 (1941) and *Am. Min.*, **24**, 351 (1939); Brace: *Am. J. Sci.*, **253**, 129, (1955).
5. Judd: *Min. Mag.*, **8**, 1 (1888) and **10**, 123 (1892).
6. Boehm: *Min. Mitt.*, **5**, 197 (1883).
7. See Ingerson and Tuttle: *Trans. Am. Geophys. Union*, **26**, 95 (1945); Fairbairn (1941); Anderson: *Bull. Geol. Soc. Amer.*, **56**, 409 (1945).
8. Quensel: *Geol. För. Förh.*, **64**, 285 (1942); Kerr: *Am. Min.*, **11**, 207 (1926).
9. Christie and Raleigh: *Am. J. Sci.*, **257**, 385 (1959).

Disorientation Phenomena on Quartz Surfaces. X-ray studies of quartz surfaces[1] that have been ground by abrasive and of dry-ground quartz powders indicate that a thin surface layer of strained and more or less disoriented material is formed by the mechanical action. The thickness of the disturbed surface layer depends on the nature of the abrasive and the polishing treatment. In fine quartz powders it has been estimated to be about 0.03 μ thick, but it is substantially thicker on heavily lapped surfaces.[4] The surface layer has been believed to be amorphous, or to be hydrous silica through the adsorption of water, but x-ray study indicates that it is in part at least crystalline. The amount of disorientation in the surface layer is not necessarily large, and values of a few degrees of arc have been measured. Freshly lapped or ground material shows an ageing effect, as if a strained or glassy surface region later broke up or recrystallized into a more or less disoriented polycrystalline mosaic. The formation of the disturbed surface layer results in a significant decrease in the intensity of reflection of x-rays from quartz powders below about 2 μ in size. This effect may seriously prejudice the quantitative measurement of the amount of finely divided quartz in mixtures.[2] With lapped quartz surfaces, etching increases the intensity of reflection.

Surface disorientation and ageing effects are markedly shown by high-frequency shear mode quartz oscillator-plates.[3] The frequency of oscillation of such plates is sensitive to changes in thickness, which is the frequency-determining dimension. In an 8 megacycle BT plate a change in thickness of about one unit cell produces a change in frequency of about 10 cycles, which can be measured. Oscillator-plates that have been lapped to final thickness entirely by abrasion show a spontaneous increase in frequency with time, as if the effective thickness of the plate had been decreased, and there is an associated decrease in the activity, or relative magnitude of oscillation, which may continue until the plate fails to oscillate in the test circuit. In a series of plates of the same frequency (thickness), both effects are more marked the coarser the abrasive employed in lapping, indicating a greater thickness of the damaged surface layer. Plates in the 6–9 megacycle range show an increase in frequency of as much as roughly 0.02 per cent of the nominal frequency, corresponding to a decrease in the effective thickness of up to approximately 0.4 μ. Comparable values for the thickness of the damaged surface layer are obtained by experiments on the stabilization of freshly lapped plates by etching the surface to various depths. The ageing progresses exponentially with time, and is very largely realized in from a few hours up to 60–100 hours. It is accelerated by heating to a few hundred degrees Centigrade or by the presence of water. Ageing phenomena are not observed if the quartz plate is adjusted to final thickness by etching rather than by abrasion.

References

1. D'Eustachio and Brody: *J. Opt. Soc. Amer.*, **35**, 544 (1945); Nagelschmidt, Gordon, and Griffin: *Nature*, **169**, 539 (1952); D'Eustachio: *Phys. Rev.*, **70**, 522 (1946) and **69**, 532 (1946); Sakisaka: *Japan. J. Phys.*, **4**, 171 (1928); Gogoberidze: *J. Exptl. Theor. Phys. USSR*, **10**, 96 (1940).
2. Brindley and Udagawa: *J. Am. Ceram. Soc.*, **42**, 643 (1959); Flörke: *Schweiz. Min. Pet. Mitt.*, **41**, 311 (1961).
3. Frondel: *Am. Min.*, **30**, 421 (1945).
4. Boyer: *Bull. soc. min.*, **77**, 1116 (1954); Gibb and Sharpe: *J. Appl. Phys.*, **3**, 213 (1953); Alexanian: *C. R.*, **242**, 2145 (1956); Finch, Lewis, and Webb: *Proc. Phys. Soc. London*, **668**, 949 (1953).

HARDNESS

The hardness of quartz is arbitrarily designated as 7 in the Mohs scale of relative scratch hardness. The scratch hardness is perceptibly greater on $\{10\bar{1}0\}$ than on $\{0001\}$. On a form other than $\{0001\}$ the scratch hardness is greater parallel to the trace of the c-axis than perpendicular thereto.[1]

Hardness is a composite character involving primarily the mechanical strength, elasticity, plasticity, and cleavage of a material. It is useful as a empirical means of comparing different substances in terms of a specified method of testing. The relative values thus obtained generally differ with the method of testing. Hardness has the aspect of a vectorial property, since in any sufficiently sensitive method of testing it is found to vary with direction in a single-crystal.

The abrasion hardness of quartz in different directions has been measured by a number of methods. Oriented quartz plates ground against a Carborundum wheel gave[2] the lap-abrasion hardness in the order $\{10\bar{1}1\} > \{10\bar{1}0\} > \{0001\}$. The impact-abrasion hardness, using a sandblast, was found[3] to be $\{10\bar{1}1\} > \{10\bar{1}0\} > \{0001\} > \{01\bar{1}1\}$. The peripheral grinding of a rotating quartz disc cut perpendicular to an a-axis gave[4] the order $\{10\bar{1}0\} > \{10\bar{1}1\} \geqslant \{0001\} > \{01\bar{1}1\}$. This disc developed an outline with two-fold symmetry. The grinding anisotropism of a disc cut perpendicular to the c-axis was less marked and could not be established. The abrasion produced by a rotating diamond tool[5] was found to be in the order $\{10\bar{1}0\} > \{0001\}$. Japanese quartz workers also have observed that the prism faces of quartz are the most resistant to cutting and grinding, whereas the direction along c is relatively soft.[6] On the other hand, spheres of natural and synthetic quartz, when tumbled in abrasive to a loss of approximately 36 per cent in weight, showed the following order of decrease in diameter on the face normals:[7] $\{0001\} < \{10\bar{1}0\} < \{01\bar{1}1\} < \{10\bar{1}1\}$. The lap-abrasion hardness of various substances has been shown[8] to vary with the grain size of the abrasive; e.g., the relative orders of abrasive hardness of quartz $\{0001\}$ and topaz parallel the c-axis, reversing with changing size. The rate of abrasion of quartz on a given surface varies with the nature of the liquid used to suspend the abrasive, and is in general less in non-polar liquids.[11] Quartz has a lower scratch hardness than topaz but a greater resistance to abrasion in a sandblast.[12]

Measurements of the hardness of quartz by an indentation method (Knoop indenter) showed[9] that $\{10\bar{1}0\}$ is harder than $\{10\bar{1}1\}$ and $\{01\bar{1}1\}$ and that the indentation hardness, like the scratch hardness, varies with direction on these planes (the indenter is a flat pyramidal diamond point with a length-to-width ratio of about 7:1). On $\{10\bar{1}0\}$ the indentation hardness is greater parallel to the c-axis than perpendicular thereto. The Vickers hardness (indentation method)[10] on polished surfaces showed $\{10\bar{1}0\}$ to be harder than $\{0001\}$.

The observed variations in the order of hardness of different planes in quartz recalls the observed variations in the cleavages in quartz. Although the directions of the cleavages in quartz are definite, the relative ease of cleavage varies, as does the hardness, with the way the disruptive force is applied. The varying hardness of quartz probably is related to the way that the potential cleavage directions are utilized in removing material. Plastic deformation does not appear to be a factor in determining the hardness of quartz.

Observations made in connection with the irradiation response and secondary twinning of quartz indicate that there is an appreciable difference in the hardness of different specimens. The hardness decreases with decreasing content of other elements in solid solution.

References

1. Berndt: *Verhl. deutsch. phys. Ges.*, **21**, 110 (1919).
2. Ingerson and Ramisch: *Am. Min.*, **27**, 595 (1942); see also Kuznetsov: *C. R. ac. sci. URSS*, **89**, 271 (1953).
3. Milligan: *J. Am. Ceram. Soc.*, **19**, 187 (1936).
4. Denning and Conrad: *Am. Min.*, **44**, 423 (1959).
5. Pfaff: *Sitber. bayer. Ak. Wiss. München, Math.-phys. Kl.*, 55, 372, 1883; 255, 1884.
6. Ichikawa: *Am. J. Sci.*, **39**, 455 (1915).
7. C. S. Hurlbut, Jr.: unpubl. data, 1961.
8. Holmquist: *Geol. För. Förh.*, **44**, 485 (1922).
9. Winchell: *Am. Min.*, **30**, 583 (1945).
10. Taylor: *Min. Mag.*, **28**, 718 (1949).
11. von Engelhardt: *Nachr. Al. Wiss. Göttingen, Math.-phys. Kl.*, no. 2, 1942.
12. Eppler: *Zbl. Min.*, 73 1941A.

DENSITY

The most probable value of the density of a randomly selected colorless crystal of natural quartz is 2.65074 $\pm$ 0.0001 in vacuum at 0° referred to water at 4°. The absolute density is about 0.00007 less. The density of colorless quartz of the ideal composition SiO_2, free from solid solution, is not known (to comparable precision). In the range from 0° to 50° the density decreases about 0.0001 per degree increase in temperature.[1]

The most probable value of the density in air at 760 mm. at 0° referred to water at 4° is 2.6528, and at room temperature (18–20°) is about 2.6510 under the same conditions. The density in air at 20° referred to water at 20° is 2.6556. A tabulation of the density of quartz in air under difference reference conditions[2] is given in Table 14.

Highly precise measurements[3] of the density of colorless quartz crystals, chiefly from vein-type occurrences, from which the most probable value given above was derived, indicate a significant variation in the range 0.0001–0.0005 in this type of quartz. It is not known whether any of the reported measurements refer to quartz of the ideal composition, SiO_2. The variation in density is accompanied by a variation in other properties of the quartz. Precise measurements are lacking of the density of quartz from other types of occurrence, such as from igneous rocks and pegmatites, which may show a greater range of variation. The reported measurements of the unit cell volume of colorless quartz show a wide and evidently systematic variation, but the analytical data are not sufficiently good to establish the numerical relations between density, cell volume, and composition. The relatively low precision to which the atomic weight of Si and the Avogadro number are known precludes the application of the x-ray method of measurement to the determination of the absolute density of quartz, since the attainable precision is less than

TABLE 14. DENSITY CONVERSION DATA
Corresponding densities in vacuum are lower by approximately 0.00200.

Temperature of Quartz	Density in Air Referred to H_2O at Stated Temperature				
	4°	10°	15°	20°	25°
0°	2.6528	2.6535	2.6551	2.6575	2.6606
5	2.6523	2.6530	2.6546	2.6570	2.6601
10	2.6518	2.6526	2.6542	2.6565	2.6596
15	2.6514	2.6521	2.6537	2.6561	2.6592
20	2.6509	2.6516	2.6532	2.6556	2.6587
25	2.6504	2.6511	2.6527	2.6551	2.6582
30	2.6499	2.6507	2.6522	2.6546	2.6577
35	2.6495	2.6502	2.6518	2.6542	2.6572

the observed variation. The density probably decreases slightly with increasing content of substitutional Al and interstitial Li, the increase in cell volume overweighing the increased molecular weight of the cell contents. The reality of the variation of the density of quartz is readily shown qualitatively by allowing quartz grains to sink through a lengthy column of heavy liquid in which there is a small density gradient, the grains sinking to different depths, or by differential sinking in a heavy liquid that is gradually and uniformly warmed. A source of error in the precise direct measurement of the density of quartz is the presence of liquid and gaseous inclusions and of solid foreign particles.

Precise measurements of the density of amethyst are lacking. The density may be slightly higher than that of average colorless quartz because of the presence of foreign pigmenting particles of iron oxide, but for all practical purposes it can be taken as 2.651 at room temperature. The density of rose quartz containing exsolved rutile was found in one instance[4] to be relatively low (Table 19). The density of smoky quartz is virtually identical with that of average colorless quartz, perhaps very slightly higher.[4]

The density of the chalcedonic varieties of quartz is less than that of crystals and is variable (usually 2.59–2.63) because of porosity, admixed impurities, and a content of non-essential water. Jasper and other pigmented micro-crystalline varieties show a wide variation, in part over 2.65, which is due to admixed material. Very finely powdered quartz shows a decrease in density[5] that has been attributed to the formation of amorphous material.

The specific gravity of quartz was measured in 1672 by Robert Boyle,[6] who said, ". . . having Hydrostatically and with a tender Ballance examin'd the weight of it, first in the Air, and then in Water, I found its weight to be to that of Water of equal bulk as two and almost two thirds to one: Which, by the way, shews us, how groundlessly many Learned Men, as well Ancient as Modern, make Crystal to be but Ice extraordinarily harden'd by a long and vehement Cold; whereas Ice is bulk for bulk lighter than Water, (and therefore swims upon it) and (to add that Objection against the vulgar error) Madagascar and other Countreys in the Torrid Zone abound with Crystal." The first precise determination of the density of quartz is that of Beudant (1828),[7] who obtained 2.6540 at an unstated temperature relative to water at the same temperature; the value corresponds to a temperature of about 18°. Fahrenheit (1726)[8] obtained 2.669.

References

1. The coefficients of thermal dilation and equations relating the various expressions of the density with regard to reference medium and temperature are given by Sosman: *Properties of Silica*, New York, 1927, p. 288, and Ahlers: *Zs. Kr.*, **59**, 293 (1924).
2. Ahlers (1924).
3. See summaries in Sosman (1927) and Ahlers (1924); Smakula and Sils: *Phys. Rev.*, **99**, 1744 (1955).
4. Frondel and Hurlbut: *J. Chem. Phys.*, **23**, 1215 (1955).
5. See Culbertson and Weber: *J. Am. Chem. Soc.*, **60**, 2695 (1938); Tammann and Moritz: *Zs. anorg. Chem.*, **218**, 267 (1934); Sosman (1927); Clelland and Ritchie: *J. Appl. Chem.*, **2**, 31 (1952).
6. Boyle: *Essay about the Origine and Virtues of Gems*, London, 1672, p. 81.
7. Beudant: *Ann. Phys.*, **14**, 474 (1828).
8. Fahrenheit: *Phil. Trans.*, **33**, 114 (1726).

MISCELLANEOUS PROPERTIES

The *thermal conductivity*[1] of quartz at room temperature parallel to [0001] is about 0.029 and perpendicular to [0001] is about 0.016 g. cal. per cm.[2] per sec., per degree C. per cm. gradient; the conductivity decreases with in-creasing temperature.

Crushing strength[2] of small pieces at room temperature is about 24,000 kg. per cm.[2] parallel to [0001] and about 10 per cent less perpendicular thereto.

Dielectric constants[3] at room temperature are about 4.6 with field parallel to [0001], and about 4.5 perpendicular thereto, decreasing with increasing temperature. The values apparently vary slightly in different specimens.

Electrical resistivity[4] at 20° is 20 × 10^{15} ohms per cm.[2] per cm. perpendic-ular to [0001] and about 0.1 × 10^{15} parallel thereto; the resistivity decreases very rapidly with increasing temperature to about 100 × 10^3 ($\perp$) and 50 × 10^3 ($\parallel$) at 1000°. The anomalously high conductivity parallel to [0001], apparently ionic in character, varies between different specimens. The conductivity at right angles to [0001] is essentially electronic.

The mean coefficients of *linear thermal expansion*[5] at 0° are $7.10 \times 10^6 \Delta m$ parallel to [0001] and $13.24 \times 10^6 \Delta m$ perpendicular thereto. The mean coefficients increase with increasing temperature to 573°, where there is a sudden increase followed by a decrease in the high-quartz region.

References

1. Knapp: *J. Am. Ceram. Soc.*, **26**, 48 (1943); Birch and Clark: *Am. J. Sci.*, **238**, 529, 613 (1940).
2. Bridgman: *Proc. Am. Ac. Arts Sci.*, **76**, 55 (1948) and **77**, 187 (1949).
3. Rao: *Proc. Indian Ac. Sci.*, **25A**, 408 (1947).
4. Verhoogen: *Am. Min.*, **37**, 637 (1952); Wenden: *Am. Min.*, **42**, 859 (1957).
5. Rosenholtz and Smith: *Am. Min.*, **26**, 103 (1941); Jay: *Proc. Roy. Soc. London*, **142**, 237 (1933).

HIGH-LOW INVERSION

The existence of an inversion in quartz was first recognized in 1889 by Le Châtelier.[1] He observed a discontinuity in the thermal dilation of a sample of clay containing admixed quartz, and by study of the dilation of bars cut from single crystals of quartz and by observation of the change in the optical rotatory power of quartz plates located the inversion at about 570°. The terms α-quartz and β-quartz were originally applied[2] to the low- and high-temperature modifications, respectively, but this usage was reversed with attendant confusion by some later authors. The terms low-quartz (or simply quartz) and high-quartz have since come into general use.

The inversion temperature had been considered by many to be a constant for all natural quartz, and the value 573° ± 1° (on rising temperature) often has been cited, but it is now known that the inversion temperature is variable[3] as a function of compositional variation in the quartz. The inversion is highly reproducible in a specific sample. The value for pure quartz is not known precisely, but is in the neighborhood of 573.5° at 1 atm. (on rising temperature). The inversion temperature in a specific sample varies slightly with the physical state of the sample. The inversion of a single piece of quartz takes place at a slightly higher temperature on heating and a slightly lower temperature on cooling, as compared to the inversion of an equivalent fine-granular sample both on heating and on cooling. In a single piece of quartz the inversion may take place 1° or 2° lower on falling than on rising temperature. The superheating and supercooling of the powdered equivalent generally is a matter of only a few tenths of a degree. Massive fine-grained samples, such as of novaculite, chert, and chalcedony, show a very weak or negligible heat effect at the inversion unless they have been ground to a particle size approaching that of the individual grains. This has been interpreted as a result of differential stresses being set up in a mosaic of randomly interlocking grains because of the variation of the coefficients of thermal expansion with crystallographic direction. Mechanical restraint has been observed to lower the inversion temperature of cristobalite.[4]

The inversion temperature of natural quartz varies over a range of at least 38°C. The inversion temperature decreases with increasing content of Al and alkalies in solid solution. When the content of these elements becomes relatively large, the transformation becomes sluggish and weak, and spreads out over a range of temperature. Over 95 per cent of all natural specimens invert within a range of 2.5° from 573°. A much larger range of inversion temperatures has been found in synthetic quartz grown under conditions permitting the entrance of Al, alkalies, and Ge into solid solution. Material prepared by sintering oxides in the proportion $Li_2O.Al_2O_3.12SiO_2$ at 1300° afforded quartz with an inversion as low as 463° ± 15° (rising temperature). In synthetic material, the substitution of Ge raises the inversion temperature, in one instance to about 640°, and the inversion become sluggish and broadened with increasing content of Ge. Increase in the content of Al and of Ge is accompanied by an increase in the volume of the unit cell (see further under *Unit Cell Dimensions*).

Single crystals of quartz that show growth zones, as revealed by differences in color or by differential etching or irradiation response, also show differences in the inversion temperatures of the various zones. The inner zones of several crystals have been found to give relatively low inversion temperatures, presumably corresponding to a higher content of Al (and a higher temperature of formation). The inversion temperature reflects the environment in which the quartz crystallized, since this determines the amount and kind of elements in solid solution, and in particular circumstances can be used to study temperature zoning in veins and similar problems.[3,5]

The high-low inversion of quartz is of the rapid or displacive type,[6] involving a change in bond angle between adjoining (SiO_4) tetrahedra rather than a complete disruption of the crystal structure during the transformation. Both high- and low-quartz can transform at the inversion point as single-crystals.

As the inversion point is approached on rising temperature, the properties of low-quartz begin to change relatively rapidly some 50° or more below the inversion point, in contrast to the behavior on cooling high-quartz down toward the inversion point. There is an abrupt change in nearly all the physical properties at the inversion point. The changes are reversible, and none of the properties are permanently altered after cycling back through the inversion, although Dauphiné twinning (which see) may be acquired on cooling down from the high-quartz region. At or very near the inversion point the chemical reactivity and the diffusivity of quartz for foreign ions may be considerably enhanced. A relatively strong, evanescent birefringence is produced as the inversion point is passed, perhaps because of large stresses set up between different parts of the (single-crystal) sample. Two apparently biaxial intermediate phases have been reported[7] in the immediate vicinity of the inversion point.

The inversion from low- to high-quartz is endothermic and is accompanied by an increase in volume of approximately 3.25 c.c. per kg. or 0.86 per cent by volume. The inversion temperature is raised approximately 1°C. by each 40 atm. increase in hydrostatic pressure,[8] to 599° at 1000 bars and 815° at 10,000 bars. No transition has been observed in low-quartz between 4.2° and 846° Kelvin.[9]

Inversion Criteria. The question arises whether a given natural quartz crystal or anhedron originally crystallized as low-quartz or as high-quartz. Various criteria have been suggested. They derive both from the properties of the two substances and from secondary characters imposed by the inversion. In general the criteria as afforded by natural material are negative in so far as the identification of high-quartz is concerned.

(a) *Twinning*. The presence of secondary Dauphiné twinning was originally suggested[10] as a criterion of high-quartz in the belief that this feature was a concomitant of the inversion. Later work, however, has shown that secondary Dauphiné twinning can be produced by strain at temperatures much below the inversion point (see *Dauphiné Twinning*). The nature of the twin boundaries, a point mentioned in the early work, is found to depend on a variety of factors, including the stress pattern and zoning in the crystals, and seems of no value in the present connection. A three-fold arrangement of Dauphiné twinning in quartz, as seen in etched basal sections, can be considered as presumptive evidence of origin as low-quartz.[11] It has been hypothesized that right-left twins should be less frequent in a magmatic environment, but evidence is lacking. The presence of the familiar triangular or lamellar types of Brazil twinning probably is indicative of crystallization below the inversion point. Crystals of a bipyramidal habit twinned on an inclined law are suggestive of high-quartz because such laws are very rare in low-quartz and are quite common in known high-quartz. Twinning on the Combined Law is a criterion of low-quartz if it can be demonstrated by morphological or other evidence that the 180° rotation part of the twin operation is not secondary.

(b) *Crystal Habit*. Crystals known or presumed from environmental evidence to have crystallized as high-quartz have a habit consistent with the hexagonal trapezohedral (6 2 2) point-symmetry of high-quartz as deduced from x-ray and etching studies. The form development that appears to be typical of the mineral is the hexagonal bipyramid $\{10\bar{1}1\}$. The occurrence of this habit, at least in an environment consistent with a high temperature of formation, can be taken as suggestive of high-quartz. It is not conclusive evidence of hexagonal trapezohedral symmetry because $\{10\bar{1}1\}$ is not a general form; the hexagonal trapezohedron is the only general form possible in high-quartz. The general form $\{21\bar{3}2\}$ has been observed on high-quartz but is very rare. An equant bipyramidal habit, with r and z more or less equal in size, is not rare in low-quartz and closely simulates in appearance and angles the $\{10\bar{1}1\}$ habit of high-quartz. General forms are of frequent occurrence on low-quartz crystals, revealing the three-fold rather than six-fold symmetry of the c-axis, and when present identify low-quartz. Vicinal growth features on the surface of the crystals also may reveal the point-symmetry and hand. The forms $\{20\bar{2}1\}$ and $\{30\bar{3}2\}$ are rather common on high-quartz, but $\{20\bar{2}1\}\{02\bar{2}1\}$ and $\{30\bar{3}2\}\{03\bar{3}2\}$ are quite rare on low-quartz.

(c) *Hand.* Both high- and low-quartz are enantiomorphous, and the hand is not affected by the inversion. Right- and left-handed crystals have essentially the same frequency of occurrence in low-quartz. Whether there are systematic differences in high-quartz is not known.

(d) *Composition.* The solid solubility of Al and alkalies is considerably greater in high-quartz, tridymite, and cristobalite than in low-quartz. The limits of compositional variation as a function of environment, however, are not well established in either low- or high-quartz. Possibly compositional criteria of origin may become available. Supplementary morphological or other evidence might be necessary to distinguish between high-quartz, tridymite, and cristobalite. Exsolution phenomena contingent on an inversion also would be useful. Since the amount of solid solution can vary between the different face-loci of a crystal, a three-fold distribution below alternate faces of the terminal pyramid would be evidence for low-quartz.

(e) *Cracking.* The development of closely spaced cracks has been suggested as a criterion of inversion, mostly on theoretical grounds as a consequence of the attendant volume change. Experimental observations show that the development of cracks during inversion is primarily a consequence of thermal gradients in the crystal. Thin plates of quartz can be repeatedly cycled through the inversion point without cracking if the heating and cooling rate is less than about 1°C. per minute.[11] At geologic rates of cooling the thermal gradients should be very small. Composition also is a factor, and quartz containing relatively large amounts of other elements in solid solution is resistant to cracking.

References

1. Le Châtelier: *C. R.*, **108**, 1046 (1889).
2. Mügge: *Jb. Min.*, Festbd., 181, 1907.
3. Keith and Tuttle: *Am. J. Sci.*, Bowen Vol., 203, 1952.
4. Plumat: abs. in *Chem. Abs.*, **42**, 7201 (1948).
5. Tuttle: *Am. Min.*, **34**, 723 (1949).
6. Buerger, in Smoluchowski *et al.: Phase Transformations in Solids*, Wiley, New York, 1951.
7. Steinwehr: *Zs. Kr.*, **99**, 292 (1938).
8. Yoder: *Trans. Am. Geophys. Union*, **31**, 827 (1950).
9. Pavlovic and Pepinsky: *J. Appl. Phys.*, **25**, 1344 (1954).
10. Mügge (1907); Wright and Larsen: *Am. J. Sci.*, **27**, 421 (1909).
11. Frondel: *Am. Min.*, **30**, 447 (1945).

PIEZOELECTRICITY AND PYROELECTRICITY

Piezoelectricity. Quartz, when submitted to mechanical stress along certain directions, develops electric charges upon its surface; an applied electric field produces an analogous mechanical strain in the crystal. Compression along an a-axis produces a negative charge on that end of the (polar) a-axis terminated by x and s faces and a positive charge on the opposite end; tension reverses the signs of the charges. Compression along the perpendicular to a $\{10\bar{1}0\}$ prism face produces charges on the ends of the a-axis identical with those produced by tension along the a-axis; tension along this direction reverses the sign of the charges. The piezoelectric effect also may be produced by properly directed flexural or torsional stresses. No effect is produced by mechanical or electrical stress directed along the c-axis, or in theory by hydrostatic pressure.

The piezoelectric and elastic properties of quartz usually are described with reference to an orthogonal axial system, XYZ, in which X is coincident with an a-axis (the positive end in front, corresponding with the sign of charge produced by tension along this axis), Z is coincident with the c-axis, and Y is perpendicular to both X and Z and coincident with $[10\bar{1}0]$. Most writers hitherto have used a right-handed axial system for both right- and left-handed quartz, with $+Y$ to the right, but this requires a change in the sign of all piezoelectric constants on passing from a right- to a left-handed crystal, and the convention has now been adopted[1] of using a right-handed axial system for right quartz and a left-handed system with $+Y$ to the left for left quartz. Angles of rotation in these axial systems are termed positive when the sense of rotation is from $+X$ to $+Y$, from $+Y$ to $+Z$, and from $+Z$ to $+X$, regardless of hand (see Fig. 6). Sometimes X and Y are called the electrical and mechanical axes, respectively.

The piezoelectric property of quartz has found important application in the control of the frequency of oscillation of electronic circuits.[2] A disc or rectangular plate of quartz cut at a particular orientation from a quartz crystal is dimensioned so that one of its natural frequencies of mechanical vibration coincides with the frequency of the circuit. When the plate is mounted between electrodes and properly introduced into the circuit, the frequency of the applied alternating field is stabilized by the reaction of the fixed mechanical vibration back on the driving circuit through the indirect piezoelectric effect. A specific orientation is selected for the quartz plate to effect compensation between the temperature coefficients of the several elastic constants involved in the particular mode of vibration utilized, so that the resonant frequency remains constant over a relatively wide range of temperature.

The values of the piezoelectric and elastic constants of quartz are tabulated below (room temperature).[3]

Piezoelectric strain constants (static method):

$$d_{11} = +6.9 \times 10^{-8} \qquad d_{14} = -2.0 \times 10^{-8}$$

Piezoelectric stress constants (calculated from the strain constants and elastic constants cited):

$$e_{11} = +5.2 \times 10^4 \qquad e_{14} = +1.2 \times 10^4$$

ADIABATIC ELASTIC CONSTANTS

(All $\times 10^{-12}$ cm.^{2} dyne^{-1})	(All $\times 10^{10}$ dyne cm.$^{-2}$)
$s_{11} = 1.269$	$c_{11} = 87$
$s_{33} = 0.97$	$c_{33} = 108$
$s_{44} = 2.00$	$c_{44} = 57$
$s_{12} = -0.17$	$c_{12} = 7.6$
$s_{13} = -0.15$	$c_{13} = 15$
$s_{14} = -0.43$	$c_{14} = 17$
$s_{66} = 2.9$	$c_{66} = 40$

The value of Young's modulus $Y = 1/s_{33}'$ for any direction Z' in quartz can be expressed as follows (where in the orthogonal axial system XYZ defined above, θ is the angle between Z' and Z and is positive when rotated from $+Z$ toward $+X'$, and ϕ is the angle of X' with $+X$ and is positive when measured toward $+Y$):

$$s_{33}' \times 10^{12} = 1269 - 841 \cos^2 \theta + 543 \cos^4 \theta - 862 \sin^3 \theta$$

$$-862 \sin^3 \theta \cos \theta \sin^3 \phi$$

Oscillator-Plates. The largest single use of quartz crystals has been in the manufacture of quartz oscillator-plates.[4] During the period 1939–1945, when quartz was in great demand for use in military radios and other devices, about 21 million pounds of quartz crystals were exported from Brazil, chiefly into the United States. The great bulk of the material comprised faced (euhedral) crystals over 150 grams in weight. Only those parts of the crystals free from twinning, cracks, inclusions, and other flaws are usable. Over 80 million oscillator-plates were made in the United States during this period, at a total cost of roughly a half-billion dollars. There also was a considerable production in other countries. The urgent demand for large production and the low level of crystallographic knowledge in the quartz oscillator-plate industry as constituted during the war years in the United States resulted in an appalling waste. Most of the quartz crystals employed came from Brazil (see further under *Occurrence*). Domestic production is negligible. Natural crystals of the size and quality needed for oscillator-plates and for optical use represent a very limited and rapidly diminishing world resource. Needs are now being partly met by synthetic quartz crystals (see *Synthesis*).

Some types of quartz oscillator-plates are oriented orthogonally with relation to the X or Y axis, but others, including most high-frequency types, have a more general orientation; many high-frequency types for radio use are rectangular plates with one edge parallel to X and the other, Z', tilted at various specific angles in a $+$ or $-$ direction relative to the c-axis or Z. Two of the most used cuts, the AT and BT, are inclined at angles of approximately $+35°15'$ and $-49°20'$, respectively. The correct orientation of inclined plates in general requires a knowledge of the hand of the quartz and the direction of polarity of the X-axis. Hand and polarity can be determined separately, by an immersion conoscope and an electrometer, or simultaneously on properly oriented surfaces by etch-figures. The angular tolerances are strict and in general require x-ray[5] goniometric control during sawing operations to better than $10'$ of arc.

A number of different cutting schemes have been employed.[6] Two methods for cutting relatively large crystals are shown in Figs. 75 and 76.

Pyroelectricity. Quartz develops an electric charge on its surface as a result of change of temperature. The effect is due to internal stresses caused by temperature gradients within the crystal and is of piezoelectric origin; although ordinarily designated as pyroelectricity, it is not the true, primary pyroelectric effect.[7] The sign and distribution of the surface charges can be demonstrated by dusting the surface of the crystal with charged powders. A mixture of red lead $(+)$ and sulfur $(-)$ is used in the Kundt method,[8] and magnesium oxide smoke[9] obtained by burning metallic magnesium in a closed vessel containing the quartz also can be employed. On cooling, the surface charges are identical in sign with the piezoelectric charge obtained by compressing the crystal in the same direction; in an untwinned crystal, regardless of hand, prism edges on which are situated the s and x faces are negatively charged and the alternate edges are positively charged. On heating, the effects are reversed. In making powder tests for pyroelectricity, care must be taken to discharge the surface after the crystal has started to cool throughout its body and then to test after a further drop in temperature. The charges set up on cooling are caused by a gradient in which the surface temperature is less than that of the interior, corresponding to compression. The distribution of twinning also may be revealed by the surface charges. In general, methods for the orientation of quartz based on pyroelectric phenomena are much less satisfactory than those based on etching.

Pyroelectricity was first observed in quartz and other minerals by Brewster (1824).[10] Following the discovery by J. and P. Curie (1880)[11] that similar effects could be obtained by pressure alone, at constant temperature, to which the name piezoelectricity was applied, the thermal effects in quartz were explained[12] as due to stress. Quartz shows a positive triboelectric effect against most metals.[13] After heating at temperatures above $350°$ and at 10^{-4} mm. pressure the quartz develops a negative charge.

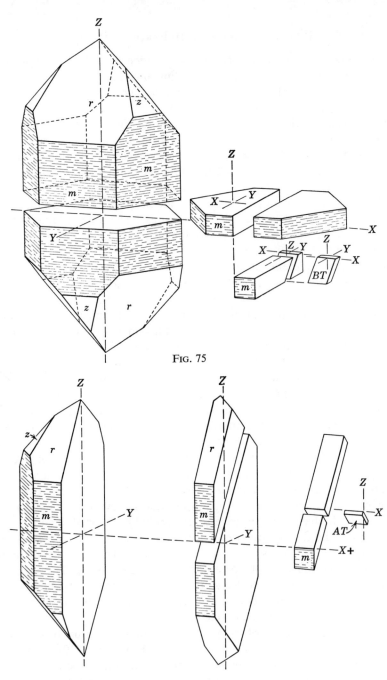

FIG. 75

FIG. 76

FIGS. 75 and 76. Cutting schemes for large quartz crystals. *Z*-slab method (above), *X*-slab method (below).

124

References

1. *Proc. Inst. Radio Eng.*, **37**, 1378 (1949).
2. For general accounts see Cady: *Piezoelectricity*, New York, 1946, and Symposium on Quartz Oscillator-Plates, *Amer. Min.*, **30**, 205–468 (1945); Scheibe: *Piezoelektrizität des Quarzes*, Dresden, 1938.
3. From Cady (1946).
4. For general review see Frondel and others: Symposium on Quartz Oscillator-Plates (1945).
5. See Parrish and Gordon: *Am. Min.*, **30**, 326 (1949).
6. See Gordon and Parrish: *Am. Min.*, **30**, 347 (1949).
7. See discussion in Cady (1946) and Sosman: *Properties of Silica*, New York, 1927.
8. Kundt: *Ann. Phys. Chem.*, **20**, 592 (1883); von Kolenko: *Zs. Kr.*, **9**, 1 (1884).
9. Maurice: *Proc. Cambridge Phil. Soc.*, **26**, 491 (1930).
10. Brewster: *Edinburgh J. Sci.*, **1**, 208 (1824).
11. J. and P. Curie: *C. R.*, **93**, 294 (1880).
12. Friedel and J. Curie: *Bull. soc. min.*, **5**, 282 (1882).
13. Mainstone: *Phil. Mag.*, **23**, 620 (1937).

OPTICAL PROPERTIES

COLOR

The color of quartz varies through a wide range and is the basis of distinction of a number of named varieties. Normally, quartz is colorless and transparent, with a vitreous luster. The intensity of the luster may be increased by heating the material to high temperatures. The luster is influenced by the nature of the pigmenting material in colored varieties, and by the grain size in aggregates; it is sometimes greasy to fatty in microcrystalline types and in white or milky granular quartz. The streak is white, faintly tinted in colored varieties.

The principal colored types of coarsely crystallized quartz are these: *amethyst*, bluish violet to reddish or purplish violet; *smoky quartz*, ranging from faint to dark smoky brown and to almost black; *citrine*, yellow to yellowish brown; *blue quartz*, pale milky blue to smoky blue and lavender-blue; *rose quartz*, rose-pink to rose-red. These varieties and the origin of their colors are described in another section (see *Varieties of Quartz*). Quartz, especially the fine-grained varieties, may be variously pigmented by admixed foreign material: yellow-brown or red by finely divided goethite or hematite; black by disseminated sulfides; white, gray, or yellowish by dispersed clayey material; green from included particles or crystals of chlorite, actinolite; etc. Many of these types, which include jasper, prase, plasma, and other named varieties, are described under the fine-grained varieties of quartz or are mentioned in the section on *Inclusions in Quartz*. Quartz may also be white, milky white, or grayish white through the scattering of light by minute cavities or flaws.

ABSORPTION

Quartz and the other polymorphs of silica are highly transparent for all visible wavelengths.[1] In the ultraviolet quartz is much more transparent than most crystalline substances, glasses, and liquids. Among ordinary optical materials it is exceeded in transparency by CaF_2 and LiF. The absorption of quartz, however, becomes appreciable in the short ultraviolet, and quartz lenses and prisms of ordinary dimensions cannot be used with radiation much below 175 mμ. In the infrared, quartz shows relatively strong absorption at about 3 μ, 4 μ, 5.3 μ, and 6–7 μ; strong metallic reflection occurs at about 8.5–9 μ and less strong reflections occur at 12–13 μ and 21–26 μ. At longer wavelengths quartz again becomes more or less transparent. In the infrared, the absorption of the ordinary ray is greater than that of the extraordinary ray. The Raman and the Rayleigh scattering in quartz have been measured.[2] The absorption of amethyst and of other colored varieties of quartz, all of which are perceptibly dichroic in the visible region with $\epsilon > \omega$, is described in another section (*Euhedral and Coarse-Grained Varieties*).

References

1. Summary in Sosman: *The Properties of Silica*, New York, 1927; see also, for infrared, Pfund: *J. Opt. Soc. Amer.*, **35**, 611 (1945), and Drummond: *Proc. Roy. Soc. London*, **153**, 318, 328 (1936); and for ultraviolet, Gillis *et al.*, *C. R.*, **229**, 816 (1949).
2. Krishnan: *Proc. Indian Ac. Sci.*, **22A**, 329 (1945); Bhagavantam and Venkateswaren: *Proc. Indian Ac. Sci.*, **19A**, 108 (1944); Saksena and Bhatnagar: *Proc. Indian Ac. Sci.*, **30A**, 308 (1949); Nedungadi: *Proc. Indian Ac. Sci.*, **11A**, 86 (1940); Michalke: *Zs. Phys.*, **108**, 748 (1938); Raman and Nedungadi: *Nature*, **145**, 147 (1940); Allen: *Nature*, **145**, 306 (1940); Gross: *C. R. ac. sc. URSS*, **26**, 757 (1940); Narayanaswamy: *Proc. Indian Ac. Sci.*, **26A**, 521 (1947).

INDICES OF REFRACTION

Quartz is an optically positive, uniaxial substance. The indices of refraction of colorless quartz may be given for ordinary purposes as:

ϵ	ω	Birefringence	(Room Temp.)
1.553	1.544	0.009	White light
1.5533	1.5442	0.0091	Na, 589.3 mμ

The refractivity of quartz has been measured by a number of different workers to a precision of 0.00001 or better, and a great many additional measurements of lower precision have been reported. High-precision measurements designed to compare the refractivities of different samples of quartz or of quartz of different colors or treatments show that a significant variation exists in the indices of refraction as a function of the material itself, aside from the errors of measurement. Differences of at least 0.00002 can be expected between any two randomly selected crystals of colorless quartz of optical grade, and the maximum variation in such material may possibly be as large as 0.0001 or 0.0002. Significant variations in refractivity also have been found between different parts of a single crystal. This intrinsic variation considerably reduces the accuracy of the measurements, in the sense of representing the refractivity of the substance quartz, below the precision with which the measurements can be made. No measurements have been reported that are known to represent liminal values, such as for stoichiometric SiO_2 free from solid solution, and very few measurements of the refractivity have yet been reported which refer to quartz in any defined or reproducible chemical composition. Measurements of the refractivity to a precision greater than about 0.0001 to be meaningful should be accompanied by determinations of the chemical composition, density, and unit cell dimensions to relevant precision. Analyses of quartz and study of other physical properties indicate that the main mechanism of compositional variation in quartz is a substitution of Al for Si, to which the variation in indices of refraction may be referred.

The indices of refraction of natural colorless quartz of optical grade are given as a function of wavelength at 18° in Table 15. The data were derived

TABLE 15. MOST PROBABLE VALUES OF THE INDICES OF
REFRACTION OF QUARTZ AT 18° FOR VARIOUS WAVELENGTHS

Radiating Element	λ in mμ	ω	ϵ
Al	185.467	1.67578	1.68997
Al	193.583	1.65999	1.67343
Au	200.06	1.64927	1.66227
Zn	202.55	1.64557	1.65842
Au	204.448	1.64288	1.65562
Au	211.07	1.63432	1.64671
Cd	214.439	1.63039	1.64262
Cd	219.462	1.62497	1.63698
Cd	226.503	1.61818	1.62992
Cd	231.288	1.61401	1.62559
Au	242.796	1.60525	1.61650
Au	250.329	1.60032	1.61139
Cd	257.304	1.59622	1.60714
Al	263.155	1.59309	1.60389
Cd	274.867	1.58752	1.59813
Au	291.358	1.58098	1.59136

TABLE 15   (cont'd)

Radiating Element	λ in mμ	ω	ϵ
Sn	303.412	1.576955	1.58720
Au	312.279	1.57433	1.584485
Cd	325.253	1.570915	1.58095
Cd	340.365	1.56747	1.577385
Al	358.68	1.563915	1.573705
Ca	396.848	1.55813	1.56772
Hg	404.656	1.557156	1.56671
H	410.174	1.556502	1.566031
H	434.047	1.553963	1.563405
Hg	435.834	1.553790	1.563225
Cd	467.815	1.551027	1.560368
Cd	479.991	1.550118	1.559428
H (F)	486.133	1.549683	1.558979
Cd	508.582	1.548229	1.557475
Mg (b)	518.362	1.547651	1.556877
Cd	533.85	1.546799	1.555996
Hg	546.072	1.546174	1.555350
Hg	579.066	1.544667	1.553791
He	587.563	1.544316	1.553428
Na (mean)	589.29	1.544246	1.553355
Au	627.82	1.542819	1.551880
Cd	643.847	1.542288	1.551332
H (C)	656.278	1.541899	1.550929
He	667.815	1.541553	1.550573
Li	670.786	1.541466	1.550483
He	706.520	1.540488	1.549472
He	728.135	1.539948	1.548913
K	766.494	1.539071	1.548005
Rb	794.763	1.538478	1.547392
O	844.67	1.537525	1.54640
	1000.00	1.53503	1.54381
Hg	1014.06	1.53483	1.54360
He	1083.03	1.53387	1.54260
	1200.00	1.53232	1.54098
	1300.00	1.53102	1.53962
	1400.00	1.52972	1.53826

by Sosman[1] from prism measurements made under superior experimental conditions and reported in the literature.[2] The precision of the original measurements is about 0.00001, or better in some instances. The data are significant to the decimal places cited, however, only in a relative sense as a function of wavelength. The indices of refraction of quartz as a function of wavelength from the ultraviolet to the infrared can be calculated to five significant figures from the following expressions:[7]

Ordinary ray (ω):

$$n^2 = 3.53445 + \frac{0.008067}{\lambda^2 - 0.0127493} + \frac{0.002682}{\lambda^2 - 0.000974} + \frac{127.2}{\lambda^2 - 108}$$

Extraordinary ray (ϵ):

$$n^2 = 3.5612557 + \frac{0.00844614}{\lambda^2 - 0.0127493} + \frac{0.00276113}{\lambda^2 - 0.000974} + \frac{127.2}{\lambda^2 - 108}$$

The indices of refraction of colorless quartz for certain wavelengths are given as a function of temperature[3] in Tables 16 and 17. The precision of the data of Table 16 is about 0.0001 at room temperature and somewhat less at higher and lower temperatures. The values decrease algebraically with increasing temperature, as with increasing wavelength, at an accelerating rate to the inversion temperature at about 573°, where there is a sudden drop followed by a decelerating rise in the values in the high-quartz region. The temperature coefficients of the refractive indices of quartz vary with temperature. Near room temperature the best values probably are -7.54×10^6 $\Delta n\epsilon$, -6.50×10^6 $\Delta n\omega$ (for Na).

The birefringence of quartz decreases with increasing wavelength, and the change is relatively rapid at small wavelengths. The birefringence also decreases slowly with increasing temperature up to 573°, where there is an abrupt discontinuity with a following increase with increasing temperature in the high-quartz region. Quartz is slightly birefringent in the direction of the optic axis (Fresnel, 1831). The difference in the two rays is about 0.000071. The effect arises in that plane-polarized light traveling in the direction of the optic axis is separated into two circularly polarized rays of opposite sign and of different speeds. The two rays reunite on emergence to form again a plane-polarized ray, rotated with respect to the original ray.

Differences ranging up to at least 0.0004 have been reported between refractivities of colorless quartz and of the colored varieties thereof (see *Euhedral and Coarse-Grained Varieties*). In smoky quartz the indices are lower than in colorless material. The indices are lower the deeper the color, and they increase when the color is discharged by heating. The indices of amethyst appear to be relatively high. Data are virtually lacking on the variation in the indices of refraction (and other properties) of blue quartz and rose quartz as a function of composition, and on the effect of irradiating quartz with x-rays. Small differences have been found in the refractivities of right-handed and left-handed quartz,[4] with right-handed quartz higher in the fifth decimal, but it is not known whether the samples studied had the same chemical composition.

TABLE 16. REFRACTIVE INDICES OF QUARTZ AT VARIOUS TEMPERATURES

ω

Solar Line	λ in mμ	Low-Quartz								High-Quartz		
		−140°C	−45°	23°	115°	212°	305°	410°	550°	580°	650°	765°
$G'(H\gamma)$	434.047	...	1.5633	1.5634	1.5629	1.5623	1.5615	1.5598	1.5551	1.5503	1.5521	1.5532
(d)	466.8	...	1.5609	1.5608	1.5603	1.5597	1.5588	1.5572	1.5526	1.5478	1.5492	1.5506
F	486.133	1.5594	1.5594	1.5593	1.5589	1.5581	1.5573	1.5558	1.5512	1.5464	1.5475	1.5490
(c)	495.75	...	1.5587	1.5587	1.5582	1.5576	1.5567	1.5552	1.5503	1.5456	1.5468	1.5481
b_2	517.27	...	1.5574	1.5574	1.5568	1.5562	1.5553	1.5538	1.5488	1.5442	1.5454	1.5469
D_2	588.997	1.5541	1.5539	1.5537	1.5532	1.5526	1.5515	1.5499	1.5451	1.5405	1.5417	1.5431
α	627.8	1.5526	1.5525	1.5522	1.5517	1.5510	1.5500	1.5486	1.5437	1.5389	1.5403	1.5416
C	656.278	...	1.5516	1.5513	1.5508	1.5502	1.5491	1.5475	1.5427	1.5380	1.5393	1.5406
B	687.2	1.5506	1.5506	1.5504	1.5499	1.5492	1.5481	1.5466	1.5419	1.5369	1.5383	1.5397
a	718.9	...	1.5499	1.5495	1.5490	1.5483	1.5472	1.5458	1.5408	1.5362	1.5375	1.5388

ϵ

Solar Line	λ in mμ	Low-Quartz								High-Quartz		
		−140°C	−45°	23°	115°	212°	305°	410°	550°	580°	650°	765°
$G'(H\gamma)$	434.047	...	1.5539	1.5540	1.5536	1.5531	1.5523	1.5510	1.5469	1.5425	1.5439	1.5454
(d)	466.8	...	1.5515	1.5514	1.5511	1.5506	1.5498	1.5483	1.5442	1.5400	1.5414	1.5429
F	486.133	1.5504	1.5501	1.5500	1.5497	1.5491	1.5483	1.5469	1.5426	1.5385	1.5399	1.5414
(c)	495.75	...	1.5494	1.5494	1.5491	1.5485	1.5477	1.5465	1.5421	1.5379	1.5393	1.5406
b_2	517.27	...	1.5481	1.5481	1.5476	1.5472	1.5463	1.5452	1.5407	1.5363	1.5377	1.5392
D_2	588.997	1.5449	1.5448	1.5446	1.5441	1.5437	1.5428	1.5414	1.5370	1.5329	1.5341	1.5356
α	627.8	1.5434	1.5434	1.5431	1.5427	1.5422	1.5413	1.5401	1.5357	1.5314	1.5328	1.5340
C	656.278	...	1.5425	1.5423	1.5418	1.5414	1.5405	1.5390	1.5349	1.5304	1.5319	1.5331
B	687.2	1.5417	1.5416	1.5414	1.5410	1.5405	1.5395	1.5382	1.5337	1.5296	1.5309	1.5321
a	718.9	...	1.5408	1.5405	1.5401	1.5396	1.5386	1.5374	1.5327	1.5288	1.5301	1.5313

TABLE 17. REFRACTIVE INDICES OF QUARTZ FOR
WAVELENGTH 5460 Å AT LOW TEMPERATURES IN VACUUM[6]

Temp.:	−200°	−150°	−100°	−50°	0°	50°
ω	1.55724	1.55694	1.55662	1.55627	1.55589	1.55552
ϵ	1.54785	1.54759	1.54732	1.54702	1.54669	1.54637

The indices of refraction of quartz were first measured by E. L. Malus in 1811, in white light. The indices were first measured[5] for particular wavelengths (solar lines) to a precision of about 0.00005 in 1828 by Rudberg, who also first measured the variation in the indices with temperature in 1832.

References

1. Sosman: *Properties of Silica*, New York, 1927.
2. Müller: *Publ. Astrophys. Obs. Potsdam*, **4**, 149 (1885); Macé de Lépinay: *J. phys.* **6**, 190 (1887); Carvallo: *C. R.*, **126**, 728 (1898); Martens: *Ann. Phys.*, **6**, 603 (1901), and **8**, 459 (1902); Trommsdorff: inaug. diss., Jena, 1901; Gifford: *Proc. Roy. Soc. London*, **100A**, 621 (1922); Paschen: *Ann. Phys.*, **35**, 1005 (1911).
3. Rinne and Kolb: *Jb. Min.*, **11**, 138 (1910), as recalculated by Sosman (1927).
4. Gifford: *Proc. Roy. Soc. London*, **70**, 329 (1902).
5. Rudberg: *Ann. Phys.*, **14**, 51 (1828).
6. Barbaron: *C. R.*, **226**, 1443 (1948).
7. Coode-Adams: *Proc. Roy. Soc. London*, **117A**, 209 (1927) and **121A**, 476 (1928).

ROTATORY POWER

The rotatory power[1] of colorless quartz measured parallel to the c-axis at 20° is given as a function of wavelength in Table 18. The absolute accuracy of the value for Na is about 0.007° per mm. Slight but real variations in the rotatory power have been reported between different specimens of colorless quartz. The rotatory power increases with temperature at an accelerating rate to the inversion at about 573°, where there is a sudden increase of 4.3 per cent, followed by a slow increase in the high-quartz region. The rate of increase[2] at room temperature is about 0.014 per cent per degree C. Quartz is optically inactive in directions inclined at 56°10′ to the c-axis.[3]

Quartz rotates the plane of polarization of light traveling parallel to the c-axis to either the right or the left. In the convention here employed, that of Biot, and correlating with the convention of identifying the morphological development as right- or left-handed, the plane of polarization is rotated clockwise in right-handed quartz and counterclockwise in left-handed quartz when the observer looks through the optical system toward the source of light and rotates the analyzing Nicol. The sense of the rotation as thus defined is opposite to the sense of the screw arrangement of the (SiO_4) tetrahedra along the [0001] direction of the structure of quartz, from which the effect stems.

TABLE 18. MOST PROBABLE VALUES OF THE ROTATORY POWER
OF QUARTZ AT 20° FOR VARIOUS WAVELENGTHS

Radiating Element	λ in mμ	ρ in degrees per mm.
Cd	226.503	201.9
Cd	231.288	190.5
Au	242.796	166.9
Au	250.329	153.9
Cd	274.867	121.10
Sn	303.412	95.02
Cd	340.365	72.45
Fe	348.534	68.585
Ca	396.848	51.115
Hg	404.656	48.945
H	410.174	47.495
H	434.047	41.924
Hg	435.834	41.548
Cd	467.815	35.601
H (F)	486.133	32.761
Cd	508.582	29.728
Hg	546.072	25.535
Na (mean)	589.29	21.724
Cd	643.847	18.023
Li	670.786	16.535
He	728.135	13.924
Rb	794.763	11.589
Hg	1014.06	6.976
	1200.00	4.889
Hg	1529.61	2.930
He	2058.20	1.527
	2500.00	0.972

References

1. Data reported before 1927 critically summarized by Sosman: *Properties of Silica* New York, 1927.
2. On the temperature coefficient of rotation in the ultraviolet region see Crook and Taylor: *Nature*, **160**, 396 (1947).
3. Szivessy: *Fortschr. Min.*, **21**, 111 (1937).

OPTICAL ANOMALIES

Undulose extinction is commonly shown by the quartz grains of rocks that have been mechanically deformed and is typically associated with cataclastic textures. Such quartz also may be biaxial with a small and variable $2V$. These features also are often shown by vein quartz. The undulose extinction may be associated with the so-called Boehm lamellae (see *Plastic Deformation*), as in certain Finnish quartzites.[4]

Undulose or patchy extinction not caused by strain may be seen in euhedral crystals with a composite or lineage structure that have grown in open spaces. Such crystals are composed of more or less separate individuals in subparallel position. In one common type of composite crystal there is a crude lamellar structure consisting of thin plate-like segments trending vertically in the crystal and radiating outward from the center (Fig. 77). The course of the lamellae can be traced by sutures on the prism faces. The internal boundaries can be seen in basal sections between crossed Nicols by the variation in interference colors, individual lamellae showing flamboyant extinction in more or less radial strips. The internal boundaries usually can be also seen in plane-polarized light as optical schlieren (or Töpler's striae)[1] because of the difference in index of refraction across the boundaries. The optical effects associated with the lamellae are best seen by examining the entire crystal or thick sawn sections thereof between crossed Nicols in an immersion tank. The distribution of the lamellae also can be studied in etched sections provided that the difference in orientation is not too small. In a prismatic crystal 2 or 3 in. thick the individual lamellae range in thickness, as measured in a horizontal direction, from roughly a millimeter up to a centimeter or so. The angular divergence between adjacent lamellae usually is approximately a degree but ranges up to 3 or 4° or more. Some of the lamellae may terminate within the body of the crystal. A well-illustrated account has been given[2] of these features in Swiss quartz crystals, and they also have been observed in quartz from Dauphiné, France,[3] and from Brazil and other localities. This type of composite structure resembles the internal structure of twisted quartz crystals and also the so-called Boehm lamellae seen in the quartz grains of metamorphic rocks, although the latter are on a much finer scale.

Quartz ordinarily is sensibly uniaxial, but an anomalous biaxial character with $2V$ small but ranging up to about 20° may be observed in strained material. A biaxial character can be produced artificially by compression in directions inclined to the optic axis, as well as by a thermal gradient, or by application of an electric field. A biaxial character is frequently shown by amethyst (which see). This variety of quartz typically shows a zonal structure composed of alternating layers of right- and left-handed quartz (Brazil twinning), and under proper circumstances the interference figure may exhibit Airy's spirals.

Very fine-grained quartz may appear isotropic because of aggregate polarization.

References

1. Töpler: *Ann. Phys.*, **127**, 556 (1866) and **131**, 33 (1867); also Kundt: *Ann. Phys. Chem.*, **20**, 688 (1883).

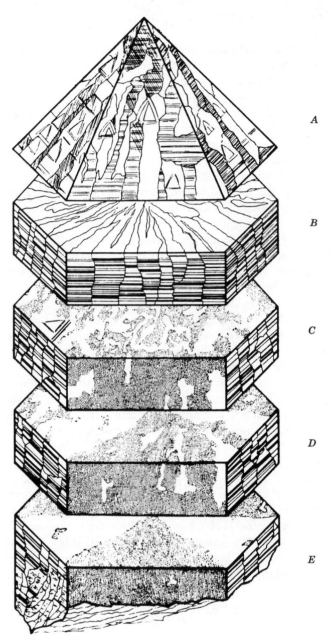

FIG. 77. Sections through a crystal of Alpine quartz, showing a crudely radial composite structure (section *B*) with associated sutures on the prism faces and the rhombohedral termination (section *A*). The variation in Dauphiné twinning (stippled) is shown in successive, etched sections (*C*, *D*, *E*). From Friedlaender (1951).

2. Friedlaender: *Beit. Geol. Schweiz, Geotech. Ser.*, no. 29, 42, 78, 89, 1951; see also Laemmlein: *Bull. Ac. Sc. URSS*, 962, 1937.
3. Weil: *C. R.*, Ire Reunion Inst. Optique, 1931.
4. Hietanen: *Bull. comm. géol. Finlande*, No. 122, 1938.

LUMINESCENCE

Single-crystals and coarse granular types of colorless quartz and of the smoky, amethyst, blue, and rose varieties do not fluoresce in either short-wave or long-wave ultraviolet radiation. Occasionally, instances of fluorescence have been observed, especially in massive granular material, but the effect is caused by inclusions or by films of foreign material along grain boundaries or cracks, often in patches or streaks, and does not arise from the quartz itself.[1] A local bluish green or yellowish white fluorescence has been noted from fluid inclusions in colorless quartz and amethyst. A milky white quartz showing parting from Vernon, British Columbia, exhibits a pinkish orange fluorescence in a long-wave ultraviolet. A bluish white fluorescence has been observed[7] under ultraviolet excitation in quartz that has been previously irradiated with x-rays. Synthetic quartz prepared by the introduction of Mn^2 and Al as coactivators into high-quartz prepared[8] in a reducing atmosphere at 1200° shows a red cathodoluminescence as a continuous band from 5400 Å to over 7000 Å with a maximum at 6500 Å. A feeble cathodoluminescence has been observed[9] in natural quartz, with bands at 3000–4900 Å and 5650–6550 Å.

Most specimens of the microgranular and fibrous varieties of quartz do not fluoresce. Occasional specimens of chalcedony, agate, and opal fluoresce a yellowish green to green, especially in short-wave ultraviolet radiation.[2] This fluorescence originates in traces of admixed or adsorbed uranyl compounds, as has been shown in the case of the chalcedonic and opaline silica associated with some sandstone-type uranium deposits on the Colorado Plateau and in the moss agates from Sweetwater, Wyoming. The thin crusts of hyalite opal found in pegmatites often show a bright uranyl fluorescence, as does some wood opal. Other specimens of chalcedony and agate have been variously reported to show a white or cream to dull yellowish orange fluorescence, best in short wave ultraviolet, and a stalactitic chalcedony from Cape Brewster, Greenland, shows a deep peach color; the activator in these instances is not known. A weak cream or yellowish response also has been noted in some specimens of chert and flint.

In x-rays, colorless quartz exhibits a weak bluish white fluorescence and a short-lived phosphorescence. The effect varies noticeably in different specimens and is less marked in quartz that is relatively weakly pigmented by radiation. Rose quartz fluoresces a pale orange.

On gradual heating, previously untreated quartz usually shows a pale bluish white or yellowish thermoluminescence[4] that in the dark becomes perceptible to the eye at about 150°, and after increasing with increasing temperature to 300–400° gradually dies out. It does not recur on reheating. The intensity of the thermoluminescence varies markedly in different samples and also in different growth zones of the same crystal. Quartz that has been irradiated with x-rays is markedly thermoluminescent (see *Irradiation Effects*). The glow is perceptible at a lower temperature and is stronger than in natural quartz. Chalcedony with a natural uranyl fluorescence also is thermoluminescent.

When heavily ground, as with a dry Carborundum wheel, or when heavily scratched, pieces of colorless, smoky, amethyst, and rose quartz show a bright yellow triboluminescence, and agate a dull reddish brown triboluminescence.[3] The effect has been known for a long time.[5] It can be observed when the test is made while the specimens are immersed in water or other liquids. The Pueblo Indians of the upper Rio Grande during their rain ceremonies beat drums to imitate thunder and rubbed together pebbles of white quartz to produce a glow simulating lightning. An instrument made for the purpose consisted of a cylinder of white vein quartz 3 in. long and 1½ in. in diameter that was rubbed in a shallow groove cut in a block of the same material.[6] A similar glow is developed when quartz is sawn with a rotating diamond blade, especially under heavy pressure and with a minimum of cooling fluid. A brilliant green glow has been observed when quartz is crushed under heavy pressure. Triboluminescence also is observed during the chipping of flint.

References

1. D. Seaman: priv. comm., 1952, from study of 1417 specimens of quartz: colorless crystals and massive (738), smoky (140), amethyst (130), rose (36), chalcedony and agate (300), jasper, chrysoprase, flint, etc. (73). See also Laurent: *Geol. För. Förh.*, 63, 59 (1941).
2. Haberlandt: *Mitt. Inst. Radiumforsch.*, no. 350, 591, 1934, and no. 391, 1, 1937; Iwase: *Bull. Chem. Soc. Japan*, **11**, 377 (1936).
3. D. Seaman: priv. comm., 1952, from study of 239 specimens of colorless quartz, amethyst, agate, etc. See also Culbertson and Weber: *J. Amer. Chem. Soc.*, **60**, 2695 (1938); Tammann and Moritz: *Zs. anorg. Chem.*, **218**, 267 (1934); Johnsen: *Cbl. Min.*, 227, 1919; Imhof: *Phys. Zs.*, **18**, 78 (1917); Lankester: *Nature*, **106**, 310 (1920) (also pp. 242, 345, 376, 409, 438, 503); Inoue, Kunitomi, and Shibata: *J. Sci. Hiroshima Univ.*, Ser. A, 129, 1939; Wick: *J. Opt. Soc. Amer.*, **27**, 275 (1937); Kratova and Karasev: *C. R. ac. sci. USSR*, **92**, 607 (1953).
4. Goldschmidt: *Forh. Vidensk. Selskr. Oslo*, no. 5, 1, 1906; Kohler and Leitmeier: *Zs. Kr.*, **87**, 146 (1934); Northup and Lee: *J. Opt. Soc. Amer.*, **30**, 206 (1940); Alt and Steinmetz: *Zs. angew. Min.*, **2**, 153 (1939); Futugami: *Proc. phys.-math. Soc. Japan*, **20**, 458 (1938); Frondel: *Am. Min.*, **30**, 432 (1945).
5. Razoumovsky: *Mém. soc. phys. Lausanne*, **2**, 13 (1787); see also refs. under Lankester (1920).
6. Ball: *Smithsonian Inst., Bur. Am. Ethnology, Anthropol.*, *Paper* **13**, 1941.
7. Audubert, Bonnemay, and Lautout: *C. R.*, **230**, 1771 (1950).
8. Claffy and Ginther: *Am. Min.*, **44**, 987 (1959).
9. Saksena and Pant: *J. Chem. Phys.*, **19**, 134 (1951); see also Peters: *Phys. Rev.*, **36**, 1631 (1930); Komovsky and Golovchiner: *Khim. Ref. Zhur.*, **4**, nos. 7–8, 19 (1941).

IRRADIATION EFFECTS

COLOR

It has been known[1] almost from the time of the discovery of radioactivity that quartz becomes smoky in color when exposed to alpha-particles, electrons, and short-wavelength x-rays proceeding from radioactive decay. Attention was first directed to this general phenomenon by the discoloration of glass vials containing radioactive preparations. Other effective radiations include the x-ray spectrum up into the long-wavelength region, and also deuterons and neutrons. A convenient and effective radiation from the point of view of absorption effects in large pieces of quartz is the general spectrum from a tungsten x-ray tube, but satisfactory results can be obtained from conventional iron- and copper-target x-ray diffraction tubes. The smoky color produced in this way can be removed by exposure to ultraviolet radiation of short wavelength and, as discussed beyond, by heat treatment.

The response of different specimens of quartz to irradiation varies markedly. The color change in colorless quartz ranges from virtually nil up to a depth comparable to that of a very dark natural, smoky quartz crystal. Rose quartz is relatively strongly affected and may be rendered black and almost opaque in small pieces. Amethyst is not affected; if the material is first decolorized by heat treatment, irradiation restores the amethystine color. Dark smoky quartz and natural citrine are relatively weakly affected. Chalcedony is weakly affected, and alternate bands may become unequally colored. Opal is not affected; tridymite strongly so. Silica glass also is colored by irradiation, the color tending to violet rather than to smoky tints, and the temperature stability of the color is considerably higher than that of both irradiated quartz and natural smoky quartz.

The artificial smoky color almost always is not uniformly distributed throughout the irradiated sample. Alternate growth bands parallel to the external crystal faces tend to be unequally affected. The zones or bands often are very thin, causing the crystal to appear uniformly colored when the bands are viewed at large angles, and are best viewed in sections cut perpendicular to the banding. The color bands or zones can be differentially etched by solvents both before and after irradiation (see *Zonal and Sectoral Structure*). The differential coloration and etching is connected with variations in the content of Al and other elements in solid solution in successive growth zones of the crystal. An unequal response also may be observed in different face-loci, as between *r* and *z*, and across the boundaries of both Brazil and Dauphiné twins. Natural smoky quartz (see *Varieties of Quartz*) generally shows a similar non-uniform distribution of the color. When a banded or zoned natural smoky quartz crystal is irradiated, the colorless or less intensely pigmented zones are more affected by the radiation and become relatively dark in color.[4] Color banding also has been observed in irradiated synthetic quartz.

The development of the smoky color is accompanied by variation in some of the properties of the quartz. The rate of solution[2] is increased, so that a small irradiated area is perceptibly more attacked by a solvent such as HF than is the adjacent non-irradiated quartz. Very precise measurements made on a quartz prism (the Arkansas quartz of column 1, Table 19) before and after irradiation showed that the indices of refraction[3] are increased by irradiation, the increase in the ϵ index being relatively large so that the birefringence also is increased. The increase in the indices of the Arkansas material is about 0.0001, the depth of irradiation color being of about average intensity. An increase in the indices of refraction also has been observed in irradiated rose quartz, amounting to approximately 0.0004.

A measurable change also may be produced in the elastic constants of quartz by irradiation. The effect can be studied and was first observed[5] in shear-mode (BT) quartz oscillator-plates in the frequency range from 5 to 9 megacycles. The frequency of oscillation decreases exponentially during irradiation, accompanying the progressive darkening of the quartz, and reaches a minimum beyond which the quartz is not further darkened. The change in frequency that can be produced varies, as does the depth of color, in different specimens of quartz over a range of about 2 kilocycles in an 8 megacycle plate. The frequency response in other types of quartz oscillator plates depends on the mode of oscillation, the results indicating that all three elastic moduli are decreased by irradiation with a relatively large decrease in C_{14}. In the so-called Y family of cuts, rotated about the X-axis, the maximum effect is obtained at about $-60°$, near the BT and BC cuts, and the minimum effect at about $+40°$ near the AT cut.[6] No specimen of quartz has been found that has not yielded a measurable change by this method, although the associated color change may be too faint to perceive by the unaided eye. The change in frequency of oscillation can be measured to a precision of a few cycles in plates in the 5–9 megacycle range. Rose quartz, which in general is relatively strongly blackened by irradiation, shows a total frequency change several times greater than that of average colorless quartz. When the smoky color is bleached by heating to temperatures over 180°, the frequency of oscillation returns to the original value. The effect is reversible. The frequency of plates made from natural smoky quartz also increases when the color is bleached thermally.[7] Conclusive direct evidence is lacking of a change in the unit cell dimensions of quartz accompanying irradiation. Theoretically the cell volume should increase on the basis of the observed changes in the elastic constants. Anomalies also have been observed in the piezoelectric constant of quartz after irradiation.[8]

BLEACHING

The smoky color produced by irradiation is unstable at temperatures over approximately 180° and bleaches out with an accompanying bluish white luminescence.[9] The spectrum is continuous[10] over the range 385–610 mμ with a maximum near 469 mμ. The elastic properties (and presumably all other properties that were affected by the irradiation) are restored to the original values if the original sample of quartz was colorless or had been heated to a temperature considerably over 180°. If the original sample had a natural smoky color, on which a further smoky color was imposed by irradiation, thermal bleaching may not restore the original properties, since the natural smoky color is itself thermally unstable. The bleaching of both the artificial and the natural smoky color is a time-temperature reaction. Using irradiation-blackened quartz cubes about 1 mm. on edge, it is found that over about 400° the bleaching takes place in a few minutes. The rate of bleaching in this temperature range apparently is primarily determined by the rate of thermal conduction in the quartz plate. At temperatures of roughly 300° the time required is a few hours, at 200° a few days, and at slightly above 180° the time of complete bleaching extends over many weeks. The time-temperature curve obtained from numerous measurements made in this way becomes asymptotic to the time axis at about 180°. The point of complete bleaching can be measured accurately by employing quartz oscillator-plates. Natural smoky quartz behaves similarly, but the temperature of stability of the color is higher, near 225°. While at fairly high temperatures smoky quartz becomes greenish in tint. Heating in an oxidizing or reducing atmosphere at relatively low temperatures has no effect on the natural color.[11] A type of smoky color (γ-centers) stable to 1000° has been introduced into quartz by electrolysis between carbon electrodes at high temperatures.[18]

THERMOLUMINESCENCE

On heating, smoky irradiated quartz shows a bluish white thermoluminescence that in part at least is associated with the decay of the color centers. The intensity of the effect is primarily a function of the depth of the irradiation color and of the rate of bleaching. Banded or zoned quartz thermoluminesces more intensely in the darker-colored parts. The thermoluminescence of irradiated quartz begins at temperatures considerably below 180°, where it apparently is not accompanied by a change in color or in elastic constants, and the thermoluminescence continues for a while after the smoky color has been completely discharged.[9] Non-irradiated natural quartz also shows thermoluminescence, in both colorless and natural smoky types, and the intensity varies markedly in different specimens and sometimes in different parts of a single crystal (see *Luminescence*). The color of the natural thermoluminescence is bluish white to yellowish. Quartz shows a bluish white fluorescence during irradiation with x-rays or electrons, and there is a brief phosphorescence. Rose quartz fluoresces a pale orange, and the color of the thermoluminescence of irradiated rose quartz is pale orange.

It has been observed that there is a change in the color of irradiated quartz in the later stages of bleaching toward brownish tints, and this is more apparent when the bleaching is effected at relatively low temperatures. There also is a suggestion of two colors with different bleaching rates or temperature dependencies in natural smoky quartz.

RELATION TO COMPOSITION

The smoky color of irradiated natural and synthetic quartz represents anisotropic color centers with absorption maxima[12] for ϵ at 460 and 625 mμ and for ω at 485 mμ. The absorption of ϵ is greater than that of ω. The color centers are related to the presence of Al in substitution for Si. The intensity of the color[13] increases with rising content of Al, or more specifically with increase in the coupled substitution of Al and Li. The different face-loci of natural crystals may be selectively colored by irradiation,[24] in general in the order $(10\bar{1}1) > (01\bar{1}1) > (10\bar{1}0)$. This order also is that of increasing content of Al and Li as found analytically in these face-loci. Similar observations have been made on synthetic quartz crystals (see *Zonal and Sectoral Structure*). Color centers apparently are not associated with Al present in interstitial positions or with Al coupled with H or OH.

Synthetic quartz containing Ge in substitution for Si also develops anisotropic color centers when irradiated.[14] These centers have an absorption maximum at about 275 mμ, and bleach more readily than the AlLi color centers. Color centers also are associated with the substitutional Ti ions of rose quartz and with Cu and As in synthetic quartz.[15] Synthetic quartz containing little or no material in solid solution is either weakly colored or not colored by radiation. Ionic diffusion in quartz is appreciable under applied electric fields at high temperatures, and both the color response of the quartz and the unit cell dimensions[16] have been altered in this way. Alkali ions can be made to migrate[25] rapidly out of quartz at temperatures roughly in the range 350–500° and applied fields of a few thousand volts per centimeter, leaving the quartz unresponsive to radiation and with altered elastic constants. The irradiation color centers also can be made to migrate out of silica glass in an applied field.[18]

A connection has been observed[17] between the tendency of quartz to darken under irradiation and the tendency to acquire secondary Dauphiné twinning by thermal strain. This also reflects compositional variation. Quartz with a high content of Al and other ions in solid solution is relatively difficult to twin in this way and is relatively responsive to radiation. Dauphiné twins that become unequally colored across the twin boundary on irradiation tend to recapture the original boundary when they are detwinned by heating over the high-low inversion at 573° and are then cooled. The relative tendencies of different natural specimens to darken under irradiation are preserved when the quartz is fused to glass.[19] There also is a correlation between the irradiation response and the temperature of formation of quartz, since the latter factor influences the extent of solid solubility of Al. Quartz formed at low temperatures, especially from non-aluminous environments such as cavities in limestone or marble, shows a very weak irradiation response and a relatively high inversion temperature corresponding to a low content of Al.

Colorless quartz may be rendered smoky in diffuse haloes up to about a millimeter in diameter surrounding tiny inclusions of radioactive minerals[20] such as monazite, uranoan microlite, zircon, and allanite; also as thin zones surrounding larger inclusions, or as dark margins on quartz pebbles set in a radioactive matrix as in the Witwatersrand conglomerate. The smoky color is produced by the hard (inhomogeneous) beta-radiation. Smoky quartz from Wölsendorf, Bavaria, contains bleached haloes and zones produced by alpha-particles from RaC' (Po^{214}) decay in radioactive inclusions.[21]

DISORDERING

Irradiation of quartz with fast neutrons results in the dislocation of ions and may ultimately convert the quartz to a glassy state.[22] Neutron irradiation is accompanied by an increase in unit cell volume, the value of a_0 increasing more rapidly than that of c_0. A sample[23] with a_0 4.903 Å, c_0 5.393, subjected to 6.6×10^{19} nvt of neutron flux, increased in dimensions to a_0 5.01 $\pm$ 0.01, c_0 5.41 $\pm$ 0.02. The greater change in a_0 is ascribed to the housing of dislocated atoms in interstitial positions. The density and mean index of refraction decrease, as do the rotatory power and the thermal conductivity; the density and mean index of refraction approach the values of low-tridymite. On annealing, the structure of quartz is restored.

References

1. Early work cited in Frondel: *Am. Min.*, **30**, 432 (1945), and Mellor: *Comp. Treatise Inorg. Chem.*, London, **6**, 1925, p. 263.
2. Kiyono: *Sci. Rpt. Tôhoku Univ.*, Ser. 1, **34**, 49 (1950); Laemmlein: *Dokl. Ak. Nauk. SSSR*, **56**, 849 (1947).
3. Frondel and Hurlbut: *J. Chem. Phys.*, **23**, 1215 (1955).
4. Early observations by Egoroff: *C. R.*, **140**, 1029 (1905); Doelter: *Das Radium und die Farben*, Dresden, 1910; Doelter and Sirk: *Sitzber. Ak. Wiss. Wien*, **119**, 1098 (1910). See also Armstrong: *Am. Min.*, **31**, 456 (1946) and Frondel: *Am. Min.*, **31**, 63 (1946) and **30**, 432 (1945).
5. Frondel (1945) and *Phys. Rev.*, **69**, 543 (1946).
6. Bottom and Nowicki: U.S. Signal Corps, Long Branch Signal Lab., *Eng. Mem. No. 1*, 1945.
7. Frondel (1946).
8. Seidl and Huber: *Zs. Phys.*, **97**, 671 (1935).
9. See Choong: *Proc. Phys. Soc. London*, **57**, 49 (1945); Holden: *Am. Min.*, **10**, 203

(1925); Frondel (1945, 1946); Chentsova and Vedeneeva: *Dokl. Ak. Nauk SSSR*, **60**, 649 (1948); Chentsova, Grechushnikov, and Batrak: *Opt. i Spektroskop.* **3**, 619 (1957); Batrak: *Kristallografiya*, **3**, 104, 626 (1958). Many earlier observations on bleaching and thermoluminescence are cited in the literature (ref. 1).

10. Futugami: *Proc. Phys.-Math. Soc. Japan*, **20**, 458 (1938).
11. Simon: *Jb. Min.*, Beil. Bd. **26**, 249 (1908); Hermann: *Zs. anorg. Chem.*, **60**, 369 (1908).
12. Cohen: *J. Chem. Phys.*, **25**, 908 (1956), with earlier literature; see also Holden (1925); Mohler: *Am. Min.*, **21**, 258 (1936); Kennard: *Am. Min.*, **20**, 392 (1935); Mitchell and Paige: *Phil. Mag.*, **46**, 1353 (1955); O'Brien: *Proc. Roy. Soc. London*, **231A**, 404 (1955).
13. See Cohen (1956), and Bambauer: *Schweiz. Min. Pet. Mitt.*, **41**, 335 (1961).
14. Cohen and Smith: *J. Chem. Phys.*, **28**, 401 (1958).
15. Brown and Thomas: *Nature*, **169**, 35 (1952).
16. Hammond, Chi, and Stanley: U.S. Signal Corps, Eng. Lab., *Eng. Rpt. E-1162*, 1955.
17. Armstrong (1946); Frondel: *Am. Min.*, **31**, 58 (1946); Zinserling: *C. R. ac. sci. URSS*, **33**, 365, 368, 419 (1941) and **94**, 1079 (1954).
18. Lietz and Wolfgang: *Glastech. Ber.*, **31**, 121 (1958).
19. Laemmlein: *Dokl. Ak. Nauk. USSR*, **43**, 247 (1944).
20. Laemmlein: *Nature*, **155**, 724 (1945); Ramdohr: *Jb. Min.*, Beil. Bd. **67**, 53 (1933) and *Abh. Ak. Wiss. Berlin, Kl. Chem.*, no. 2, 3 (1958).
21. Ramdohr (1933).
22. Primak, Fuchs, and Day: *J. Am. Ceram. Soc.*, **38**, 135 (1955); Simon: *Phys. Rev.*, **103**, 1587 (1956); Wittels and Sherrill: *Phys. Rev.*, **93**, 1117 (1954); Primak: *Phys. Rev.*, **110**, 1240 (1958); Wittels: *Phil. Mag.*, **2**, 1445 (1957).
23. Wittels: *Phys. Rev.*, **89**, 656 (1953); see also Johnson and Pease: *Phil. Mag.*, **45**, 651 (1954); Crawford and Wittels: *Proc. Intern. Conf. Peaceful Uses Atomic Energy*, Geneva, **7**, 654 (1956).
24. Bambauer (1961).
25. Vedeneeva and Chentsova: *Dokl. Ak. Nauk. SSSR*, **87**, 197 (1952); Lietz and Hänisch: *Naturwiss.*, **46**, 67 (1959); Wondratschek, Brunner, and Laves: *Naturwiss.*, **47**, 275 (1960); Wood: *J. Phys. Chem. Solids*, **13**, 326 (1960).
26. Cohen: *J. Phys. Chem. Solids*, **13**, 321 (1960).

CHEMICAL COMPOSITION

Quartz is silicon dioxide, SiO_2, containing Si 46.751 (for atomic weight 28.095) and O 53.249, total 100.000 weight per cent. Quartz shows a significant but very small range of variation in chemical composition. The principal elements that enter solid solution in natural quartz are Li, Na, Al, and Ti. Mg and (OH) also may be present in significant amounts, and a number of other elements, in part of uncertain role, have been reported in smaller quantities. Although the nature of the chemical variation is known qualitatively, there is not sufficient information based on quantitative analyses of the requisite precision to establish the mechanisms and the extent of the solid solution. The quantitative relations between the variation in composition and the dependent variation in properties such as the indices of refraction, density, and unit cell dimensions also are not known. The principal difficulty has been the precise experimental measurement of the very small variations involved. The existing information as to compositional variation has been obtained chiefly from qualitative and semi-quantitative analyses obtained by optical spectrography and from studies of the variation of composition-dependent properties that have been interpreted on crystallochemical grounds. A selection of the reported analyses[1] is given in Tables 19–27.

The most important mechanism of compositional variation in quartz involves the substitution of Al for Si. Valence compensation apparently is effected by the concomitant entrance of a small alkali ion, either Li or Na or both, into interstitial positions. The coupled entrance of one interstitial Al for three substitutional Al ions has been indicated, and Mg also may enter similarly into interstitial positions. The coupled entrance of Al and alkalies into solid solution occurs on a much larger scale in tridymite. The maximum extent of the substitution by Al in quartz is not clearly established. The amount of Al present usually is in the range $0.000X$–$0.00X$, but amounts up to 0.01–0.05 per cent have been reported in both synthetic and natural hydrothermal quartz. A higher content may be found in magmatic quartz and in quartz formed by inversion from high-quartz. The general level of Al content reported in different analytical studies varies considerably, and the precision of some of this work is questionable. The analytical work and studies of properties that have been correlated with the Al content, such as the high-low inversion temperature and the irradiation blackening, indicate that virtually all colorless quartz contains more or less Al. In general, the unit cell volume increases and the high-low inversion temperature decreases with increasing content of Al. The amount of Al present in solid solution increases with increasing temperature of formation of the quartz, as is indicated by the variation in the high-low inversion temperature of samples taken from different environments[2] and by the variation in the unit cell dimensions of certain samples of synthetic quartz.[3] The content of Al is greater in the quartz of certain granites than in the quartz of associated pegmatites (Table 21). The content of Al and of other elements present in solid solution generally is not uniform throughout a single-crystal of quartz but varies in a zonal fashion (see *Zonal and Sectoral Structure*). The content of Al, Li, and H in quartz crystals from Alpine localities is larger in colorless than in smoky quartz and is largest in imperfect crystals with a birefringent zonal structure. The content of Al shows a positive correlation with that of Li and H but a less marked correlation with that of Na. Analyses[24] of over forty samples gave the following range and averages (in parentheses) in parts per million Si atoms:

	Al	Li	Na	H
Smoky	13–60 (41)	2.5–37 (22)	~1–25 (<9)	3–22 (8)
Colorless	30–140 (82)	10–105 (45)	<2–30 (<6)	<1–55 (16)
Zonal	300–1700 (1138)	105–1250 (516)	20–40 (25)	35–1200 (464)

TABLE 19. PRECISION ANALYSES OF QUARTZ

Analyses by J. A. Maxwell and R. B. Ellestad; physical constants by Frondel and Hurlbut (1955)[15]. Alkalies determined by flame photometer; Ti and Mn colorimetric; Fe by thioglycolic acid, and Al by aurintricarboxylic acid method, using photoelectric colorimeter. Valence of Mn, Ti, and Fe not determined. Analyses on large acid-washed crushed samples dissolved in HF and H_2SO_4, corrected for blanks on reagents. Density in grams per cubic centimeter in vacuum at 25°, precision ±0.0001. Precision of indices of refraction ±0.00001. Precision of cell dimensions ±0.0001 (Arkansas); ±0.00015 (others).

	Colorless Hot Springs, Ark.	Colorless Palermo, N.H.	Smoky Brazil	Smoky Amelia C.H., Va.	Rose Auburn, Me.	Rose Brazil	Rose Rossing, S.W. Africa	Rose Albany, Me.	Amethyst Delaware Co., Pa.	Amethyst Uruguay
Li	0.0005		0.0004		0.0038		0.0038			
Na	.0004		.0000		.0011		.0104			
K	.0002		.0000		.0001		.0010			
Al	.0008	0.0002	.0008	0.0003	.0001	0.0002	.0004	0.0004	0.0000	0.0004
Fe	.0000	.0007	.0005	.0019	.0003	.0000	.0006	.0003	.0008	.0216
Mn	.00002	.00000	.00002	.00004	.00005	.0000	.00004	.00005	.00000	.00000
Ti	.0001	.0004	.0002	.0002	.0015	.0026	.0011	.0013	.0000	.0004
$d(25°)$	2.64847		2.64848		2.64797					
ϵ (25°, Na)	1.544258		1.544214		1.544226	1.544241				
ω	1.553380		1.553369		1.553357	1.553368				
a_0 (25°, Å)	4.91350		4.91289		4.91320	4.91376				
c_0	5.40507		5.40442		5.40474	5.40534				
Inv. temp. Heat	573.3°		573.4°		572.2°	570.5°				
Cool	573.3°		573.2°		572.1°	570.3°				

TABLE 20. SPECTROGRAPHIC ANALYSIS OF QUARTZ
(Cohen and Sumner, 1958)[14]

Absolute values believed to be within one-third and three times the values given. Precision of cell dimensions ±0.0002 kX.
N.F. = not found.

	Colorless Synthetic	Colorless Synthetic	Colorless Synthetic	Colorless Synthetic	Colorless Ign. Rock	Smoky Calif.	Smoky Brazil(?)	Amethyst Brazil	Amethyst Brazil	Rose Local. Unknown
Ag	0.0001	0.0001	0.001	0.0001	0.0001	N.F.	0.0001	0.0001	<0.0005	0.0001
Al	.003	.003	.01	.03	>.01	0.01	~.01	>.01	.005	≫.01
Ca	.0003	~.001	~.001	~.001	~.01	.003	.003	.03	>.0005	.003
Cr	.0003	.0003	.0003	.001	.003	.0003	N.F.	N.F.	<.002	N.F.
Cu	.001	.001	.001	.003	~.0001	.01	<.0001	>.0001	<.0005	~.0001
Fe	.003	~.001	~.001	.01	~.01	.001	>.01	.01	>.005	>.01
Li	.0003	.0001	.0003	.01	Not det.	.0003	Not det.	Not det.	Not det.	Not det.
Mg	<.001	<.001	<.001	<.003	.01	~.003	~.001	~.001	<.0005	~.001
Na	.003	.003	.003	.003	Not det.	.003	Not det.	Not det.	Not det.	Not det.
Ti	N.F.	N.F.	N.F.	N.F.	.003	N.F.	Tr.(?)	Tr.(?)	N.F.	.003
See note		a	b	c	d				e	
a_0 (25°, Å)	4.9134	4.9133	4.9132	4.9139	4.9138	4.9132	4.9131	4.9134	4.9131	4.9141
c_0	5.4049	5.4051	5.4052	5.4052	5.4047	5.4050	5.4051	5.4047	5.4047	5.4052

[a] Also Ge 0.01.
[b] Also Bi 0.001(?), Pb 0.003, Rb 0.03.
[c] Also Ge 0.1.
[d] Also Ni 0.001, Pb 0.003.
[e] Also Ni 0.001, Pb < 0.05

Alkalies are usually found to be present, generally in amounts in the range $0.000X$–$0.00X$ per cent. The precision of the determinations usually is not such as to establish the stoichiometric relations with the Al or other cations present. The alkalies, chiefly Li and Na, are believed present in inter-

TABLE 21. SPECTROGRAPHIC ANALYSES OF QUARTZ FROM GRANITE AND PEGMATITE (Tatekawa, 1954)[16]

Relative intensities on scale of 10. Below detection limit by method employed: Ni, Co, Cr, Ag, Zn, Cd, Sb, Mo, Cb, Ta, Li, Rb, Ge, Ga, Tl, Sc, Cs, rare earths.

	Ti	Al	Mg	Na	Sn	Ca
Granite	6	9	10	5	...	8
	7	7	9	3	...	10
	5	9	7	5	1	...
	4	9	7	8	1	9
	5	9	7	8	1	...
	7	7	6	5	1	6
	6	8	6	3	1	...
	5	8	6	3	1	...
	5	9	5	3	2	...
	5	8	5	5	7	...
	5	9	4	5	...	...
	4	9	4	5	...	...
	5	9	4	3	1	...
	3	9	2	3	1	...
Pegmatite	5	8	3	3	1	...
	3	7	3	3	1	...
	6	7	2	3	1	...
	3	6	2	3	1	1
	3	7	2	3	1	...
Transparent	1	7	2	3	...	4
White	3	6	2	3	1	...
Smoky	5	7	2	3	1	...
White	5	6	2	3	1	...(Ag 4)
Gray	6	9	2	3	1	...
Milky	4	8	2	3	...	...
Smoky	4	7	2	3	1	...

stitial solid solution in compensation for the Al. The coupled entrance of Li and Al is clearly indicated by a synthetic quartz grown at 890° from a solution saturated with $LiAlSiO_4$. This material[4] had the relatively large unit cell dimensions a_0 4.904, c_0 5.3950 (kX at 25°) and had a very sluggish high-low inversion at 556° (heating) and 565° (cooling). In some analyses, as of rose quartz, Li and Na are present in amounts considerably greater than needed

TABLE 22. ANALYSES OF SMOKY QUARTZ
(Holden, 1925)[17]

Color	Fe_2O_3	TiO_2	MnO	Locality
Dark	0.003	0.001	0.0001	Colorado
blackish	.000	.001	.0000	Maine
brown	.000	.001	.0000	Maine
	.03	.002	.0005	Colorado
	.001	.002	.0003	Ontario
	.001 +	.001 +	.0005	Ontario
	.000	.001	.0004	Nova Scotia
Smoky	.001	.001	.0000	Siberia
brown	.008	.008	.0000	Maine
	.001	Tr.	.0002	New York
	.003 +	.001	.0001	North Carolina
	.004	Tr.	.0002	New York
Pale grayish	.003 +	.001	.0000	Connecticut
brown	.005	Tr.	.0002	Pennsylvania
	.000	Tr.	.0001	Herkimer, N.Y.
	...	.001	.0001	Hot Springs, Ark.
Almost colorless	.003	Tr.	.0000	Switzerland

TABLE 23. SPECTROGRAPHIC ANALYSES OF SYNTHETIC QUARTZ
(Stanley and Theokritoff, 1956)[18]

Element	Amount in Quartz in %	Amount in Solution	Temp. Gradient in Autoclave
Ge	0.3	$0.2N\,Ge^4$	339–381°
	.1	$.2N\,Ge^4$	344–385°
	.3	$.5N\,Ge^4$	325–358°
	.1	$.2N\,Ge^4$	287–344°
Al	.01	$.015N\,Al$	330–370°
	.02	$.015N\,Al$	351–399°
	.005	$.03N\,Al$	320–376°
	.01	$.015N\,Al$	332–348°
Pb	.1	$.06N\,Pb^4$	315–395°
	.02	$.05N\,Pb^4$	345–378°
Sn	.03	$.018N\,Sn^4$	311–341°
Ag	0.2	$.05N\,Ag$	350–365°

TABLE 24. ANALYSES OF AMETHYST
(Holden, 1925)[17]

Color	Fe$_2$O$_3$	TiO$_2$	MnO	Locality
Dark violet	0.35	Tr.	0.0002	Unknown
	.24	0.0005	.0001	Uruguay
	.06	Tr.	.0000	Lake Superior
	.05	.001	.0001	Unknown
	.02	Tr.	.0000	Rhode Island
Violet	.07	.001	.0080	Schemnitz, Hungary
	.03	.001	.0011	Korea
	.02	Tr.	.0015	Guanajuato, Mexico
	.006	Tr.	.0002	Aspen, Colo.
	.005	Tr.	.0002	North Carolina
Pale violet	.011	Tr.	.0004	Madagascar
	.007	Tr.	.0002	North Carolina
	.006	Tr.	.0000	Delaware Co., Pa.
	.005	Tr.	.0001	Jefferson Co., Mont.
Very pale violet	.015	Tr.	.0002	Japan
	.004	.001	.0021	Guanajuato
	.004	Tr.	.0002	Iredell Co., N.C.

TABLE 25. ANALYSES OF AMETHYST AND ROSE QUARTZ
(Watson and Beard, 1917)[19]

	TiO$_2$	MnO	Fe$_2$O$_3$	Locality
Amethyst	0.00199	0.00036	0.0775	Virginia
	.00166	.00068	.0382	Montana
	.00145	.00085	.0404	Montana
	.00521	.00029	.0935	Brazil
	...	.0002	...	Colorado
Rose	.00577	.00052	.0679	Colorado
	.00139	.00068	.0399	South Dakota
	.00269	.00017	.0578	Greenland
	.00288	.00074	.0351	South Dakota
	.00288	.00036	.0042	South Dakota
	.00405	.00018	.0067	Paris, Maine
	.00289	.00057	.0032	Connecticut
	.00293	.00017	.0040	New York
	...	.00012	...	Unknown

TABLE 26. ANALYSES OF ROSE QUARTZ

(Holden, 1924)[20]

Li, Co, and Al determined only on specimens indicated; Li by visual flame test, Mn and Ti colorometric. All samples showed an ignition loss between 0.1 and 0.2 per cent.

Color	MnO	TiO$_2$	Fe$_2$O$_3$	CoO	Li$_2$O	Al$_2$O$_3$	Locality
Very deep rose	0.0005	0.004+	0.016				Rossing
Deep rose	.0006	.003	.008	0.0001	0.00X		Madagascar
	.0005	.004	.022	.0000+	.0X-.00X		South Dakota
	.0002+	.003	.006				Bohemia
	.0003	.004	.010				Rabenstein
Rose	.0004	.003	.007	.0001	.00X	0.06	Brazil
	.0004	.006	.007				California
	.0003	.003	.012				South Dakota
	.0002+	.002+	.006	.0000+		.16	Zwiesel, Bohemia
	.0002	.003	.006				California
Pale rose	.0002+	.003+	.006				Maine
	.0002+	.003	.006	.0000+	.0X-.00X	.12	New Hampshire
	.0002	.003	.008				Connecticut
	.0001+	.003+	.005				New Hampshire
	.0002+	.003	.007				Japan
	.0003	.003	.007				Connecticut
	.0002	.003+	.004				Connecticut
	.0002+	.003	.013				California
Very pale rose	.0002	.003	.006	.0001	.0X-.00X	.12	Bedford, N.Y.
	.0002+	.002+	.012				California
	.0002	.003	.012				Maine
White	.0005	.003	.008	.0001	.00X	.13	Madagascar
	.0002+	.005					New Hampshire
	.0003+	.006	.007				California
	.0002	.003					New Hampshire
	.0003+	.003	.008	None	.00X	.12	New Hampshire
	.0002+	.003+					New Hampshire

TABLE 27. ANALYSES OF BLUE QUARTZ

TiO$_2$	Fe$_2$O$_3$	Locality	Ref.
0.069	0.539	Virginia	10
0.05–0.07	0.65–0.70	Sweden	9
0.03	0.011	India, charnockite	4
0.024	0.006	India, charnockite	
0.020	0.005	India, charnockite	
0.018	0.004	India, charnockite	
0.014	0.004	India, charnockite	
0.002	0.001	India, charnockite	
0.025	0.010	Mysore, gneiss	
0.024	0.012	Mysore, gneiss	
0.028	0.005	Mysore, gneiss	
0.017	0.015	Mysore, gneiss	
0.025	0.005	Mysore, granite	
0.024	0.004	Mysore, granite	

for valence compensation of the Al present. Lithium and also Na and K diffuse readily through quartz parallel to the c-axis in the range 300–500° under potential differences of a few hundred volts per mm.[5] The diffusion coefficients of Mg, Ca, Fe^2, and Al are considerably smaller. Solid inclusions and bubbles or cavities containing entrapped liquid are common in quartz (see *Inclusions in Quartz*) and may contribute to the apparent content of alkalies and other cations. Microscopic inclusions of other minerals are typical of the quartz grains of metamorphic and sedimentary rocks, and their constituents show up on chemical analysis.[22] The liquid inclusions presumably are the source of the small amounts of CO_2, N, and H_2O frequently reported in quartz in amounts up to about 0.0X per cent. Other substances found in quartz and known or presumed to be of the same origin include NH_3, SO_2, H_2S, F, Cl, and organic substances. The halogens probably are present as alkali halides in entrapped saline solutions. Ferric thiocyanate also has been reported in quartz, but not verified.[6]

The presence of (OH) in quartz has been established by infrared absorption study.[7] It presumably substitutes for O ions, and would provide valence compensation for the entrance of Al or other trivalent ions. The content of (OH) is not known. It varies in successive growth zones of the crystal. The bulk of the water found on ignition probably comes from liquid inclusions and from films of moisture adsorbed from the air on the surface of powdered samples.[8]

In the rose-quartz of pegmatites and in the granular quartz of certain metamorphic and igneous rocks Ti^4 enters solid solution in relatively large amounts (see *Varieties of Quartz*). Reported analyses of these varieties generally show 0.00X per cent TiO_2. The temperature of formation of such quartz is believed to be considerably higher than that of the euhedral crystals of hydrothermal veins. The level of Ti in colorless and smoky quartz crystals appears to be usually about 0.000X or below. Amethyst sometimes contains rutile needles, but whether these are the result of exsolution or are mechanical inclusions is not known (see also Tables 20, 24, and 25). The Ti presumably substitutes for Si in four-coordination and usually has in large part exsolved as microscopic needles of rutile in oriented position (see *Oriented Growths*), giving rise to asterism or chatoyance in cut spheres or properly oriented cabochon stones. Rutile frequently occurs as macroscopic hair-like or acicular inclusions in quartz crystals, which have not formed by exsolution but have been enclosed during the growth of the crystal (see *Inclusions in Quartz*).

Magnesium appears to be a significant minor constituent in the quartz of some igneous rocks (Tables 20 and 21), but ordinarily it is present only in amounts of 0.000X or below. It is not always sought, however, by analysts. Normally Mg goes into six-coordination with oxygen, and this element may enter substitutional solid solution in quartz in interstitial positions rather than in substitution for Si in four coordination. The coupled solid solution of Mg and Al up to about 8 per cent $MgAl_2O_4$ has been observed in low-quartz formed by inversion from high-quartz solid solutions. Generally Mn is present in quartz in relatively small amounts, about $0.000X$ or below. It does not appear to be characteristic of any particular type or color of quartz such as amethyst or rose quartz. In amethyst Fe in the trivalent state is present in relatively large amounts, up to $0.0X–0.X$ per cent (Tables 24 and 25), and it also occurs in relatively large amounts in some rose quartz. The Fe in amethyst probably is not in solid solution but is present as colloidally dispersed particles of an iron oxide (see *Amethyst*, under varieties of quartz). The Fe content of colorless and smoky quartz is very low in comparison, although variable, and the manner in which it is contained is not known. A number of other elements have been found in relatively small amounts or in traces in quartz. These include Rb, Cs, Ca, Ba, Pb, Ag, Sn, Cu, Zn, V, Cr, Zr, and U. It is not known whether these elements are present in solid solution or as impurities representing solution entrapped in cavities or solid inclusions.

The extent of compositional variation in colorless quartz is very small, less than that of almost any other mineral. Among common natural phases, only diamond, graphite, and ice appear to be less variable in composition. The reasons for this behavior in quartz are found in a complex of crystallochemical and geochemical factors. The small size of the Si ion precludes extensive substitution by quadrivalent ions such as Ti, Zr, Sn, Th, and U, all of which are considerably larger in size and normally go into higher coordination with oxygen. The smallest of these ions, Ti^4, occurs in small amounts in substitution for Si in rose quartz and at high temperatures enters into other SiO_2 polymorphs in larger amounts. Quadrivalent Ge is relatively close in size to Si, and one of the polymorphs of GeO_2 is isostructural with quartz. Although Ge^4 has been introduced into solid solution in synthetic quartz (Table 23),[18] this element ordinarily is lacking or is present only in minute traces in natural quartz. A much larger substitution of Ge for Si is found in certain silicates, especially in other than tectosilicates.[13] The very great disparity in the crustal abundances of Si and Ge and the relatively high solubility of Ge compounds are factors.

The entrance of trivalent ions into solid solution in quartz involves both the matter of ionic size and the necessity of effecting a concomitant substitution or omission in other structural sites to provide valence compensation. Of the common trivalent ions, Al approaches Si relatively closely and has a radius ratio with oxygen close to the critical value between four- and six-coordination. The substitution of Al for Si in four-coordination is a dominant theme in the crystal chemistry of the silicates. In quartz, however, with its simple composition and relatively tightly packed structure the opportunities for valence compensation of trivalent ions are more limited. The principal opportunities here are in the coupled entrance of relatively small monovalent ions into interstitial positions, these positions being too small to house K or larger ions, and in the coupled substitution of (OH) for O. The latter mechanism, however, disrupts the tetrahedral framework, since an (OH) cannot link two Si atoms as does the divalent O. Omission mechanisms of compensation in the oxygen positions have not been identified in SiO_2 and appear unlikely in view of the high binding energy of Si for O. Valence compensation by the coupled entrance of a pentavalent cation such as P or As in place of Si has not been clearly established, although the compounds $AlPO_4$ and $AlAsO_4$ have polymorphs isostructural with quartz. The substitution of Al for Si becomes more important in the relatively open structures of the high-temperature polymorphs tridymite and cristobalite. Numerous alumino-silicates are structurally derived in a more or less idealized way from these polymorphs, but not from quartz. The substitution of Fe^3 for Si also becomes more important in these polymorphs. Other trivalent ions that come into consideration with regard to quartz include Mn^3 and B. Boron, much smaller in size than Si, occurs in four- as well as in three-coordination with oxygen. It enters various silicates as tetrahedral (BO_4) or (BO_3OH) groups in place of (SiO_4), although generally in systematic rather than serial fashion. The B content of quartz has been reported[23] to range between ~0.5 and 34 p.p.m. with most analyses showing under 2 p.p.m. Manganese is commonly present, but in relatively small amounts, and the valence state is not known.

Historically, quartz has been known as a distinct entity since ancient times. The composition of quartz and the chemistry of silicon in general were not established, however, until the nineteenth century. The element Si was prepared by Berzelius in 1823, by reduction of K_2SiF_6 with K, and its chief properties were determined. Berzelius (1811) showed that silica unites with bases in definite proportions and regarded silicate minerals as salts of silicic acid. Silica was synthesized by Berzelius and others in this general period, by the oxidation of Si and in other ways, and the combining weights of O with Si were approximately determined. The quadrivalence of Si was not satisfactorily established[9] until the middle 1800's, the formula of quartz earlier being variously represented as SiO, SiO_2, and SiO_3. Before this, quartz was regarded by many as a primitive earth or element, primarily because of its infusibility and resistance to chemical attack. The earliest chemical analysis of quartz is that of Bergman (1779),[10] who obtained SiO_2 93, Al_2O_3 6, and CaO 1, and additional analyses of quartz and its varieties were reported by Klaproth (1795) and others before 1810. Much of the analytical work on quartz has been stimulated by efforts to determine the nature of the pigmenting material in amethyst, smoky quartz, and other colored varieties. The presence of Fe and Mn in amethyst was reported by H. Rose (1800), and Ti was found in rose quartz from Rabenstein by Fuchs (1831). Lithium was recognized spectrographically in colorless quartz[11] in 1908. Rock-crystal, quartz sand, and flint were shown[21] to be of the same qualitative composition by J. H. Pott in 1753.

References

1. Six approximate analyses reported before 1822 are cited by Haüy: *Traité de minéralogie*, 1822, **2**, 228; and most of the analytical work on quartz and its varieties and on opal reported before 1915 is cited by Rammelsberg: *Mineralchemie*, Leipzig, second ed., 1875, p. 162; Doelter: *Handbuch der Mineralchemie*, **2**, 1914, Pt. 1, p. 118; and Hintze: *Handbuch der Mineralogie*, **1**, 1915, Abt. 2, p. 1266. Many additional partial analyses have been reported, among which may be mentioned: Dale: *Trans. Engl. Ceram. Soc.*, **23**, 211 (1924); Kennard: *Am. Min.*, **20**, 392 (1935); Keith and Tuttle: *Am. J. Sci.*, Bowen Vol., 203, 1952; Petrun: *Zap. Vsesoy. Min. Obsh.*, **84**, 191 (1955); Hurlbut: *Am. Min.*, **31**, 443 (1946); Becker: *Min. Mitt.*, **6**, 158 (1885); Weinschenk: *Zs. Kr.*, **26**, 396 (1896); Konta: *Rozpravy Československ. Ak. Věd.*, **64**, no. 4 (1954); Bray: *Bull. Geol. Soc. Amer.*, **53**, 765 (1942).

2. Keith and Tuttle (1952); Tuttle: *Am. Min.*, **34**, 723 (1949).

3. Frondel and Hurlbut: *J. Chem. Phys.*, **23**, 1215 (1955); Keith and Tuttle (1952).

4. Frondel and Hurlbut (1955).

5. Verhoogen: *Am. Min.*, **37**, 637 (1952); Wenden: *Am. Min.*, **42**, 859 (1957); Sosman: *Properties of Silica*, New York, 1929.

6. Nabl: *Min. Mitt.*, **19**, 273 (1900); Wöhler and Kraatz-Koschlau: *Min. Mitt.*, **18**, 304 (1899); Holden: *Am. Min.*, **10**, 203 (1921).

7. Brunner, Wondratschek, and Laves: *Naturwiss.*, **24**, 664 (1959); Kats and Haven: *Phys. Chem. Glass*, **1**, 99 (1960); Bambauer: *Schweiz. min. pet. Mitt.*, **41**, 335 (1961). On the infrared absorption spectrography of the SiO_2 polymorphs see Dachille and Roy: *Zs. Kr.*, **111**, 462 (1959).

8. Koenigsberger: *Min. Mitt.*, **19**, 149 (1900).

9. See Wurtz: *Ann. chim. phys.*, **69**, 355 (1863), and *Chem. Jahresber.*, 144, 1860, and 211, 1864.

10. Bergman: *De terra silicea*, Upsala, 1779.

11. Curie and Gleditsch: *C. R.*, **147**, 345 (1908).

12. Keith and Tuttle (1952); Stanley and Theokritoff: *Am. Min.*, **41**, 527 (1956).

13. Wickman: *Geol. För. Förh.*, **65**, 371 (1943).

14. Cohen and Sumner: *Am. Min.*, **43**, 58 (1958).

15. Frondel and Hurlbut (1955), with hitherto unpublished data.

16. Tatekawa: *Mem. Coll. Soi. Univ. Kyoto*, **21**, ser. B, 183 (1954).

17. Holden (1921).

18. Stanley and Theokritoff (1956); see also Cohen and Smith: *J. Chem Phys.*, **28**, 401 (1958), and Augustine and Hale: *J. Chem. Phys.*, **29**, 685 (1958).

19. Watson and Beard: *Proc. U.S. Nat. Mus.*, **53**, 553 (1917).

20. Holden: *Am. Min.*, **9**, 75, 101 (1924).

21. Pott: *Lithogéognosie*, Paris, 1753.

22. Keller and Littlefield: *J. Sed. Petrol.*, **20**, 74 (1950).

23. Harder: *Nachr. Ak. Wiss. Gottingen, III, Math.-phys. Kl.*, **5**, 67 (1959).

24. Bambauer: *Schweiz. min. pet. Mitt.*, **41**, 335 (1961).

SOLUBILITY

Fragments of quartz are not appreciably attacked by HCl, HNO_3, or H_2SO_4, but are readily dissolved by HF and by warm concentrated solutions of NH_4HF (see further under *Etching Phenomena*). Quartz is slowly acted upon by alkaline solutions at room temperature, the microcrystalline varieties more readily, and is strongly attacked by alkaline solutions, as of NaOH, Na_2CO_3, K_2CO_3, Na_2SiO_3, or $Na_2B_4O_7$, in sealed vessels at elevated temperatures and pressures. Quartz also dissolves readily in fusions of borax, NaOH, or Na_2CO_3, and in $KHSO_4$ containing added fluorides.

The equilibrium solubility of quartz in water at room temperature has been reported[1] to be in the range from 6 to 30 parts per million and probably is in the neighborhood of 7 p.p.m. The laboratory determination of the solubility is difficult because of the very long time needed to achieve equilibrium. The solubility of so-called amorphous silica in water at room temperature is considerably higher than that of quartz and is in the range 100–140 p.p.m. at 25°. The term amorphous silica here includes silica gel either wet or dried, silica sols, silica glass, the opaline silica of the skeletal material of silica-secreting organisms such as diatoms, and opal in more or less part. This material is not truly amorphous, but in general contains regions of local order, or crystalline particles of extremely small size, which from x-ray diffraction study are found to be of the cristobalite type of structure. The cristobalite structure probably is more or less disordered and may contain H_2O or other material in the interstices of the structure as stabilizing agents.

The solubility of amorphous silica at ordinary temperatures is essentially independent of *p*H over the range from 1 to 9. The silica is present in true solution, principally as monomeric silicic acid, or $Si(OH)_4$, which probably is further hydrated in dilute solutions. At *p*H values over about 9 the solubility increases rapidly because of ionization of the H_4SiO_4. The value of the *p*H at which the solubility increases is independent of temperature up to at least 100°. The solubility is reduced by Al^3 ions, and presumably also by other ions that would react to form relatively insoluble silicates, but is not materially affected by NaCl in the concentration range of sea water. The solubility increases with increasing temperature[2] up to a value of about 400 p.p.m. at 100° and 800 p.p.m. at 200°C. The rate of attainment of solubility equilibrium also increases with increasing temperature.

In the silica-rich portion[3] of the system SiO_2-H_2O, the liquidus of cristobalite is lowered by water vapor, and at pressures above approximately 400 kg. per cm.^{2} tridymite melts directly to a hydrous liquid. At water vapor pressures above approximately 1400 kg. per cm.^{2} tridymite is not formed as a stable phase, and quartz melts directly to a hydrous liquid (Fig. 78).

The solubilities of silica in water against normal cristobalite or tridymite as the solid phase are not known but presumably are greater than the solubility of quartz in its region of stability. The solubility of cristobalite may be close to that of amorphous silica. Water solutions saturated with respect to these polymorphs would be supersaturated with respect to quartz. A saturated silica solution obtained from amorphous silica, however, does not deposit

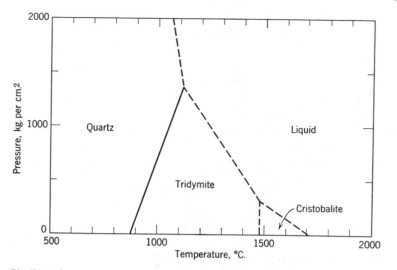

FIG. 78. Tentative pressure-temperature diagram for the silica-rich portion of the system SiO$_2$–H$_2$O.

quartz at an appreciable rate under low-temperature laboratory conditions. In general, silica precipitated by the cooling of saturated solutions at relatively low temperatures deposits colloidal silica or silica gel, which consists of or ages to cristobalite as a metastable phase. This is analogous to the crystallization of metastable cristobalite instead of the stable phase when silica glass is cooled to temperatures below the stability region of cristobalite. At relatively high temperatures the cristobalite-like phase in the types of amorphous silica earlier mentioned rapidly recrystallizes in water to quartz. A solution saturated with silica and containing excess solid amorphous silica at such a temperature should with time convert the cristobalite to quartz, and crystallize quartz in addition to reach the equilibrium solubility against quartz at that temperature. The conversion of metastable cristobalite to quartz in hydrothermal environments at relatively high temperatures is illustrated by the natural formation of moss agate and of agate geodes in igneous rocks and tuffs, in which chalcedonic quartz forms from the original silica gel or opal.

References

1. Reviews in Alexander, Heston, and Iler: *J. Phys. Chem.*, **58**, 453 (1954); Iler: *Colloid Chemistry of Silica and Silicates*, Ithaca, N.Y., 1955; Krauskopf: *Soc. Econ. Paleont. Min. Spec. Publ.* **7**, 1959, p. 4; Siever: *Amer. Min.*, **42**, 821 (1957). See also Meier and Schuster: *Zs. anorg. Chem.*, **196**, 220 (1931); Wyart and Sabatier: *C. R.*, **238**, 702 (1954).
2. Kennedy: *Econ. Geol.*, **45**, 629 (1950); Morey and Hesselgesser: *Econ. Geol.*, **46**, 821 (1951); Wasserburg: *J. Geol.*, **66**, 559 (1958).
3. Tuttle and England: *Bull. Geol. Soc. Amer.*, **66**, 149 (1955).

SYNTHESIS

Quartz can be synthesized[1] in a variety of ways, but crystals larger than a millimeter or so in size are obtained only under special circumstances.[6] The first published account of the synthesis of quartz crystals is that of Schafhäutl[2] (1845), who heated freshly precipitated silica gel in water in an autoclave at about 100–120° for 8 days. The first synthetic crystals of appreciable size and good quality were grown hydrothermally by Spezia[3] in 1905. He was the first to use cut seed crystals and a temperature gradient (actually a solubility gradient) in the pressure vessel. In one experiment, using a water solution containing 1.9 per cent Na_2SiO_3 and 12.7 per cent $NaCl$, and a temperature gradient between 165–180° and 320–350°, a cut seed about 6 mm. in diameter was grown into a clear euhedral crystal about 13 mm. in size. Primarily as a consequence of the shortage and difficulties of supply of natural quartz of piezoelectric grade during World War II, efforts were undertaken both in the United States and abroad to synthesize large quartz crystals of high quality. Quartz crystals are now grown on a mass-production scale, using Spezia's basic method. The reported syntheses of quartz may be grouped according to the method employed.

From Solutions at Elevated Temperatures and Pressures. The most successful syntheses of quartz are those in which silica is recrystallized or is produced by reaction in alkaline solutions confined in steel pressure vessels held at temperatures ordinarily in the range from 200 to 500° and at pressures up to 25,000 p.s.i. The various methods include heating fragments of soda lime or other glasses in water, heating silica gel or dried "silicic acid" with water alone or in alkaline solutions, and heating silica glass or fragments of quartz in solutions of $NaOH$, KOH, or alkali carbonates. The size and quality of the quartz crystals in such experiments are rather sensitive to the presence of certain cosolutes, such as fluorides, various metallic ions, and organic substances, e.g., oleic acid.

Flawless quartz crystals of large size (Fig. 79) are grown commercially by a technique in which a temperature gradient is maintained along a steel pressure vessel. A nutrient supply of quartz fragments is kept at the lower, relatively hot end of the vessel, and this is slowly dissolved and transported by convection through the contained solution of $NaOH$ or Na_2CO_3 and redeposited because of decreased solubility in the upper, relatively cool part of the vessel upon seed crystals or sawn plates of quartz suspended in the solution. Coarse fragments of quartz are employed rather than silica gel or silica glass because the latter give rise to uncontrolled nucleation on the walls of the vessel. The crystallographic orientation of the sawn seed plate and the rate of growth are important factors. The best quality is obtained at relatively slow growth rates, about 0.08 in. or less per 24 hours on {0001}. At high growth

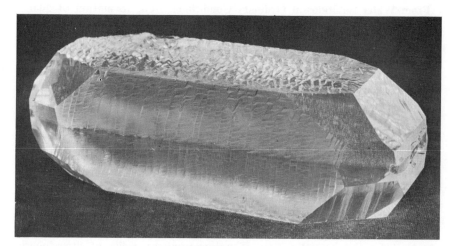

FIG. 79. Synthetic crystal of quartz grown hydrothermally. The crystal is 5½ in. in length and weighs 1.65 pounds. The face (0001) is at top, with (2$\overline{1}$10) in front and faces of {10$\overline{1}$1}, {01$\overline{1}$1}, and {10$\overline{1}$0} at right and left. Part of a metal strap used to support the seed plate can be seen at right.

rates the crystals become flawed or cloudy. The presence of Al in the solution tends to decrease the quality of the crystals. Solutions containing NaOH or Na$_2$CO$_3$ may form a protective coating of acmite on the walls of steel bombs. With K salts the steel is attacked and the quartz crystals may become green or yellow through included Fe ions. Foreign ions such as Al present in the solution in general are taken up in solid solution in the quartz crystals in different amounts in different face-loci (see section on *Zonal and Sectoral Structure*).

The temperature gradients employed may range up to 100°, depending on the size and shape of the containing vessel, the average temperature usually being near 350°; and the pressures may range up to 15,000–25,000 p.s.i. Of the several forms bounding the crystal, the highest growth rate is shown by {0001}, if this surface is present on the seed plate. The growth surface of this form is not smooth but hummocky or uneven; if growth of the seed is continued sufficiently long the {0001} surfaces will disappear through the lateral extension of *r* and *z* faces. The growth rates of *r* and *z* are much greater than that of the prism *m*, the slowest growing of the bounding forms. Since the *a*-axis of quartz is polar, the growth rates of parallel faces such as (2$\overline{1}$10) and ($\overline{2}$110) present on a sawn seed are different; in right quartz +X is the faster direction. Synthetic crystals have been grown up to about 10 lb. in weight; larger crystals can be obtained, depending primarily on the size of the containing vessel. A cut sphere grown by this technique[4] developed into a polyhedron bounded by *r*, *z*, and *m*, with small faces of {20$\overline{2}$1} and {51$\overline{6}$1}.

From Water Solutions at Ordinary Conditions. The formation of quartz crystals by the evaporation of silica solutions or sols in water at room temperature has been claimed, but not verified, and it is probable that the products were only hardened gels (opal). Quartz crystals can form at relatively low temperatures and pressures in nature, however, and distinct crystals have been observed in plant cells. Quartz also has been reported by the action of dilute H_2SO_4 on granulated basalt at 32–52° (analogous to the natural alteration of igneous rocks by acid, solfataric action).

From Vapors. Crystalline silica has been obtained by the thermal decomposition of $SiCl_4$, and by passing the mixed vapors of NaCl and KCl over silica glass. It is also produced by the action of superheated steam on silica or silicates, and by the detonation of silica and hexogene with some water and alkalies at 450–650° and very high pressures in a sealed vessel.[5] Silica is soluble in superheated steam.[7]

From Dry Melts. High-quartz, tridymite, and cristobalite have been obtained by the crystallization of polycomponent melts at atmospheric pressure and without water, but the temperatures necessary are too high for the direct crystallization of low-quartz itself. The syntheses include the crystallization of molten rhyolite or other siliceous rocks in the presence of a flux such as molybdic oxide, halogen compounds, or tungstic oxide; fusion of silica in sodium or lithium tungstate; and crystallization from silicate glasses or silica glass when annealed at a high temperature.

References

1. A summary of the older literature is given by Kerr and Armstrong: *Bull. Geol. Soc. Amer.*, **54**, Suppl. 1, 1943.
2. Schafhäutl: *Gelehrte Anzeigen Bayer. Ak.*, **20**, 557 (1845).
3. Spezia: *Atti accad. sci. Torino*, **33**, 289 (1898), **40**, 254, 1905, **41**, 158, 1905, **44**, 95, 1908.
4. Hale and Hurlbut: *Am. Min.*, **34**, 596, 1949; see also Lincio: *Beitr. Kryst. Min.*, **1**, 87 (1916) and Nacken: *Jb. Min.*, **1**, 71 (1916).
5. Michel-Lévy and Wyart: *Bull. soc. min.*, **69**, 156 (1946); *C. R.*, **212**, 89 (1941), **210**, 733 (1940), **208**, 1594, 1939.
6. Buehler and Walker: *Sci. Monthly*, **69**, 148 (1949) and *Ind. Eng. Chem.*, **42**, 1369 (1950); Praagh: *Geol. Mag.*, **84**, 98 (1947); Hale: *Science*, **107**, 393 (1948); Wooster and Wooster: *Nature*, **157**, 297 (1946); Barrer: *Nature*, **157**, 734 (1946); Swinnerton, Owen, and Corwin: *Sympos. Faraday Soc.*, no 5, 172 1949; Brown *et al.*: *Min. Mag.*, **29**, 858 (1952); Franke: *Bull. soc. chim. France*, 1950, 454; Friedman: *Am. Min.*, **34**, 583 1949; Corwin and Swinnerton: *J. Am. Chem. Soc.*, **73**, 3598 (1951); Franke: *Bull. soc. min.*, **75**, 591 (1952); Walker: *Ind. Eng. Chem.*, **46**, 1670 (1954); Laudise: *J. Am. Chem. Soc.*, **81**, 562 (1959).
7. Morey and Hesselgesser: *Econ. Geol.*, **46**, 821 (1951); Wasserburg: *J. Geol.*, **66**, 559 (1958).

ETCHING PHENOMENA

The etch pits produced by natural[1] or artificial[2] solvents on crystal faces or on sawn surfaces or cut spheres of quartz conform to trigonal trapezohedral

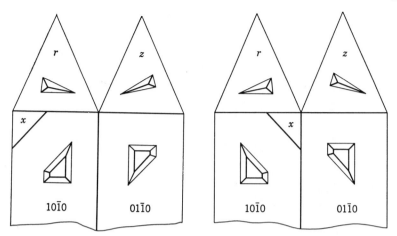

FIG. 80. Idealized appearance of etch pits produced on the $r(10\overline{1}1)$, $z(01\overline{1}1)$, and $m(10\overline{1}0)$ faces of right and left quartz.

(32) symmetry. The details of the etch pits vary with the kind and concentration of the solvent, with the temperature, and, to a certain extent, with the duration of the attack. Etch pits with anomalous symmetry have been reported.[3]

Etch pits are useful in distinguishing between right- and left-handed crystals and between positive and negative forms. Natural and artificial etching also is useful in revealing the presence and kind of twinning that may occur in quartz. The problem of orienting a euhedral quartz crystal for morphological or technical purposes in the absence of general forms usually resolves to the separate identification of the terminal rhombohedra. Etching effects are particularly useful for this purpose. The shapes of the etch pits on the positive $r\{10\overline{1}1\}$ and negative $z\{01\overline{1}1\}$ rhombohedra and on $m\{10\overline{1}0\}$ are distinctive, and their attitudes reveal the hand of the quartz (Fig. 80). The details of these figures vary considerably in different solvents and with the etching conditions. The rate of solution is greater on $z\{01\overline{1}1\}$ than on $r\{10\overline{1}1\}$. Etched surfaces of z usually are matt or dull in luster and are composed of relatively small and closely spaced etch pits, whereas the surfaces of r tend to be lustrous, because of the relatively weak attack, and show widely scattered although rather distinct etch pits. The faces of r generally are larger than those of z (see *Crystal Habit*), but this is not a reliable criterion, especially when the crystals are distorted. The boundaries of Dauphiné twins on the rhombohedral faces may be revealed by natural or artificial etching. In sawn and lapped sections cut parallel or nearly parallel to r or z, long-continued etching produces characteristically patterned solution surfaces, with a shingled appearance on r and a rippled appearance on z. Figure 81 shows these appearances on an etched section cut across a Dauphiné twin boundary, r and z being coplanar.

FIG. 81. Dauphiné twin in etched section of quartz. Bottom part nearly parallel to r, top part nearly parallel to z. Photomicrograph.

The $\{10\bar{1}0\}$ prism faces are less readily etched than the rhombohedral faces. They show characteristic pits which, when sufficiently distinct, lead to the identification of hand and of polarity of the a-axes (Figs. 80 and 82). The etching phenomena shown by $\{11\bar{2}0\}$ sawn sections usually are studied not by direct examination of the etch pits, since these are rather indistinct, but by the light-figures derived therefrom (see *Light-Figures*).

On $\{0001\}$, usually studied in sawn sections because this form is extremely rare on quartz crystals, etching produces three-sided pyramids with the apical edges approximately perpendicular to the a-axes and parallel to Y (Fig. 82). Each edge points to $+Y$, i.e., to the alternate prism faces over which would be located the faces of $r\{10\bar{1}1\}$. The triangular base of the pyramids usually is indistinct, because of overlap, but is approximately parallel to the a-axes. The ρ angle of the sides of a pyramid generally is in the range 25–35°. The faces and edges of the pyramids are more or less rounded. With long-continued etching, the ρ angle of the side faces decreases and other solution forms are developed in addition, including general forms. The latter forms cannot be clearly seen on direct inspection under a microscope. Reflections from them, however, appear in light-figures obtained from the bulk etch surface and give a screw appearance to the figure which identifies the hand of the quartz. The details of the etch pyramids on $\{0001\}$ and of the associated light-figures, and to some extent the orientation of the pyramids to the a-axes, vary with the conditions of etching, especially time and concentration. The nature of the abrasive finish on the surface also is important. Etching while the section is irradiated with ultraviolet radiation tends to flatten and to rotate the pyramids. The rotation of the pyramids is in a sense corresponding to the hand of the quartz. Under some conditions of etching, especially with weak acid and lapping with fine abrasive, hexagonal pyramids or three-sided pyramids modified by secondary facets at 60° may develop. These

(11$\bar{2}$0)

(10$\bar{1}$0)

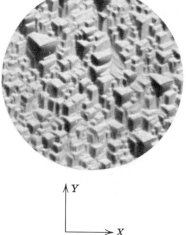

(0001)

FIG. 82

developments correlate with changes in the associated light-figures, but the exact relations have not been described.

Deep etch tubes more than a centimeter in length have been observed in natural and synthetic quartz.[17]

ETCHED SPHERES

Cut and polished spheres of quartz have been applied to the study of the etch phenomena and rates of solution of quartz.[4] The rate of solution is greatest in the direction of the c-axis and is at a minimum along an a-axis (X) in the direction from that end of this polar axis which becomes positively charged on compression. The rate of solution along the opposite direction of this axis is much greater. A deeply dissolved sphere thus takes on a triangular, lenticular appearance (Fig. 83). A slightly etched sphere viewed in a beam of light, so that the surface etch pits reflect in aggregate, shows two major, large-scale features (Fig. 84): (a) two equivalent triangular areas with curved projections, the direction of curvature indicating the hand of the quartz, present on the surface at opposite parts of the sphere at the points of emergence of the c-axis; and (b) three equivalent parallelogram-shaped patterns centering about the points of emergence of those ends of the a-axes that become positively charged on compression. The four corners of the parallelogram are defined by poles of m and r faces. One side of the parallelogram, connecting the poles of m and r, is vertical and parallel to the trace of the c-axis (Z). The direction of slope of the long side of the parallelogram, marking a great circle through the poles of m' and r, indicates the hand of the quartz. The equatorial corners of adjacent parallelograms are connected by luminous lines. The position of the reflecting patterns can be measured by

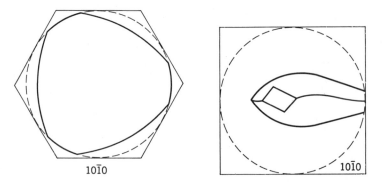

FIG. 83. Rates of solution on a sphere of left quartz. The dotted line shows the outline of the original sphere; the inner lines the outline of the solution-body remaining after long-continued attack by HF. Left: view along the c-axis, showing the crystal outline; right: view perpendicular to a prism face.

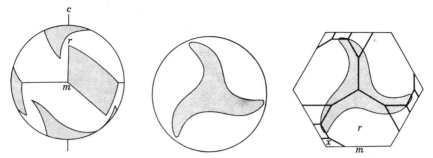

FIG. 84. Etching effects produced on a polished sphere of left quartz. Left: equatorial view, the poles of m, r, and c indicated; center: view along the c-axis; right: relation to morphology.

mounting the sphere on an optical goniometer.[8] These triangular and parallelogram areas are seen more distinctly as pinhole light-figures on etched sections cut perpendicular to the c-axis (Z) or an a-axis (X). The microscopic etch pits on the surface of the sphere that give rise to these effects have been figured photographically.[5] They conform, according to their position, to the etch effects seen on plane faces or sawn sections. The etch phenomena on quartz spheres are rendered more distinct if the surface of the sphere has been roughened with abrasive before etching.

The variation in rate of solution with direction in quartz also can be investigated in plane sections and on euhedral crystals. The rate decreases[6] in the order $\{0001\} \gg \{01\bar{1}1\} > \{10\bar{1}1\} \gg \{10\bar{1}0\}$. The difference in rate between $\{0001\}$ and $\{10\bar{1}0\}$ is over 100-fold. The trigonal trapezohedral zone is attacked relatively rapidly. On euhedral crystals,[7] the six prism faces are equally attacked, but alternate prism edges, corresponding to the positive ends of the a-axes, on which the s and x faces would be located, are much more rapidly rounded. Grooves may develop on these edges with long etching. The rounding of the rr and zz edges by solution is symmetrical, but rz edges are asymmetrically rounded and reveal the hand of the crystal. The rounded edges in general are not smooth and uniform but are composed of small facets of solution forms. More or less definite indices can be assigned to such facets, the optical goniometer signals showing a continuous band with superposed broad, bright reflections. In the prism zone the rounded edges give signals approximating to $\{11\bar{2}0\}$ or to ditrigonal prisms, and solution forms, usually vicinal in nature, can be recognized on edges in other zones. If the solution of a euhedral crystal is long continued, the rhombohedral termination may be largely dissolved away, leaving a low irregular surface terminating the hexagonal prism. Natural crystals, such as from the Alpine veins of Switzerland, sometimes are found deeply corroded into irregular cavemous bodies without traces remaining of the original faces.

LIGHT-FIGURES

The so-called light-figures[8] are best obtained on sawn and lapped sections of quartz that have been etched in a solvent. In the contact pinhole method, a sawn section with the top surface etched is placed upon a pinhole in a thin sheet of opaque material and is illuminated with a widely divergent beam of light. The light in passing through the upper etched surface is refracted by the facets bounding the etch pits, giving a luminous geometrical figure. In photographing the light-figure it is necessary to focus on the virtual image, which lies between the top and bottom surfaces of the section. With the camera lens directly over the section, the elevation E of the virtual image from the bottom surface is given by

$$\frac{E}{T} = 1 - \sqrt{1 + \frac{R^2/T^2}{n}}$$

where T is the thickness of the section, n the mean index of refraction, and R the radial displacement of the virtual image of the pinhole P from the actual position P, which can be measured experimentally. The ρ angle of the reflecting etch face can be calculated from T, R and n from the relation $R/T = \tan{[\rho - \sin^{-1}(\sin^{-1}\rho/n)]}$. Since E depends on R, an extended virtual image is not in a single plane and cannot be brought into exact focus.

Light-figures also can be obtained by transmission by using a focusing lens between the light source and the section, the incident surface again being rendered optically flat by polishing or by covering with an index oil and cover glass. Simple reflection in a parallel beam of light can be used to examine etched sections for twinning by tilting and rotating the section during examination. Geometrical light-figures also can be obtained by reflection if a vertically incident beam is thrown on the etched section through a hole in a screen positioned perpendicular to the beam, the light reflected back from the facets of the etch pits affording a visual pattern on the screen. In these and other methods in which the light-figure is viewed directly or through the aid of reflecting mirrors, the correct correlation of the figure to the microscopic image of the etch pits and to the quartz itself requires a knowledge of all the image reversals in the optical path.

The pinhole method is useful to determine the hand or polarity of the a-axes in etched quartz sections, particularly on $X = \{11\bar{2}0\}$ and $Z = \{0001\}$ sections. The light-figures obtained on such sections are illustrated in Figs. 85, 86, 87, and 88. Complicated light-figures which vary in their nature with rotation of the section are obtained from regularly arranged groups of pinholes or of groups of triangular or other apertures. In $\{11\bar{2}0\}$ or X sections that have been sawn, lapped, and etched, the light-figures differ on opposite sides of the section because the a-axis is polar. On viewing the negative (on

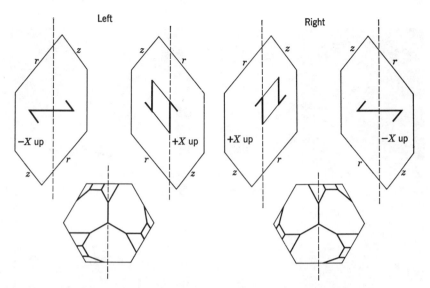

FIG. 85. Idealized appearance of pinhole light-figures on etched facing (11$\bar{2}$0) sections through right and left quartz.

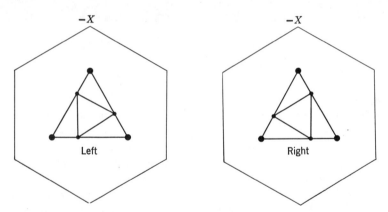

FIG. 86. Idealized appearance of pinhole light-figures on etched (0001) surfaces of right and left quartz.

compression) side when laid uppermost on a pinhole source of light an arrow-shaped figure is seen (Fig. 85). The horizontal, middle part of the figure is parallel to the Y-axis in the plane of the section. When the other or positive side of the section is laid uppermost, a parallelogram figure is obtained. The short sides of the parallelogram are parallel to the c-axis (Z), and the long sides are parallel to the intersection edge of $r\{10\bar{1}1\}$ and the plane of the section. This figure correlates with the parallelogram figure seen on etched spheres. A Dauphiné twin in a $\{11\bar{2}0\}$ section shows a parallelogram figure on

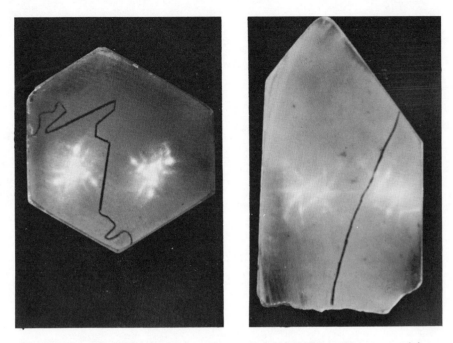

FIG. 87. Left: pinhole light-figures through an etched (0001) section of quartz containing a Dauphiné twin. Right: pinhole light-figures through an etched (11$\bar{2}$0) section of quartz containing a Dauphiné twin. Dark line marks twin boundary.

one side of the twin boundary and an arrow-shaped figure on the other side, on both surfaces of the section (Figs. 87 and 89). The details of the light-figures seen on {11$\bar{2}$0} sections vary somewhat with the etching conditions.

In a {0001} or Z section, the light-figures are identical in appearance on both sides of the section. The details of the figure vary considerably with the conditions of etching, especially with time. With long etching time, the light-figure appears as a group of three equal curved rays. The direction of curvature indicates the hand of the quartz: clockwise in right-handed quartz, and counterclockwise in left-handed quartz (Fig. 88). The tip of each ray points roughly to the negative end of an a-axis. Under other etching conditions an equilateral triangle may be obtained with the corners curved as above. With shorter etching

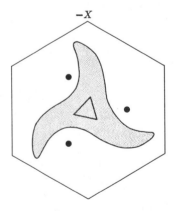

FIG. 88. Idealized pinhole light-figure on an etched (0001) surface of left quartz.

times, a group of three bright spots may be obtained, which if connected by lines define an equilateral triangle with the sides approximately parallel to Y. Three additional bright spots may be seen on the sides of this triangle, defining another triangle oriented at 90° to the first. The central part of the inner triangle may show a bright blur or a faint small triangle with curved corners (Fig. 87). All these effects vary markedly with the etching conditions, and consistent results are obtained only if the experimental conditions are carefully standardized. If a Dauphiné twin is present in the {0001} section, the light-figures obtained across the twin boundary are oriented at 180° to each other. A light-figure consisting of a simple hexagon is obtained from unetched {0001} surfaces that have been ground with coarse abrasive. It is caused by the development of cleavage facets on r and z planes.

Light-figures are a convenient means of establishing both the presence and the identity of twinning on the parallel-axis laws in sawn and etched sections.[10] The recognition of the twin law is based on the geometrical relation of the light-figures on opposite sides of the twin boundary in oriented sections. The effect of Dauphiné twinning on the light-figures in {0001} or Z sections and {11$\bar{2}$0} or X sections has already been mentioned. Brazil twinning cannot ordinarily be investigated by this method because of the very thin nature of the twin lamellae. The effect of twinning on the Combined Law on light-figures in {11$\bar{2}$0} sections is shown in Fig. 89. Systematic drawings showing the effect of Dauphiné, Brazil, and Combined twinning on the mutual orientation of the etch pits as seen in X, Y, and Z sections have been published.[15]

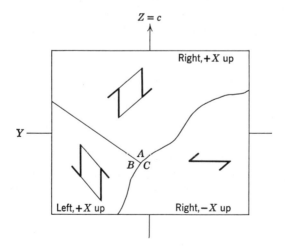

FIG. 89. Idealized appearance of pinhole light-figures on an etched (11$\bar{2}$0) section through a twin on the Combined Law. C is twinned to B by the Brazil Law, A to C by the Dauphiné Law, and B to A by the Combined Law.

SOLVENTS

A number of different solvents can be used to etch quartz, including HF, NH_4HF_2, and hot solutions of NaOH, KOH, and substances such as alkali carbonates and borates that hydrolyze in water to give alkaline solutions. At high temperatures in closed bombs water and dilute water solutions with a pH slightly over 7 are effective etchants. The attack in fused NaOH, NH_4HF_2, etc., is very rapid and in general produces transparent, more or less regularly undulating, hummocky or grooved surfaces without distinct etch pits. The best results, in the sense of the most sharply defined etch pits or light-figures derived therefrom, are obtained with hydrofluoric acid in the concentration range from about 50 to 20 per cent. Acid stronger than about 50 per cent reacts too vigorously to give good figures. The final product of the reaction of SiO_2 with HF is fluosilicic acid, H_2SiF_6, in solution. Care should be used with HF, since it produces serious skin burns and the fumes cause fluorine poisoning. Satisfactory results also can be obtained with a warm saturated solution of NH_4HF_2, especially if a small amount of HF is added, and the solution is relatively safe to use. A finer-scale etching effect, leading to matt surfaces, can be obtained if fluoboric acid or alkali borates are added to the solution. Sugar also may be added to keep the solution from creeping if stored in an open tank. The quality and to a certain extent the nature of the etch-figures or light-figures vary with etching time, as well as with the nature and concentration of the solvent and with temperature. With long-continued etching or with rapid attack in strong solvents, so that a considerable amount of quartz is dissolved from the surface, the initially pitted and dull etched surface may become smooth with a glazed, transparent appearance. In general the best light-figures are obtained by long etching of sawn surfaces that have first been lapped with a coarse abrasive. Relatively large and perfect, isolated etch pits may be obtained by the long-continued action of dilute solutions on unlapped natural faces.

The rate of solution may be increased and the nature of the etch pits altered by placing the quartz in ultraviolet light below 2800 Å or in an a-c or d-c field during the etching.[13] The sensitivity of the etching, in the sense of developing the zonal growth structure commonly present in quartz crystals, also is increased. The etching also is affected[14] by exposing the quartz to a high-frequency electrical discharge at low pressure, and to the vapors of alkali metals at 350°. Condensation of moisture on a polished surface stimulated by an electrical discharge also reveals a zonal pattern.[14]

HISTORICAL

The first experimental etching of quartz or of any single-crystal was done in 1816 by the English chemist J. F. Daniell,[11] best known to crystallographers for his work, extending that of Haüy and Wollaston, on the interrelation of the morphology and the structure of crystals. Daniell produced rectangular and rhomboidal etch pits on the prism and rhombohedral faces of quartz crystals, by HF, and remarked ". . . the surface of a body is never equally acted upon by a solvent . . . . This new process of dissection admits of more extensive application than might at first be imagined, and we are thus furnished with a method of analyzing crystalline arrangements, which promises to lead to important results." Etching was first applied to the study of the twinning and the hand of quartz by Leydolt[12] in 1855. His indication of the type of twinning in quartz now known as the Combined Law was not fully described, again by the etch method, until about 90 years later. Another important contribution of this period was made by Des Cloizeaux.[16] A number of experimental studies of the etching of quartz appeared in the scientific and technical literature in the latter nineteenth and early twentieth centuries. Not all of these (and of later) studies correctly related the symmetry of the etch effects to the hand of the quartz. The literature of the subject increased greatly in the period 1938–1945. The examination of quartz objects for optical and piezoelectric purposes by both optical and etch tests to reveal twinning is a relatively recent practice, and some of the early measurements of the piezoelectric properties of quartz were prejudiced by the presence of unrecognized twinning. In the quartz oscillator-plate industry the sawn blocks or wafers produced at various stages of manufacture are etched to reveal twinning and to permit the correct orientation of the quartz as to hand and polarity of the a-axes by means of pinhole light-figures or similar tests. The light-figures, known to Leydolt (1855), were first described in detail by de Gramont (1931) and Gaudefroy (1933). Some of the light-figure methods of orientation have been granted patents.

References

1. In particular see Ichikawa: *Am. J. Sci.*, **39**, 455 (1915); Mügge: *Cbl. Min.*, 609, 641, 1921; Cavinato: *Rend. accad. nazl. Lincei*, **1**, 246 (1925); Molengraaff: *Zs. Kr.*, **14**, 186 (1888) and **17**, 138 (1889); Trommsdorf: *Jb. Min.*, Beil. Bd. **72**, 464 (1937); Kalb: *Zs. Kr.*, **86**, 439 (1933).
2. The literature is very large; in particular see Molengraaff (1888, 1889); Ichikawa (1915); Friedlaender: *Beitr. Geol. Schweiz, Geotech. Ser.*, no. 29, 45, 1951; Bömer: *Jb. Min.*, Beil. Bd. **7**, 516, 1891; Parrish and Gordon: *Am. Min.*, **30**, 296 (1945); Honess: *Nature, Origin and Interpretation of Etch Figures*, New York, 1927; Frederickson and Cox: *Am. Min.*, **39**, 886 (1953); Weil: *C. R.*, **191**, 935 (1930). See also refs. 4 and 5.
3. Beckenkamp: *Zs. Kr.*, **32**, 18 (1899).
4. Meyer and Penfield: *Trans. Conn. Ac. Sci.*, **8**, 158 (1889) [the hand of the crystal is erroneously described as right]; Bond: *Zs. Kr.*, **99**, 488 (1938) [hand of crystal not stated, but right]; Nacken: *Jb. Min.*, I, 71 (1916); Van Dyke: *Proc. Inst. Radio Eng.*, **28**, 403 (1940); Gill: *Zs. Kr.*, **22**, 97 (1893); Gaudefroy: *Bull. soc. min.*, **56**, 5 (1933); Lincio: *Beitr. Kryst. Min.*, **1**, 87 (1916); Ichikawa (1915).
5. Bond (1938) and others.

6. Mügge: *Festschr. H. Rosenbusch*, 1906, p. 96.
7. Molengraaff (1888, 1889); Meyer and Penfield (1889); Leydolt: *Sitzber. AK. Wiss. Wien, Math. Nat. Kl.*, **15**, 39 (1855); Ichikawa (1915).
8. See Nacken (1916); Gaudefroy (1933).
9. Willard: *Bell Syst. Tech. J.*, **23**, 11 (1944); Parrish and Gordon (1945); Nomoto: *Proc. Phys. Soc. Japan*, **3**, 136 (1948); de Gramont: *Rev. opt. théor. instrum.*, **10**, 213 (1931) and *Recherches sur le quartz piezoel.*, Paris, 1935; Gaudefroy (1933); Booth: *J. Inst. Elect. Eng.*, **88**, Pt. 3, 140 (1941); Schubnikov: *Trudy Lab. Krist. Ak. Nauk SSSR*, no. 1, 41, 1941; Laemmlein: *Dokl. Ak. Nauk SSSR*, **72**, 775 (1950).
10. The technical literature on the manufacture of quartz oscillator-plates contains many descriptions; see Parrish and Gordon (1945); Gordon: *Am. Min.*, **30**, 269 (1945); Gaudefroy (1933).
11. Daniell: *Quart. J. Sci.*, **1**, 26 (1816).
12. Leydolt: (1855).
13. Choong Shin-Piaw: *Nature*, **154**, 464, 516 (1944) and *J. Opt. Soc. Amer.*, **35**, 552 (1945); Kleber and Koch: *Naturwiss.*, **39**, 19 (1952).
14. Choong Shin-Piaw: *Am. Min.*, **37**, 791 (1952).
15. Booth and Sayers: *Post Office Elect. Eng. J.*, **31**, 245 (1939); see also ref. 10, etc.
16. Des Cloizeaux: *Ann. chim. phys.*, **45**, 129 (1855) and *Mém. ac. sci. (inst. imp. France)*, **15**, 404 (1858).
17. Nielsen and Foster: *Am. Min.*, **45**, 299 (1960).

VARIETIES OF QUARTZ

The varieties of quartz may be divided into two broad categories:

(1) *Coarse-crystallized varieties*, including euhedral crystals and anhedral aggregated types with the individual grains visible to the unaided eye. This material ordinarily is transparent or strongly translucent, with a vitreous luster, and affords values for the specific gravity and indices of refraction that vary only within very narrow limits. The chief varieties in this category are based on color, such as amethyst and smoky quartz, and are described in this section. In addition, there are varieties of more or less significance, sometimes designated by specific names, that are based on features of crystal habit or aggregation or on the presence of inclusions. The latter varieties, including rutilated quartz and aventurine and asteriated types, are described in a separate section on *Inclusions in Quartz*.

(2) *Fine-crystalline to dense varieties*, with the individual grains or fibers distinctly seen only under magnification. This material usually has a sub-vitreous to waxy or dull luster and is weakly translucent to virtually opaque. The fibrous material generally affords variable and relatively low values for the indices of refraction and, if free from admixture, for the specific gravity. In general there are no abrupt transitions or absolute distinctions of class between the fine-grained varieties of quartz. Many of the recognized types stem from times of antiquity, and almost all antedate the application of the petrographic microscope and of x-rays as descriptive tools. Their distinction

is based on secondary, gross characters, particularly the color, as determined by physical admixture, and on the mode and scale of their geological occurrence. Additional factors are the distribution of the color, the fracture and texture, and visible inclusions of other materials.

Following general and long-established usage, the name chalcedony is restricted, as the main variety, to fibrous quartzthat is relatively light colored, without appreciable admixture of foreign material and without gross color banding, and that occurs in small masses as crusts, cavity fillings, and the like. Most of the fine-grained types of quartz that have been distinguished by given names, including agate, carnelian, flint, and chert, basically are aggregates of fibrous quartz, for which the term chalcedonic silica is appropriate, and in a sense are variants of chalcedony. In addition, some of the fine-grained types of quartz are found on microscopic examination to be composed of granular rather than fibrous quartz. Novaculite and much of what is called jasper and prase are of this nature, as is at times some of the normally fibrous material classed under specific names on the basis of color or other non-textural factors.

EUHEDRAL AND COARSE-GRAINED VARIETIES

AMETHYST

The color of amethyst ranges through shades of violet, bluish violet, reddish violet, and purplish violet. The depth of color varies widely, from almost colorless to deep rich tones of pure violet or purplish violet. The color rarely is distributed uniformly in the crystal, although it may appear to be in a well-cut gem stone, and ordinarily is disposed in very thin layers or bands parallel to the external crystal faces. It also occurs in irregular patches. The color frequently is darkest and internal flaws are less frequent toward the apical part of the crystal. The color may be selectively distributed in the growth-loci of the terminal r faces, giving rise to a three-fold sectoral appearance in basal sections or on the table of cut gem stones.[1] It also may be distributed beneath the faces of both r and z but not beneath the prism faces. The effect is caused by the selective adsorption and taking up of the pigmenting substance by the r faces. Instances also are known in which these faces of quartz crystals are selectively filmed[1] by finely divided iron oxide. Sceptre crystals occur of amethyst upon colorless, milky, or smoky quartz, and parallel overgrowths of amethyst crystals upon other varieties of quartz are common. Amethystine phantoms occur within colorless crystals of quartz, and alternate zones of colorless, smoky, and amethystine quartz have been observed in single crystals. More commonly the outer zone is amethyst, and colorless and smoky zones alternate inwardly. Close-spaced or extremely

thin zones are best seen in polished sections cut perpendicular to a rhombohedral face. Specimens from Coos County, New Hampshire, have been
found with as many as ten narrow, alternating bands of amethyst and smoky
quartz. Coarsely zoned amethystine and milky quartz has been found at
Chambelève, Puy-de-Dôme, France,[34] and other localities.

Amethyst crystals almost always have a very simple habit, with r and z
(the former usually dominant or present to the exclusion of z) either alone or
in combination with m. Faces of s and x are lacking. The crystals often are
densely aggregated into crusts, and then show only the terminal rhombohedral
faces. When individual crystals are broken from such crusts, they taper
irregularly toward the point of attachment and are striated laterally because
of interference with adjoining crystals. Usually the lower parts of such
crystals are flawed and turbid with a milky appearance.

Amethyst virtually always shows polysynthetic twinning on the Brazil Law.
Untwinned crystals have been noted.[32] The twin lamellae, a fraction of a
millimeter thick, are remarkably uniform and are arranged parallel to the
terminal r or r and z faces. The lamellae are alternately right- and left-
handed. They may give rise to sets of delicate striations or of open polygonal
markings on the rhombohedral faces, and cause a rippled or finger-print
appearance on fracture surfaces. The twinning can be studied in etched sections or, more conveniently, by optical means. Electron microscopy has
revealed[38] very fine transverse striations down to about 35 Å on the twin
lamellae as exposed on fracture surfaces. The amethyst color may be equally
developed in successive twin lamellae, but often alternate lamellae of the same
hand, either right or left, are selectively pigmented. Some descriptions note
the pigment only in the right-handed twins. In other instances the pigment
appears to be concentrated between the lamellae. When color banding is
marked parallel to the rhombohedral faces, the colorless or relatively faintly
pigmented bands also show the twin structure. The liquid inclusions in
amethyst also may be rhythmically arranged, like the twin lamellae.

The optical structure of amethyst was first described by Brewster (1823) and
has been the subject of many later investigations.[2] In polarized light under
crossed Nicols, basal sections of amethyst (preferably 2–4 mm. thick, so that
there is an appreciable amount of optical rotation) show the twinning as a
series of uniform, alternating light and dark stripes representing extinction
bands (Fig. 90). Both double refraction and optical rotation are involved in
the formation of the black extinction bands by the equal and opposite re-
tardation of laminae of opposite hand and equal thickness that are inclined
to the plane of the section.

The distribution of the lamellae in basal section depends on the habit of the
crystal in question. In a crystal terminated wholly by the positive rhombohedron r, the stripes radiate in sets 120° apart from the center (apex) of the
crystal, being perpendicular to the $\{10\bar{1}0\}$ prism faces and parallel to the

terminating rhombohedral edges. If both r and z faces are present, only the growth-loci of r being pigmented, as is usually the case, the polysynthetic twinning is lacking in the growth-loci of z, as exposed in section, and the stripes in the growth-loci of r have a polygonal arrangement parallel to the traces of the terminal rr and rz edges if these forms are of unequal size. If the crystal is of distorted habit (the r faces, e.g., not all meeting at the apex), the distribution of the twinned areas as seen in basal section varies accordingly, and isolated, usually triangular areas of twinning may appear.

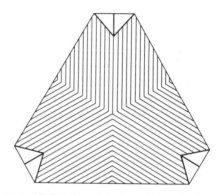

In convergent polarized light, the interlamination of right- and left-handed layers of equal thickness produces an interference figure with the usual black cross and circular rings of an ordinary, optically inactive, uniaxial substance. Other parts of a basal section may consist wholly of right- or left-handed quartz, and then show in thick section the typical uniaxial figure of quartz consisting of rings without isogyres in the central part of the field. At the juncture of

FIG. 90. Basal section of amethyst, showing a system of extinction bands observed between crossed Nicols in polarized light and caused by lamellar Brazil twinning.

two such areas of right- and left-handed quartz the overlapping parts may show Airy's spirals. This optical effect, discovered in quartz,[35] is best observed when two equally thick basal plates of right and left quartz are superposed and examined in convergent monochromatic polarized light. Spiral isogyres are obtained that rotate outward from the center in the rotation sense of the quartz plate in which the light from the polarizer enters. In white light the spirals appear red on the concave side and blue on the convex side. Amethyst may show an anomalous biaxial character,[21] with $2V$ ranging up to $35°$ and with the axial plane inclined at small angles, sometimes at about $90°$, to the $\{10\bar{1}0\}$ edges. The effect is associated with twinning or strain, and other specimens show uniaxial interference figures.

Dauphiné twinning also occurs in amethyst,[3] but seems to be relatively uncommon. A (secondary?) Dauphiné twin boundary that transects the polysynthetic twinning causes the Brazil laminae as counted in alternate sets back and forth across the twin boundary to stand in the relation of the Combined Law.

Chemically, amethyst differs from other varieties of quartz in containing a relatively large amount of Fe_2O_3, in the range $0.0X$ per cent (Tables 24 and 25, under *Chemical Composition*). The depth of color is proportional to the iron content. Precision determinations of the content of alkalies, Al, etc., are lacking, as are highly precise measurements of the indices of refraction and the unit cell dimensions. Measurements of these physical properties of amethyst of moderate or low precision are identical within the experimental error for the properties of colorless quartz; slightly higher values for the indices of refraction of amethyst have been reported.[4] Amethyst was reported[5] to contain a coupled solid solution of B and P (BPO$_4$ is isostructural with quartz), in the range 0.001–0.0001 B$_2$O$_3$, with an attendant variation in the density and the indices of refraction, but this work has not been confirmed.[6] Numerous spectrographic analyses of amethyst have been made;[7] the principal minor constituents reported are Fe, Al, Li, Ca, Mg, Cr, Mn, Ti, and Cu.

Color and Heat Treatment. Amethyst can be decolorized or changed in color to citrine by heat.[8] The decolorization has been known for hundreds of years. The lower limit of stability of the color is not clearly established, but appears to be in the range 230–260° for most specimens. At lower temperatures amethyst becomes gray-violet when hot, but takes on the original color when cooled. The rate of decolorization is very slow below about 300°, but increases with increasing temperature; at about 400° several hours may be required to render the material completely colorless when cold, and less at 550°. After the colorless stage has been reached, further heating at higher temperatures, roughly in the range 500–600°, generally produces a citrine or brownish yellow to reddish brown color in the darker amethysts, shading off to pale yellow as the depth of the original color decreases. Most of the commercial gem citrine is made in this way. At still higher temperatures, the citrine color may be bleached out, leaving the material colorless when cold, and the material shows a milky turbidity or opalescent shimmer.[36] Colorless and smoky quartz does not develop this milky turbidity at high temperatures.

There is a considerable variation in the response of different specimens to heat treatment. Many amethysts do not show a colorless stage; the material from Madagascar is said to turn colorless but not citrine. Some specimens phosphoresce on heating. The nature of the atmosphere during heating, whether oxidizing or reducing, has no effect.[9] Amethyst, like most quartz, shows a weight loss in the range 0.00X–0.0X when ignited. A grayish green or grass green color also may be developed, apparently in place of the citrine color, on heat treatment. The tendency to become green has been known for many years; it seems to be marked only in amethyst from certain localities. Amethyst from Montezuma, Minas Gerais, Brazil, was found to become green when heated in air at 510° and cooled, but to become colorless when heated in hydrogen at 550°; measurements of the absorption spectrum have been made.[10] The amethyst color of bleached or of citrine material can be restored by irradiation with x-rays, an effect first noted in 1906 by irradiating heated amethyst with radioactive material;[11] and the color of natural amethyst sometimes can be slightly darkened in this way. The irradiation of untreated or of heat-treated amethyst with x-rays also produces more or less of a smoky tint, which is superposed on the amethystine color. An amethystine color also has been produced[39] by the irradiation of synthetic quartz crystals containing Fe^2 and Fe^3.

The color of amethyst is believed to be produced by a compound of ferric iron. The main evidence[12] in support of this is as follows: (a) the amethystine variety of quartz is characterized by a relatively high content of Fe_2O_3, in comparison to other quartz, and the depth of color is proportional to the amount of iron found on analysis; (b) the absorption spectrum is very near that of some ferric compounds, certain of which have a violet color like amethyst; (c) on heating, amethyst become identical in color and absorption spectrum with natural citrine, which appears to be colored by a ferric compound; and (d) amethyst typically contains goethite and hematite as macroscopic inclusions. The inclusions often occur in an outer colorless zone overlying amethystine material.

Other theories[12] have attributed the color of amethyst to hydrocarbons, Ti, and Mn, and some early observations of historical interest are mentioned in a following section. One early view, based on an analysis of Brazilian amethyst by H. Rose[33] in 1800 that showed 0.25 per cent manganese oxide and 0.5 per cent iron oxide, attributed the color to Mn. This theory later gained wider support when it was observed[13] that radiation produced a violet color in glasses containing a trace of Mn. The content of Mn-varies over about the same range in all kinds of quartz regardless of color, and is very low or lacking in some amethyst.

The state in which the Fe is present in amethyst is not known. It may be in substitution for Si, or in interstitial positions, but there is no evidence of the accompanying variation in physical properties and in cell dimensions. The identity of the coupling ion also is not apparent, although the coupled entrance of Fe^3 and OH is a possibility. The amount of Fe present in amethyst is considerably greater than that of Al in quartz in general. The Fe in amethyst also may be present as an adsorbed or admixed phase, perhaps as colloidally dispersed goethite or hematite, analogous to the adsorption of large dye molecules[14] by crystals such as K_2SO_4.

Several coloring mechanisms also may be involved. The color of amethyst is perceptibly different in material from certain localities, ranging from a bluish violet to a reddish violet, and at times to smoky shades, at the same general intensity of coloration, and the response to heat treatment and to irradiation also is not wholly uniform. The range in color of amethyst is apparent on a side-to-side comparison in sunlight of crystals from Madagascar (in part) and from Porkura in Hungary, which tend toward a blue-lavender or bluish violet color, with dark amethyst from Iredell County, North Carolina, or from occurrences such as Guanajuato and Creede, which incline toward a reddish violet. The amethyst from Uruguay and most other localities stands between these extremes. The amethyst from Bahia, Brazil, tends towards reddish or brownish red tones as compared to that from Uruguay. A smoky cast of color is shown by the amethyst from Upper Providence township, Pennsylvania, and from some pegmatitic occurrences. The Madagascar amethyst also tends to be smoky. A superposition of an amethystine color on a natural citrine or smoky color may be involved, and the amethyst color itself may be variable, perhaps as a function of particle size and orientation if the color derives from a dispersed phase. Amethyst, however, unlike smoky quartz, does not scatter light[31] to a greater extent than does colorless quartz. It does show a relatively strong absorption in the infrared[32] corresponding to (OH). Another associated problem is the polysynthetic Brazil twinning, which seems to be genetically associated with the color.

The absorption spectrum[15] of amethyst shows characteristic maxima at 340, 540, and 950 mμ. Certain specimens show additional maxima at 225 and 266 mμ. All the absorption bands can be enhanced by irradiation with x-rays.[16] Amethyst shows a more or less strong dichroism, with ϵ reddish violet and ω pale blue and absorption $\epsilon > \omega$. A weak anomalous dichroism can be seen in (0001) sections; it has been established[7] that the absorption indicatrices for the 540 mμ and 340 mμ bands are differently oriented. The indicatrix for 340 mμ conforms in position to the symmetry of quartz, but the indicatrix for 540 mμ has the rotation axis of the ellipsoid, ϵ, perpendicular to a face of $\{10\bar{1}0\}$ in each of the rhombohedral growth-loci.

Occurrence. Amethyst crystals generally are small in size, rarely ranging over 4 or 5 in., and giant crystals such as are occasionally found in smoky and colorless quartz are not known. An unusually large and uniformly colored amethyst crystal from Brazil, found in 1946, shows a rhombohedral termination with edges about 5 in. in length and weighs almost 8 lb. Crystals weighing as much as 50 lb. have been found, but they are for the most part milky and translucent, or badly flawed, and only local areas are transparent and of good color. Amethyst forms at relatively low temperatures and pressures, below those generally obtaining for smoky quartz and rose quartz. When associated with smoky quartz, amethyst follows the crystallization of that variety.

Amethyst occurs in deposits of varied nature. It is widespread, although generally of quite pale color, in hydrothermal veins formed at relatively low temperatures (epithermal type), where it is associated with barite, calcite, fluorite, sulfides, and, at times, zeolites. A well-known occurrence of this type is at Guanajuato, Mexico, where drusy crusts of amethyst occur with calcite and apophyllite in the silver veins. Also widespread, although in relatively small amount, in veins of the Alpine type and much less typical than smoky quartz. In cavities in pegmatites and granitic rocks. Also in deposits of hematite formed in the meteoric circulation or by low-temperature hypogene solutions.

The most important manner of occurrence of amethyst is in basic igneous flow rocks, chiefly basalts, where it is found in cavities and is often associated with agate and zeolites. Commercial amethyst is chiefly obtained from Uruguay and the state of Rio Grande do Sul in southern Brazil, where it occurs with agate in large amygdaloidal cavities on a weathered melaphyre. Typical specimens show concentric layers of agate lining the walls with an inner cavity partly or completely filled with crystals of amethyst or of colorless quartz with amethystine tips radiating inward. The amethyst also occurs lining large open geodes and other cavities. At Serro do Mar, Rio Grande do Sul, Brazil, a single cavity 33 ft. in length, 5½ ft. in width, and 3 ft. in height was entirely lined with brilliant, deep violet amethyst crystals averaging about 4 cm. across. Large specimens from this occurrence are preserved in many museums. Amethyst also is obtained commercially in the states of Minas Gerais, Goiaz, and especially Bahia, in Brazil. The export of amethyst from Brazil (including Uruguay) is said to have begun in 1727. Amethyst, in part of the finest quality, is mined in the Urals, in the Reshev and Alapayev districts around Mursinsk, where it occurs in quartz veins in granite, and fine stones are also obtained from the gem gravels of Ceylon. The term Siberian amethyst is now generally applied to any cut stone of rich, deep color regardless of locality. Fine amethyst specimens are found in hydrothermal sulfide veins at Porkura and in an iron deposit at Macskamezö, Transylvania. Massive vein amethyst with a cockscomb structure has been worked at localities in Puy-de-Dôme, France, and along the Bay of Fundy, Nova Scotia, to provide slabs for ornamental purposes. A locality important in past centuries was at Idar, Oberstein, and nearby places, Germany, where amethyst occurred with agate (which see) in melaphyre.

Amethyst occurs abundantly in the Thunder Bay region on the north shore of Lake Superior in Ontario. Here crystals up to 5 or 6 in. across occur as groups and crusts associated with fluorite, calcite, and barite in sulfide veins. Also in Canada as light-colored crystals and aggregates with chalcedony and zeolites in basalt along the shores of the Bay of Fundy, Nova Scotia. Only a very small selection of the localities for amethyst in the United States can be mentioned. In Maine in pockets in pegmatite on Deer Hill north of Stowe, Oxford County, and in New Hampshire at numerous small localities in pegmatite in Stark and Milan townships in Coos County, especially on the Diamond Ledges on Long Mountain. Drusy crusts of small amethyst crystals are found at various of the zeolite localities in the diabase sills of northern New Jersey. Fine crystals, in part of gem quality, occur with colorless and smoky quartz at numerous minor localities in pegmatite or loose in the soil in Delaware and Chester counties, Pennsylvania, notably in the neighborhood of Media in Upper Providence township, Delaware County. Crystals and cut stones from the latter area sometimes show a marked sectoral distribution of the color, with alternating amethystine and colorless or smoky regions beneath the terminal rhombohedral faces. Fine gem material and exhibit specimens have been obtained in North Carolina in Iredell, Macon, Alexander and Lincoln counties; in Nelson County, Virginia; and at various places in Georgia. Large amethyst crystals from Clayton, Rabun County, Georgia, have been found, containing liquid-filled cavities almost an inch in size. Amethyst occurs in the hollow trunks (or casts) of petrified trees on Amethyst Mountain in Yellowstone National Park, Wyoming. Fine composite crystals of amethyst were found at Cripple Creek, Colorado, and pale amethystine quartz is a gangue mineral in the silver veins of the Creede district, Mineral County, Colorado. Amethyst, in part of gem quality, occurs with hematite and apatite in cavities in a mineralized zone in quartzite at Four Peaks, Maricopa County, Arizona.

Historical. Amethyst has been known since ancient times. Theophrastus, a pupil of Aristotle, in his work *On Stones*, written in the fourth century B.C., mentioned both rock-crystal and amethyst, and said that these substances and *sardion* (sard), a brownish red chalcedony, were found when certain rocks were cut through. This is perhaps a reference to the occurrence of amethystine-tipped quartz crystals in the central cavity of chalcedony geodes. Amethyst was used for engraved seals in Helenistic and early Roman times, although red and brownish red types of chalcedony were more commonly employed. The name is derived from the Greek ἀμέθυστ-ος, *not drunken*, and according to Pliny expresses the mistaken belief that the wearer is protected against the intoxicating effect of wine or means that the color approaches but does not quite reach that of red wine. The ancient term *amethystus* included not only the true amethyst, now also called occidental amethyst, but also purple or violet corundum, or oriental amethyst, together with the purple garnets. Amethystus was one of the twelve gem stones mentioned in the Bible as used to adorn the linen bag worn by the High-Priest when ministering to Jehovah; the stones were severally engraved with the names of the twelve tribes of Israel. Some of the other stones also were varieties of quartz, including agate, sard, and jasper. In ancient times amethyst and other colored stones were fashioned by roughly rounding and polishing the surface, or were used for engraved seals and other carved objects, and the faceting of gem stones to deliberately enhance the internal color and brilliance did not develop until later centuries. Amethysts generally are faceted as step- or emerald-cuts; stones of a fine and uniform color over about 20 carats in weight are rare. Formerly costly, amethysts depreciated greatly in value after the early 1800's with the export of large amounts of material from Uruguay and Brazil. Pliny and other writers up to Agricola stated that the best amethysts were obtained from India.

The identification of amethyst as a variety of ordinary quartz was indicated by the Swiss natural scientist J. J. Scheuchzer[17] (1672–1733) in 1708 and later was explicitly stated by Rome de l'Isle and R. J. Haüy. Scheuchzer is given qualified priority for this discovery by von Kobell,[19] but Agricola,[18] in his *De natura fossilium*, published in 1546, said that amethyst is found in large crystals having a hexagonal base and terminated in a point similar to quartz. As late as 1817, James Sowerby in his work, *British Mineralogy, or Coloured Figures Intended to Elucidate the Mineralogy of Great Britain*, stated,[20] "Few would have thought that the Amethyst, so long well known as a precious stone was Quartz or Rock-crystal coloured by Manganese." The anomalous dichroism in basal sections of amethyst has led to numerous suggestions that the symmetry is lower than that of quartz, and perhaps monoclinic,[22] but the x-ray Laue symmetry of untwinned amethyst is identical with that of colorless quartz. It also has been proposed, first by Brewster (1819), that amethyst be classed as a species separate from quartz on the basis of the twinning.

The color of amethyst has been a matter of interest for many years. J. F. Henckel,[23] in 1725, thought that it might be due to colloidal gold: "In particular the violet cast of amethist, and the red blush of the jasper, whereby it comes to resemble coral, would seem to suggest an enquiry, whether their tint be not owing to the share of metal they contain. I know not, whether the amethist does not hold gold; as there is no experiment extant for giving such a color to a stone, or stony glass-flux, except by means of gold, with the addition of tin, especially as I have a method of imparting this color to spring water, without making it less sweet and potable . . . ." Henckel is referring to the Purple of Cassius. Silica gel with an amethyst, yellow, or rose color has been produced[24] by precipitating silicic ether mixed with an alcoholic solution of gold chloride and exposing the gel to light. Colloidally dispersed gold is produced. Quartz colored rose by colloidal gold has been produced[25] by passing a current through a gold-plated section held at a high temperature.

Many early authors took the view that iron is the cause of the color. In 1729 John Woodward[26] spoke of an occurrence of quartz as ". . . some of the crystals Red, others Amethystine; concreted on a crust of Iron-ore, growing on the side of a perpendicular fissure in St. Vincent's Rock, Bristol. These different colors are owing to the different Proportions of ferreous corpuscles uniting with the Crystalline in the Concretion." Haüy (1817)[27] thought that the color was caused by iron oxide, and Carl M. Marx (1831)[28] spoke of hydrated iron oxide as the pigment. Later work on the role of iron in amethyst is reviewed in the classical study by E. F. Holden[12] in 1925. Titanium also has been suggested as the pigment, and a statement[29] appears in the 1813 volume of the short-lived *American Mineralogical Journal* that the color of the amethyst from Delaware County, Pennsylvania, is caused by this element. The amethyst from this locality sometimes contains inclusions of rutile. Brewster (1853)[30] spoke of amethyst as having grown from amethystine solutions.

References

1. Frondel: *Am. Min.*, **19**, 318 (1934), and *Am. Mus. Nat. Hist. Novit.* **758**, 1934; Laemmlein: *Trudy Inst. Krist. Ak. Nauk. SSSR*, no. 6, 255, 1951.
2. Brewster: *Trans. Roy. Soc. Edinburgh*, **9**, 139 (1823) (for 1819); Dove: *Ann. Phys.*,

180 QUARTZ

40, 607 (1837) and *Jb. Min.,* 550, 1838; Haidinger: *Sitzber. Ak. Wiss. Wien, Math.-nat. Kl.,* **12,** 401, 1854; Böklen: *Jb. Min.,* I, 62 (1883); Liebisch: *Sitzber. Ak. Wiss. Berlin,* **36,** 870 (1916); Des Cloizeaux: *Mém. ac. sci. France,* **15,** 1858; Judd: *Min. Mag.,* **8,** 1 (1888) and **10,** 123 (1893); Lacroix: *Min. de France,* **3,** 43 (1901); Tutton: *Cryst. and Practical Cryst. Meas.,* London, 1922, **1,** p. 508, and *The Natural History of Crystals,* London, 1924, p. 206; Brauns: *Zbl. Min.,* **97,** 289 (1932); Mügge: *Zs. Kr.,* **83,** 460 (1932); Trommsdorf: *Jb. Min.,* Beil. Bd. **72,** 464 (1937).

3. See Mügge (1932).

4. Dufet: *Bull. soc. min.,* **13,** 271 (1890); Trommsdorf (1937).

5. Trommsdorf (1937).

6. See discussion in Lietz and Münchberg: *Jb. Min., Monatsh.,* **10,** 217 (1956).

7. Cohen: *Am. Min.,* **41,** 874 (1956); Leutwine: *Urania,* **14,** 373 (1951); Trommsdorf (1937); Hoffman: *Zs. anorg. Chem.,* **196,** 225 (1931); Wild: *Zbl. Min.,* 428, 1930; Klemm and Wild: *Zbl. Min.,* 270, 1925; Wild and Liesegang: *Zbl. Min.,* 737, 1923; Chudoba: *Der Aufschluss,* **9,** 233 (1961).

8. Holden: *Am. Min.,* **10,** 203 (1925); Simon: *Jb. Min.,* Beil. Bd. **26,** 249 (1908); Klemm and Wild (1925); Wild in Bauer (Schlossmacher, ed.); *Edelsteinkunde,* Leipzig, third ed., 1932, p. 664.

9. Simon (1908); Hoffman (1931).

10. Rose and Lietz: *Naturwiss.,* **19,** 448 (1954); Cohen (1956).

11. Berthelot: *C. R.,* **143,** 477 (1906); Simon (1908).

12. Holden (1925); Vedeneeva: *Trav. lab. crist. ac. sci. URSS,* no. 2, 107 (1940).

13. Berthelot (1906).

14. Buckley: *Crystal Growth,* Wiley, New York, 1951.

15. Lietz and Münchberg (1956); Cohen (1956); Vedeneeva: *Trav. lab. crist. ac. sc. URSS,* no. 2, 107, 1940.

16. Cohen (1956).

17. Scheuchzer: *Helveticus, sive itinera per Helvetiae Alpinus regiones,* 1708, **1,** p. 233.

18. Agricola: *De natura fossilium,* 1546 (*Geol. Soc. Amer. Spec. Paper 63,* 1955, p. 131, transl. by M. C. and J. A. Bandy).

19. von Kobell: *Geschichte der Mineralogie,* Munich, 1864, pp. 22, 427.

20. Sowerby: *British Mineralogy, or Coloured Figures Intended to Elucidate the Mineralogy of Great Britain,* London, 1817, **5,** p. 49.

21. Trommsdorf (1937); Liebisch (1916); Lietz and Münchberg (1956); Böklen (1883); Wyrouboff: *Ann. chim. phys.,* **8,** 355 (1886) and *Bull. soc. min.,* **13,** 231 (1890).

22. Dove: *Jb. Min.,* 550, 1838; Pancharatnam: *Proc. Indian Ac. Sci.,* **40A,** 196 (1954); Trommsdorf (1937).

23. Henckel: *Pyritologia,* Leipzig, 1725 (Engl. transl., London, 1757).

24. Ebelmen: *C. R.,* **25,** 854 (1847).

25. Doelter: *Kolloid Zs.,* **4,** 188 (1909).

26. Woodward: *An Attempt towards a Natural History of the Fossils of England,* London 1729, **1,** pp. 157, 161.

27. Haüy: *Traité des char. phys. des pierres precieuses,* Paris, 1817.

28. Marx: *Jb. Chem. Phys.,* **1,** 13 (1831).

29. Bruce: *Am. Min. J.,* **1,** no. 4, 242 (1813).

30. Brewster: *Trans. Roy. Soc. Edinburgh,* **20,** 55 (1853).

31. Raman and Jayaraman: *Proc. Indian Ac. Sci.,* **40A,** 189 (1954); Holden (1925).

32. Brunner, Wondratschek, and Laves: *Naturwiss.,* **24,** 664 (1959).

33. H. Rose cited by Karsten: *Min. Tabellen,* Berlin, 1800, p. 23.

34. Lacroix (1901), p. 75.

35. Airy: *Phil. Trans.,* **4,** 79, 198 (1831); Neumann: *Vorles. Theor. Optik,* Leipzig, 1885, p. 251.

36. Mackowsky: *Jb. Min.*, Ref. I, 579, 1939; Chudoba (1961).
37. Lietz and Münchberg: *Jb. Min.*, *Monatsh.*, 17, 1958; Martini: *Jb. Min.*, II, 73, 1905.
38. Rice and Cohen: *Am. Min.*, **43**, 25 (1958).
39. Tsinober and Chentsova: *Kristallografiya*, **4**, 593 (1960).

CITRINE

The name citrine, from the French *citrin*, *lemon*, is applied to the transparent varieties of quartz that range in color from yellow and yellowish brown to, less commonly, saffron, honey, or golden yellow. Citrine also may have more or less of a smoky cast of color; the smoky component may be removed by heat, leaving the material lighter and brighter in color. Citrine quartz occurs as such in nature, but the material that ordinarily passes by this name in the gem trade is made by the heat treatment of amethyst and smoky quartz. Citrine is often sold under names that confuse it with true topaz, which it closely resembles in color (Spanish topaz, Madeira topaz, Indian topaz, occidental topaz, etc.). The confusion is sometimes intensified by selling the yellow quartz under the name topaz itself, the true topaz being called Brazilian topaz. Citrine made from amethyst can be identified by the polysynthetic Brazil twinning and by the rippled fracture surfaces. Natural citrine generally lacks the reddish tints seen in much heat-treated amethyst.

The color of natural citrine[1] apparently is caused by colloidally dispersed particles of hydrous ferric oxide. The dispersed material is not visible under the microscope. The transmission spectrum of citrine is the same as that of sols of hydrous ferric oxide. Analysis of three specimens from Brazil gave 0.026 per cent Fe_2O_3 for a dark amber sample and 0.011 and 0.008 for two lighter-colored samples. Citrine made from amethyst also contains ferric iron. Citrine is faintly dichroic in yellow and pale yellow shades. On heating, citrine may become lighter in color and at high temperatures becomes milky. Irradiation with x-rays causes natural citrine to become browner in color, because of the superposition of an irradiation smokiness on the original yellow color. It does not become amethystine, as does amethyst that has been decolorized or turned citrine by heat treatment. Studies of "citrine" should clearly specify whether the material is of natural origin and untreated or has been made by heat treatment of amethyst or smoky quartz. Some of the published observations on citrine are of uncertain relevance.

Natural citrine is quite rare as compared to smoky quartz and amethyst. It occurs chiefly at localities for amethyst: Madagascar; the Isle of Arran, Scotland; Salamanca province, Spain; the states of Minas Gerais, Goiaz, and Rio Grande do Sul in Brazil; Uruguay; Dauphiné, France; the neighborhood of Mursinsk in the Ural Mountains, Russia; the United States in North Carolina and at a few other places.

References

1. Holden: *Am. Min.*, **10**, 127 (1925) and **8**, 117 (1923); Laemmlein: *Trudy Inst. Krist. Ak. Nauk SSSR*, no. 6, 269, 1951; Nabl: *Min. Mitt.*, **19**, 273 (1900).

SMOKY QUARTZ

Smoky quartz grades indefinably into colorless quartz. The color ranges from a barely perceptible smokiness through smoky brown to dark smoky brown and almost black. Deeply colored crystals are translucent to nearly opaque in large masses (*morion*) and are perceptibly dichroic.[16] The dichroic colors are ϵ yellowish to reddish brown and ω pure brown, with absorption $\epsilon > \omega$; in thick sections of deep-colored crystals ϵ is nearly black and ω dark brown. The color in general is not uniformly distributed throughout the crystal, but tends to be concentrated in thin bands or growth zones parallel to the external faces. Very thin bands are best observed in sections cut perpendicular to the crystal faces, especially to a terminal rhombohedron. The smoky color also may be localized in face-loci, especially beneath the terminal *r* faces, then giving a three-fold segmental appearance in basal sections; growth banding also is present in the smoky parts. Thick zones of smoky quartz, themselves banded on a fine scale, may alternate with colorless or milky zones, and smoky quartz also occurs interbanded with amethystine quartz in single-crystals (see *Amethyst*). In some instances a smoky or yellowish color is apparent in alternate face-loci of amethyst crystals, as in specimens from Delaware County, Pennsylvania. The color of smoky quartz may tend to yellowish or yellowish brown to reddish brown shades and then seems to grade into or become admixed with a citrine color.

Smoky quartz can be decolorized by heat.[1] The bleaching is a time-temperature reaction. The color is stable below about 225°. At slightly higher temperatures bleaching requires a heating period of many days; however, the rate of bleaching increases rapidly with increasing temperature, and over about 450° small pieces are bleached in a matter of minutes. Rapid decolorization is accompanied by a bluish white thermoluminescence. At relatively low temperatures some material becomes clove brown to golden brown in color before decolorization is complete. The effect is more marked in smoky quartz that has a yellowish or brownish tone. Part of the gem material sold as citrine is made in this way. The smoky quartz from Hinojosa, Cordoba, Spain, turns to a fine yellow on heating. Many quartz crystals classed as colorless on ordinary examination actually are slightly smoky, as can be demonstrated by heating a sawn piece of the material to a high temperature and then making a direct comparison with the original against a white background. Much of the seemingly colorless quartz from Brazil and Switzerland has a faint smoky cast of color that can be recognized by this test or by comparison to the Carrara, Marmaros, Dauphiné, or Herkimer material, which generally is colorless.

The natural smoky color of quartz has a higher temperature stability than that of irradiated quartz[2] (see *Irradiation Effects*). Decolorization is accompanied by an extremely small decrease in the unit cell dimensions[3] and by a small increase in the indices of refraction. There also is a change in the elastic constants of the material.[17]

Composition and Properties. The chemical composition of smoky quartz appears to be within the range of variation of colorless quartz. The material, so far as known, is not characterized by the presence of relatively large amounts of another element, such as the Fe content of amethyst and citrine or the Ti content of blue quartz and rose quartz. Smoky quartz does contain Al, which apparently is a prerequisite for the color, but comparable or larger amounts of Al are present in both colorless quartz[19] and in other colored varieties of quartz. Spectrographic[4] and other analyses of natural smoky quartz (see *Chemical Composition*) show Li, Al, Ca, Mg, etc., and some analyses show little or no Fe, Mn, or Na. Smoky quartz generally contains fluid inclusions, sometimes very abundantly (see *Inclusions in Quartz*). Smoky quartz has been said to contain a hydrocarbon, but this has been disproved,[15] other than that it may be present in fluid inclusions. Chemical tests have been interpreted[5] as indicating that smoky quartz contains free silicon, decreasing in amount on heating. The amounts of Si reported in dark smoky material are of the order of 0.01 per cent. The smoky color has been attributed[5] to free Si formed by natural irradiation (see beyond).

Precise measurements indicate that the indices of refraction of smoky quartz are slightly less than those of colorless quartz, decreasing with increasing depth of color and varying also within a single crystal. The difference, however, is very small, in the fifth significant figure or beyond, and is comparable with the variation found in different crystals of colorless quartz, or within single-crystals, as a function of variation in chemical composition. The validity of the observations is supported by the fact that the indices of refraction of smoky quartz increase when the smoky color is removed by heat treatment:

Color	ϵ	ω	Ref.[6]
Light-colored zone	1.55299	1.54403	7
Dark-colored zone	1.55289	1.54387	
Same, bleached	1.55344	1.54436	
Very dark color	1.553369	1.544214	8
Same, bleached	1.553413	1.544284	

It has not been shown experimentally that there is a difference in the densities of smoky and colorless quartz that is associated with the color. The same difficulty associated with the problem of the indices of refraction of smoky quartz exists here, that the variation in the density of colorless quartz, by mechanisms whose relation to the color is not known, is as large or larger than the effect sought in smoky quartz. The variation in density of colorless quartz is in the range 0.0001–0.0005. Precision measurements[8] by the same method of dark smoky and colorless quartz gave values for the density of 2.64849 and 2.64847 at 25°, identical within the errors of measurement. The values 2.65027 and 2.65022 (both ±0.00009 at 0° referred to H_2O at 4°) for natural smoky quartz before and after bleaching also have been reported.[13]

The origin of the color of smoky quartz has been the subject of much investigation.[9] The color has been attributed to pigmentation by hydrocarbons or by various inorganic impurities, but it appears to be caused by a disturbance of the internal electron configuration accompanying compositional variation. Smoky quartz has a characteristic absorption[10] at 460 mμ that is caused by the interaction of an unpaired electron on an oxygen ion or in an anion vacancy with an Al ion in substitution for Si. Other absorption maxima also have been reported. The presence of Al in tetrahedral coordination appears to be a necessary condition for the smoky color, but not all quartz containing Al is smoky. The difference is in the way that the Al is housed, whether in substitution for Al or in an interstitial position, or in the way in which valence compensation is effected (see *Irradiation Effects*). Although the absorption spectrums of natural and of irradiated smoky quartz apparently are identical, the temperature stabilities of the two colors definitely are different. Natural smoky quartz scatters light[11] more strongly than colorless quartz or amethyst; the scattering is diminished when the color is bleached by heat.

It has been suggested[5] that the natural smoky color is secondary, produced by the natural irradiation of an initially colorless crystal. It is well known that a related if not identical color can be produced by irradiation of colorless quartz by x-rays. In any case, the disposing factor, known to be Al, was acquired by the crystal during its initial growth. It is this that produces the distribution of the color in growth bands and face-loci. The lack of linear absorption phenomena in very large crystals, and the fact that the colorless parts of zoned crystals are more strongly pigmented than the natural smoky parts by x-rays and particle radiation, suggest that the crystals were pigmented from the beginning by a mechanism not involving irradiation. Linear absorption effects, however, would not be present if the crystals grew from a radioactive solution, since the growth surface would be continuously irradiated. Inclusions of radioactive minerals are surrounded by diffuse smoky haloes a millimeter or so thick. Dispersed microscopic inclusions of carbon, hydrocarbons, or sulfides may produce a dark color resembling that of true smoky quartz. The quartz crystals from Herkimer County, New York, often contain black specks of an anthracite hydrocarbon either floating in cavities containing liquid or embedded in the quartz.

Occurrence. Smoky quartz is very common. It typically occurs in deposits formed at relatively high temperatures, including pegmatites, where it

is found chiefly in cavities formed during the hydrothermal stage of mineralization, and in miarolytic cavities and drusy pegmatitic segregations in granitic
igneous rocks. It is found also in veins of the Alpine type. The crystals are of
the ordinary prismatic habit, and apparently tend to be more highly modified
by general forms than does colorless low-temperature quartz. Vicinal
development is common. The crystals often are relatively imperfect because
of the development of lineage and composite structure. Twisted crystals are
more common in smoky than in colorless quartz. Brazil twinning is less
common in smoky quartz, at least from some localities, but Dauphiné twinning is usually present.

Probably the best-known occurrences are in the Alpine veins of Switzerland.[12] Fine specimens are abundant, and very large crystals have been
found. A vein cavity discovered about 1719 on the Zinkenstock near
Grimsel afforded over 50 tons of crystals, one weighing 800 lb., and a cavity
in the Vieschthal in Upper Valais yielded over 1700 crystals ranging in weight
from 50 to 1400 lb. Another famous find,[13] in 1868, near the Tiefen glacier in
Uri gave over 10 tons of dark smoky, transparent quartz suitable for ornamental working, together with another 5 tons of museum material, including
a perfectly developed doubly terminated crystal weighing 148 lb. This
crystal and other fine specimens from the locality are preserved in Berne.
A magnificent group of smoky crystals is exhibited in the Kunstgewerbe
Museum in Zurich, and outstanding single crystals are to be seen in many of
the great mineral collections of the world. In the central Alps the intensity of
the smoky color is said to increase with increasing altitude of the occurrence.[14]
Another well-known locality is on Cairngorm, a mountain on the border of
Inverness-shire and Banffshire in Scotland, where smoky quartz occurs in
pegmatites and in weathered surface debris overlying granite. It is especially
valued as an ornamental stone in Scotland, under the name cairngorm or
Scottish topaz. Fine specimens also are found in Russia, at numerous places
in Brazil, and in Madagascar, as well as in veins in granite at San Piero and
nearby places on the island of Elba, and in pegmatite in the Alto Lighona
district, Mozambique.

In the United States, smoky quartz crystals up to several feet in length have
been found in pegmatites in New England and in various other places.
Numerous pale smoky and milky crystals up to 6 ft. in length were found in
cavities in pegmatite at Buckfield, Maine. Lustrous black crystals, in part as
overgrowths on large milky quartz crystals, were found at Mt. Apatite near
Auburn, Maine, and several good localities are known in Coos County, New
Hampshire. Pale smoky quartz occurs in some of the vein-quartz localities in
Arkansas. Handsome specimens of dark smoky quartz associated with green
microcline are obtained from pegmatitic pockets in the granite of the Pikes
Peak region, and on Mt. Antero, Colorado. Masses of dark smoky quartz
and euhedral crystals reaching 1000 lb. or more were found in the rare-earth

pegmatite at Baringer Hill, Llano County, Texas. Smoky quartz having a marked zonal structure and containing enclosures of tourmaline occurs in Jefferson County, Montana, and also in pegmatites near Milford, Beaver County, Utah, and in San Diego County, California. Complexly modified pale smoky crystals of small size have been found at numerous localities in Burke and Alexander counties, North Carolina,[18] and elsewhere in pockets in pegmatites or quartz veins in the metamorphic rocks of the Piedmont region. The anhedral quartz grains of some igneous and metamorphic rocks have a slightly smoky appearance, often difficult to judge because of their small size and the presence of internal flaws. The quartz of graphic granite also usually is more or less smoky.

References

1. Holden: *Am. Min.*, **10**, 203 (1925); Simon: *Jb. Min.*, Beil. Bd. **26**, 249 (1908); Koenigsberger: *Abh. bayer. Ak. Wiss.*, **28**, 31, 117 (1918–19).
2. Frondel: *Phys. Rev.*, **69**, 543 (1956); Chentsova and Vedeneeva: *Dokl. Ak. Nauk SSSR*, **60**, 649 (1948); Vedeneeva and Chentsova: *C. R. ac. sci. URSS*, **55**, 437 (1947) and *Trudy Inst. Krist. Ak. Nauk SSSR*, **7**, 159 (1952).
3. Hammond, Chi, and Stanley: U.S. Signal Corps Eng. Labs., Fort Monmouth, N.J., *Eng. Rpt. E1162*, 1955.
4. Kennard: *Am. Min.*, **20**, 392 (1935), with literature; Bambauer: *Schweiz. min. pet. Mitt.*, **41**, 335 (1961).
5. Holden (1925).
6. See also Dufet: *Bull. soc. min.*, **13**, 271 (1890), and Forster: *Ann. Phys.*, **143**, 408 (1871).
7. Hlawatsch: *Zs. Kr.*, **27**, 605 (1897).
8. Frondel and Hurlbut: *J. Chem. Phys.*, **23**, 1215 (1955).
9. Holden (1925) with older literature.
10. Cohen: *J. Chem. Phys.*, **25**, 908 (1956); O'Brien; *Proc. Roy. Soc. London*, **231A**, 404 (1955); Marshall: *Am. Min.*, **40**, 535 (1955); Griffiths, Owen, and Ward: *Nature*, **173**, 439 (1954); Holden (1925); Mohler: *Am. Min.*, **21**, 258 (1936). See also ref. 16.
11. Sur: *Proc. Indian Assoc. Cultiv. Sci.*, **8**, 271 (1923); Holden (1925); Raman and Jayaraman: *Proc. Indian Ac. Sci.*, **40A**, 189 (1954).
12. Niggli, Koenigsberger, and Parker: *Die Mineralien der Schweizeralpen*, Basel, 1940; Friedlaender: *Beitr. Geol. Schweiz, Geotech. Ser.*, no. 29, 1951.
13. Forster: *Ann. Phys.*, **143**, 173 (1871).
14. Koenigsberger: *Jb. Min.*, Beil. Bd. **14**, 43 (1901).
15. Weinschenk: *Min. Mitt.*, **19**, 144 (1900); Koenigsberger: *Min. Mitt.*, **19**, 148 (1900); Schneider: *Ann. Phys.*, **117**, 653 (1862).
16. Vedeneeva: *Trav. lab. crist. ac. sc. URSS*, no. 2, 87, 1940; Vedeneeva and Rudnitskaya: *C. R. ac. sci. URSS*, **87**, 361 (1952).
17. Frondel (1956).
18. See Kunz: *North Carolina Geol. Surv. Bull.*, **12**, 30 (1907).
19. Bambauer (1961).

ROSE QUARTZ

Rose quartz ranges in color from very pale shades of pink, rose-pink, and rose to deep rose-red. Some rose quartz has a faint purplish or lavender cast of color. It always is more or less turbid and much cracked. Unflawed pieces larger than an inch or two in size are unusual. Rose quartz occurs in pegmatites, generally in the central quartz core, and may form very large masses. These masses have a very coarse granular structure, as may be seen by etching sawn sections, or by the development of rhombohedral cleavage surfaces, and crystal faces are not developed. Growth zones may be seen by transmitted light in cut sections, or are developed by etching, and the lack of euhedral crystals is due to the complete in-filling of the available space. Distinct euhedral crystals of rose quartz also occur as a great rarity. The best-known locality is in granite pegmatite at Newry, Oxford County, Maine, where ill-formed crystals up to about 1 cm. in size are found as drusy crusts or small groups. Some specimens show overgrowths of rose quartz crystals in parallel position upon earlier crystals of slightly smoky quartz. The color sometimes is bright pink or rose-pink, and the crystals may be transparent, but they generally are turbid to milky and of poor color. Specimens are highly prized by collectors. At Newry and at the few other known occurrences in pegmatite in New England the rose quartz crystals are found in small cavities associated with fairfieldite and other late-stage phosphates and silicates as a hydrothermal product. Rose quartz is found similarly in pegmatite at Pala and Mesa Grande in southern California; fine specimens come from the Arassuahy-Jequitinhonha district, Minas Gerais, Brazil. Some so-called rose quartz, especially that from non-pegmatitic occurrences, is ordinary quartz pigmented by inclusions of iron oxide or pinkish clayey or other foreign material. Such quartz generally is opaque or very turbid, and the inclusions can be seen in thin section under the microscope. Quartz has been observed with a rose or rose-pink color caused by microscopic inclusions of manganese silicates (Langban, Sweden) or of dumortierite.

Rose quartz typically contains needle-like oriented inclusions of rutile, which have formed by exsolution. The rutile needles occur in a number of positions of orientation[1] (see Table 13, under *Oriented Growths*). Spheres cut from such material may show marked asterism,[2] with light rays extending along various directions in the quartz, together with bright spots or nodes, conforming in their distribution to the symmetry of quartz. The effect is best seen in sunlight or some other point source of light. The relative intensity and number of the light rays and nodes vary in material from different localities and even within material from the same locality. Usually only one symmetrical set of light rays is developed or is greatly predominant; this is six-rayed in the direction $[1\bar{2}10]$ and equivalents, but a weaker six-rayed set on $[1\bar{1}00]$ and equivalents is often observed, and a number of other positions have been observed (Table 13).[3] The rutile inclusions in blue quartz have identical orientations. The light nodes have been observed on $\{0001\}$, $\{10\bar{1}0\}$, $\{10\bar{1}1\}$, $\{10\bar{1}2\}$, $\{11\bar{2}1\}$, and $\{11\bar{2}0\}$, including both positive and negative and right and left equivalents. Cabochon stones usually are cut perpendicular to the c-axis to display the six-rayed star from $[1\bar{2}10]$ or the twelve-rayed star from

[1$\bar{1}$00] and [1$\bar{2}$10] together. The asterism is caused by the scattering of light from the oriented inclusions, each light ray being perpendicular to the direction of elongation of the rutile needles. The intensity of the asterism depends on the number of inclusions per unit volume in each position of orientation and on the thickness of the inclusions. The effect in rose quartz never is as sharp or intense as in a superior natural or synthetic star sapphire or ruby, and many specimens of rose quartz do not show distinct asterism. Plates of asteriated rose quartz have been employed in inexpensive artificial doublets intended to simulate star sapphires and rubies. Good examples of asteriated rose quartz have been found in the United States in pegmatites at the Kinkel quarry, Bedford, Westchester County, New York; in Maine at localities in Oxford County; in Georgia near La Grange in Troup County.

The rutile needles usually are 0.01–0.2 μ in thickness and may range from a fraction of a millimeter up to a centimeter or more in length. Sometimes the needles are so thick and long as to be visible to the unaided eye; in such instances asterism is weak or lacking, and the quartz is relatively turbid with a milky appearance. Some of the rutile needles observed in rose quartz, such as long curved crystals and divergent aggregates, appear to be mechanical enclosures. Bulk analyses of rose quartz are relatively high in TiO_2, usually in the range 0.00X per cent TiO_2 (see *Chemical Composition*). An analysis[3] made on a 241.9 gram sample of Brazilian rose quartz gave: SiO_2 99.79, TiO_2 0.048, Al_2O_3 0.042, MgO 0.008, CaO 0.010, MnO 0.009, Fe_2O 0.007, PbO 0.006, total 99.92, remainder chiefly SO_4, Cl, and alkalies. The content of TiO_2 and Al_2O_3 in this analysis is unusually large. Lithium and Fe are often present in relatively large amounts (see Tables 25 and 26 under *Chemical Composition*).

Rose quartz is very much more responsive to x-ray irradiation than colorless quartz, or than any other colored variety, and usually becomes quite black and virtually opaque. There is a correspondingly large change in the elastic constants and in the rate of solution, and on heating after irradiation there is a brilliant thermoluminescence. Rose quartz also is relatively resistant to secondary Dauphiné twinning by thermal strain.

The color of rose quartz, like that of the other colored varieties of quartz, has been the subject of much investigation. The color has been attributed to Mn in solid solution,[4] and there is an apparent correlation between the depth of color and the Mn content (Table 26 under *Chemical Composition*). Comparable amounts of Mn, however, are found in smoky quartz and amethyst, and some rose quartz is reported to be very low or lacking in Mn. The absorption spectrum[5] of rose quartz shows a maximum in ϵ at about 500 mμ, ω varying uniformly; both Ti^4 and Mn^3 show an absorption at about this wavelength. Blue quartz contains somewhat larger amounts of Ti than rose quartz, but analyses for Mn are lacking. The color of blue quartz is caused by Tyndall scattering from the (relatively large) rutile needles. The lavender and purplish shades of some rose quartz suggest a transition to blue quartz, with a blue color caused by scattering superposed on a rose color originating in absorption by an ion, perhaps Ti^4, still in solid solution. Rose quartz also may tend toward a white cloudy appearance or contain white streaks due to the presence of large numbers of liquid inclusions.

Rose quartz bleaches when heated in air at about 575°. There are conflicting statements about the stability of the color of rose quartz in sunlight, and different specimens may vary in this regard. The change in color, if any, is small. In pegmatites, a freshly quarried surface may whiten slightly on exposure because of loss of moisture from minute cracks and flaws.

The unit cell dimensions of rose quartz are significantly larger than those of colorless or smoky quartz, by as much as 0.01 Å in some instances (see Tables 6, 19, and 20). The observed values presumably are less than those originally obtaining, since the Ti, which substitutes for Si and enlarges the cell volume by virtue of its larger size, has in part exsolved as TiO_2. The increase in the cell volume of rose quartz also reflects the content of Al, Fe, Li, etc., which is large in some samples. The indices of refraction and the density of rose quartz apparently are not markedly different from those of colorless quartz. Rose quartz is faintly dichroic, as seen in sections about 1 cm. thick, with ω pink and ϵ nearly colorless.

Rose quartz occurs in pegmatites, sometimes in large amounts, and has been mined for use as an ornamental or carving material. Localities are very numerous, but material that is of deep color or is relatively translucent and free from cracks is not abundant. Rose quartz occurs in many pegmatites in New England, as in Maine at the Bumpus mine near Stoneham and at Albany and Paris. It is obtained in large amounts in pegmatite in the Black Hills, South Dakota, as at the Scott quarry[6] 6 miles southeast of Custer, Custer County, and from California and Colorado. Rose quartz also occurs at Rabenstein, Bavaria; Rossing, South West Africa; at numerous places in Brazil; etc.

References

1. von Vultée: *Zs. Kr.*, **107**, 1 (1956) and *Jb. Min.*, Abh. **87**, 389 (1955).
2. Goldschmidt and Brauns: *Jb. Min.*, Beil. Bd. **31**, 220 (1911); von Vultée (1955); Kadokura: *Beitr. Min. Japan*, no. 5, 269, 1915.
3. von Vultée (1955).
4. Holden: *Am. Min.*, **9**, 75, 101 (1924).
5. von Vultée and Lietz: *Jb. Min.*, *Monatsh.*, **3**, 49 (1956).
6. Scott: *Rocks and Min.*, **16**, 360 (1941).

BLUE QUARTZ

The so-called blue quartz found as grains in some metamorphic and igneous rocks is turbid and has a rather greasy luster. The color is pale and ranges from a faint, soft blue or milky blue to smoky blue, plum-blue, and lavender-blue. The color resembles that of some chalcedony. Most blue quartz is crowded with needle-like inclusions of rutile about 0.02–1 μ thick and 1–500 μ in length.[1] The average thickness and length vary in material from different localities. Some specimens have been shown to contain from 0.4 to 2 million inclusions per cubic centimeter. The amount of rutile per unit volume that is visible under the microscope, as determined by measurement of the number and size of the inclusions, has been found in some instances to be much less than that calculated from the TiO_2 found on chemical analysis, and it is believed that colloidally dispersed scattering material is present below the visible range of size. The depth of color is not always related to the amount of visible rutile. The inclusions usually are rather uniformly distributed, but instances have been observed in which they are more concentrated in particular growth zones with accompanying deepening of the color.

The blue color in reflected light is owing to the selective scattering of the blue wavelengths of ordinary light by the inclusions (Tyndall effect).[2] In transmitted light the color is complementary, reddish to yellowish red and yellowish brown; the yellowish tones presumably are contributed by absorption in the inclusions themselves. The rutile needles are oriented to the quartz and have formed by precipitation from solid solution. The inclusions mostly fall into a few positions of orientation with the length (c-axis) of the rutile needles parallel to [0001], [12̄10], or [11̄00] of the quartz, but a total of eleven different orientations has been recognized[3] (see Table 13 under *Oriented Growths*), and additional positions probably would be found on statistical study. Blue quartz commonly is chatoyant and shows asterism. The asterism is difficult to observe because of the small size, turbidity, and flawed nature of the quartz grains. Minute dust-like particles of unidentified substances sometimes are present in addition to rutile, and a blue color may be produced by the scattering of light from such inclusions in the absence of rutile.[4] The scattering is primarily a function of the dimensions of the included material and of the difference in the mean index of refraction of the inclusions and of the quartz.

Chemical analyses of blue quartz show TiO_2 in the range $0.0X$ (Table 27), much over the usual range in rose quartz. Determinations of Al, Mn, and alkalies are lacking. The Fe_2O_3 reported in analyses[4] is said to be largely removed by washing with acid; it originates in impurities on the surface of the grains or in cracks. The depth of the color is proportional to the amount of TiO_2 found on analysis.

On heating,[4] some specimens of blue quartz, such as those from Nelson County, Virginia, maintain their color to temperature as high as 1000° and 1100°. Other material, such as that from the charnockites of India, loses its color completely at 300–400°, and some specimens are said to bleach on gentle heating or on exposure to sunlight. In another study,[1] some specimens heated at 400° for 24 hours altered in color toward violet, while others became colorless or remained unchanged. The color of rose quartz is bleached at about 575°. Both blue quartz and rose quartz differ from other varieties of quartz in containing a relatively large amount of TiO_2, but so far as known do not differ chemically between themselves except in the amount of Ti present. A relatively large amount of Ti apparently remains in solid solution in rose quartz. The difference in color of the two varieties is due to the mechanism by which the color is produced, by scattering from rutile needles in one case and by absorption by the Ti^4 ion in the other. Precision measurements of the unit cell dimensions and of the indices of refraction of blue quartz (both before and after heating) are lacking.

Blue quartz is widespread in igneous and metamorphic rocks, and only a few localities can be mentioned. It is very common in the quartz-containing rocks of the charnockite series in India[4] and is found in quartz porphyry and in the Champion gneiss of Mysore. Blue quartz phenocrysts occur in dikes of a reddish porphyry containing microcline with some albite and biotite near Llano, Llano County, Texas.[5] Deep blue quartz, the color apparent in thin sections, occurs abundantly in the titanium-bearing rocks of Amherst and Nelson counties, Virginia.[6] It also occurs in Stafford County, Virginia,[7] in a quartz diorite and with tourmaline, hornblende, and andesine in a pegmatitic facies of this rock. Blue quartz is typically present in the Milford granite[8] of Rhode Island and Massachusetts and occurs in other igneous and metamorphic rocks in New England, as well as in the Baltimore gneiss in Pennsylvania. It has been described from granites in Sweden, in rapakivi from Finland,[9] and in albitite in Australia.[11]

References

1. von Vultée: *Jb. Min.*, Abh. **87**, 389 (1955).
2. Summary in von Vultée and Lietz: *Jb. Min.*, *Monatsh.*, **3**, 49 (1956); see also Postelmann: *Jb. Min.*, Beil. Bd. **72A**, 401 (1937); Jayaraman: *Proc. Indian Ac. Sci.*, Sect. A, **9**, 265 (1939) and *J. Indian Inst. Sci.*, **23A**, 28 (1940); Kalkowsky: *Zs. Kr.*, **55**, 23 (1915–20).
3. von Vultée: *Zs. Kr.*, **107**, 1 (1956) and ref. 1.
4. Jayaraman (1939).
5. Iddings: *J. Geol.*, **12**, 227 (1904).
6. Watson and Taber: *Virginia Geol. Surv. Bull. III-A*, 1913, p. 214.
7. Lonsdale: *Am. J. Sci.*, **11**, 505 (1926).
8. Emerson and Perry: *U.S. Geol. Surv. Bull.* **311**, 1907, p. 46.
9. Robertson cited in Watson and Beard: *Proc. U.S. Nat. Mus.*, **53**, 553 (1917).
10. Postelmann (1937).
11. Tilley: *Trans. Roy. Soc. South Australia*, **43**, 325 (1919).

OTHER COLORED TYPES

Milky and white quartz owes its color to the scattering of light by great numbers of minute cavities or flaws. The massive granular milky quartz of .

pegmatites and hydrothermal veins is of this nature. The color generally is unevenly distributed, such as along irregular bands or zones which may transect adjacent grains in granular material. The white color of granular massive quartz tends to become more uniform as the grain size decreases, and in pure white material there is a tendency toward a fatty appearance. The material may become translucent to nearly opaque in thin splinters. Euhedral crystals may have a uniform milky or white color, but more often the color is localized toward the attached end of the crystal or is distributed along irregular or wavy zones or healed cracks. Fine groups of uniformly white crystals have been obtained at Ouray, Colorado. A white to gray color also may be imparted by included, finely divided clayey material. The quartz may then be turbid to virtually opaque with a pearly to metalloidal luster. Only a thin film may be present on the crystal faces, giving a glazed or porcelanous appearance.

As the foreign pigmenting material varies in kind, or in degree of admixture with other foreign substances, the color may vary from white or gray to pink, brown, or other shades. Ferruginous quartz crystals are red- or yellow-brown from the presence of dispersed hematite or goethite. The crystals usually are small, up to a centimeter in size, and are coarsely developed and opaque, sometimes uniformly tinted and translucent. Well-known localities for ferruginous quartz crystals are at Sundwig, Westphalia; in Czechoslovakia;[1] and in Cumberland and Lancashire, England. Transparent crystals faintly pigmented by hematite may resemble rose quartz. Growth zones rich in finely divided particles of iron oxide or containing annular rings of iron oxide a fraction of a millimeter in diameter are sometimes observed in amethyst, such as that from Thunder Bay, Ontario; similar annular rings are formed by the evaporation of drops of water containing suspended particles of iron oxide or other matter. Dark gray to almost black quartz pigmented by finely divided sulfides has been observed in hydrothermal veins; also black because of finely divided carbon.[2]

References

1. Tuček: *Věstnik Stát. geol. ústavu Českoslov. rep.*, **21**, 335 (1946); Slavíková and Slavík: *Časopis Mus. Českeho, Prague*, **93**, 105 (1919).
2. Boyle: *Am. Min.*, **38**, 528 (1953).

ROCK-CRYSTAL

Although not properly a variety, much less a colored variety, colorless and transparent crystals of quartz are regarded with a special interest, both aesthetic and practical. In Grecian and Roman times and extending up through the Middle Ages large quartz crystals were employed to carve vases, bowls, and other types of vessels, frequently elaborately engraved and mounted with silver or gold ornamentation. This art reached a peak during the Italian Renaissance, centering in Milan but with other lapidary centers in Turin, Prague, and Vienna. A remarkable collection of these vessels is preserved in the Kunsthistorisches Museum[1] in Vienna. The quartz used was obtained chiefly from Madagascar and Switzerland. Similar objects were carved in China, especially during the Ch'ien Lung period (1736–1796), from quartz probably obtained in upper Burma. This art, especially the polishing of spheres, also was practiced in Japan. Rock-crystal formerly was carved at Delhi, India. Another center for the carving of rock-crystal and especially of the colored fine-grained varieties of quartz has been Sverdlovsk (Ekaterinburg), Russia, where lapidary works were established by Catherine the Great in 1721.

A flawless polished sphere of rock-crystal 12⅞ in. in diameter and weighing 107 lb. is exhibited in the United States National Museum in Washington. This unique sphere was cut in China from rock-crystal believed to have been obtained at Sakangyi in the Katha district of Burma. At this locality a large, deeply weathered pegmatite has afforded euhedral crystals of transparent quartz weighing over 1500 lb., together with fine crystals of topaz. Most of this quartz of carving grade was exported to China. Flawless spheres over 6 or 7 in. in diameter are rare and costly.

Several fine collections of carved quartz objects of Oriental or European origin are preserved in the United States, as in the Philadelphia Museum of Fine Arts[2] and the Metropolitan Museum in New York City. The carving of rock-crystal and its use in small utilitarian objects such as paper weights died out in Europe during the eighteenth century with the development of the manufacture of perfectly clear and colorless glass. Fine glassware, however, still carries the name crystal, which itself was first applied to quartz.

Rock-crystal, especially in the past, has been used as a gem stone but is now almost wholly displaced by glass and by plastics. It usually is cut as brilliants. Rhinestone ornaments gained their name from quartz pebbles, used for the purpose, found in the Rhine (*Germ., Rhinekiesel*) and carried from the Alps through the Aar tributary. The formerly wide use of colorless quartz in jewelry and the rather distant resemblance to diamond are indicated by the application of the word diamond as an affix to the locality at which the quartz was found. Thus the limpid crystals from Herkimer County and Lake George in New York, are known as "Herkimer diamonds" and "Lake George diamonds," and the fine clear crystals from sandstone and slate in Marmaros Comitat, Hungary, are called "Marmaros diamonds." Other names of this nature are the "Cape May diamonds" and "Arkansas diamonds" of the United States; the "Bristol diamonds" and "Cornish diamonds" of England; the "Briançon diamonds" and "Alençon diamonds" of France; the "Schaumberg diamonds" of Germany; etc.

The chemical and physical stability of quartz early led to the use of large quartz weights cut from rock-crystal for national standards of mass. Measurements of the volume of these objects have afforded extremely precise, varying values of the density of quartz.[3] The practical uses of rock-crystal have been numerous. A lens of rock-crystal found[4] at Nimrud has been dated at 3800 B.C. Brazilian quartz was extensively used in Europe in the nineteenth century for lenses of eyeglasses and for hand mirrors. The first extensive scientific use for rock-crystal was as prisms in optical spectrographs. In recent years very large amounts have been employed for piezoelectric applications.

Among the best-known localities for large, colorless, and flawless crystals or masses of quartz may be mentioned Madagascar, which supplied material to Europe for centuries; Goiaz and other states in Brazil, which today is the principal source of supply; Switzerland, most of the large crystals being more or less smoky; and Burma. In 1938, at Itamarandiba in the Diamantina district, Minas Gerais, Brazil, an almost completely doubly terminated quartz crystal was found that weighed $5\frac{1}{2}$ tons. It was broken with sledgehammers and a flawless piece weighing 992 lb. was obtained. It was sold in England in 1951 to a manufacturer of quartz prisms for over $60,000. Only a few localities for very large transparent crystals of quartz are known in the United States. An unusual occurrence was found in the Green Mountain and other mines near Mokelumne Hill, Calaveras County, California. Here more or less abraded crystals and boulders of transparent quartz were found in Tertiary stream sands and gravels, buried under lava flows, that were mined for gold by underground methods. The crystals ranged in weight up to 2000 lb., and most of them weighed over 50 lb. About 12 tons were sold in the 1890's for carving purposes, and the commercial production up to 1942 was over 30 tons. Large, fine crystals weighing up to 300 lb. were found at a minor locality in Chestnut Hill township, Ashe County, North Carolina;[5] also occasional large crystals from the vein deposits in Arkansas (see *Occurrence*). At the present time, transparent quartz crystals free from twinning, inclusions, and other imperfections are in demand chiefly for optical and piezoelectric applications (see *Piezoelectricity*). Very large imperfect crystals of quartz are commonplace (see *Occurrence*, and *Smoky Quartz* under *Varieties of Quartz*).

The origin of the traditional term rock-crystal is indicated in the section on *History of the Names Quartz and Crystal*.

References

1. Strohmer: *Prunkgefässe aus Bergkristall*, Wulfrum, Vienna, 1947.
2. Crozier Collection, Pt. 1, Rock Crystals, *Philadelphia Mus. Bull.* **40**, no. 203, 1944.
3. See summary in Sosman: *Properties of Silica*, New York, 1927.
4. Layard: *Discoveries in the Ruins of Ninevah and Babylon*, New York, 1853, p. 197.
5. See Kunz: *North Carolina Geol. Surv. Bull.* **12**, 29, 1907.

FINE-GRAINED VARIETIES

CHALCEDONY

CHALCEDONY. Iaspis pt. *Theophrastus; Pliny* (*Nat. Hist.*, **37**, A.D. 77). Chalcedon, Achates vix pellucida, nebulosa, colore griseo mixto *Wallerius* (*Mineralogia*, 83, 1747). Chalcedon *Germ.* Calcédoine *Fr.*
Beekite, Beckite *Dufrénoy* (*Traité de Min.*, **3**, 750, 1847; Hughes: *Min. Mag.*, **8**, 265, 1889). Chalcedonite *Becker* (*U.S. Geol. Surv. Mon.* **13**, 390, 1888). Quartzin *Michel-Lévy and Munier-Chalmas* (*Bull. soc. min.*, **15**, 161, 1892; *C. R.*, **110**, 649, 1890). Lutécite, Lutécin *Michel-Lévy and Munier-Chalmas* (*op. cit.*). Calcédonite *Lacroix* (*Min. de France*, **3**, 121, 1901). Pseudocalcédonite *Lacroix* (*C. R.*, **130**, 430, 1900). Pseudoquartzin *Braitsch* (*Heidelberg. Beitr. Min. Pet.*, **5**, 331, 1957).

Structure. Chalcedony occurs as crusts with an irregularly rounded, botryoidal, or warty surface (Fig. 91); as pendant masses in openings, stalactitic, and grading into grape-like forms or irregular sheets; as isolated reniform, irregular, or rounded isolated masses, either concretionary in origin or yielded by the weathering-out of cavity fillings; sinter-like; as veinlets and as an interstitial cement. Chalcedony typically occurs as geodes and amygdule-fillings in altered basic igneous rocks; these are often hollow, and may contain entrapped liquid (*enhydros*). Chalcedony often occurs as incrustation pseudomorphs after other minerals, especially calcite and fluorite, and as a replacement of these and other minerals and of fossil shells. The term chalcedonic silica is of broader meaning than chalcedony, which chiefly includes the crustiform types and cavity fillings, and is applied to any fine-grained type of quartz with a fibrous microstructure. It extends to massive and nodular material of sedimentary or diagenetic origin, including chert and flint, and to replacements of limestone and other rocks.

Chalcedony shows a more or less distinct banding parallel to the free surface or parallel to the walls of the cavity. The banding is not always apparent to the unaided eye, but can be recognized under the microscope in sections cut perpendicular to the free surface or can be brought out by staining or etching techniques.

Optical Properties. Under the microscope, chalcedony and its subvarieties show a fibrous structure[1] with the fiber direction perpendicular to the layering and to the free surface. The individual fibers are not physically separable, and interstitial material is not visible in section at high magnification or on fracture surfaces examined under the electron microscope. The fibers are of variable thickness and length, usually only a few microns in diameter and up to several hundred microns in length, but range up to sizes visible under a hand lens or to the unaided eye. The fibers occur as parallel or subparallel aggregates or bundles, sometimes sheaf-like or divergent, and also as laths

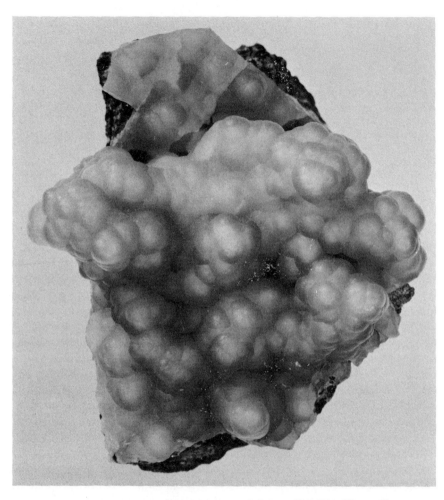

FIG. 91. Chalcedony, botryoidal crust. Arizona. U.S. Nat. Mus. coll.

or leafy aggregates, or spherulitic; as fibers carrying smaller fibers attached at acute angles which in turn splinter and diverge. Certain bands in chalcedony may not show fibers under the highest magnification, but such material exhibits optical extinction parallel and perpendicular to the banding. Light transmitted through sections of chalcedony cut parallel to the fibrosity is plane polarized.[2]

The fibers usually are elongated in a direction at right angles to the c-axis, with the fiber axis [11$\bar{2}$0] or less commonly [10$\bar{1}$0] as shown by x-ray fiber diagrams;[3] these fibers show negative optical elongation. The fiber axis sometimes is [0001], with positive optical elongation. (See further under *Lussatite*.)

Inclined extinction also may be observed, usually at small angles; also a type[4] with an extinction angle of about 30°, perhaps representing elongation along a rhombohedral edge. Several types of elongation may occur in the same specimen. The aggregates often show a biaxial character, with $2V$ variable and usually positive. It is commonly observed that adjacent fibers, or adjacent bundles of fibers in each of which the extinction is parallel, will extinguish at slightly different angles (since the fibers usually are perpendicular to a

TABLE 28. INDICES OF REFRACTION OF CHALCEDONY

ω (Na)	ϵ (Na)	Locality	Ref.
1.526	...	Austria	9
1.528	...	Austria	9
1.533–1.537[a]	...	Austria	9
1.532–1.535[b]	...	Various	10
1.531	1.539	Holtenau	11
1.531	1.539	...	12
1.5325	1.5435	France	13
1.531	1.5375	Iceland	14
1.532	1.538	Iceland	14
1.534	1.541	Olomuczan	14
1.534	1.540	Uruguay	14
1.5345	1.5405	Ferrol	14
1.535	1.541	Rochlitz	14
1.5350	1.5430	Oregon	15

[a] Range of five specimens.
[b] Range of ten specimens.

surface or band which itself is curved). The fibers also may be helically twisted about the axis of elongation,[5] giving a periodic variation in birefringence along the fiber length (see further under *Iris Agate*). This feature apparently is very common in chalcedonic silica and recalls the macroscopic, euhedral crystals of quartz that are twisted about an *a*-axis of elongation (see *Twisted Crystals*). The (apparent) birefringence of the fibers generally is 0.005–0.008, lower than in coarsely crystallized quartz, and the (apparent) indices of refraction also are relatively low. The value for the ϵ index generally cannot be measured with precision; reported values range upward from about 1.538. The ω index usually is in the range 1.530–1.539, usually near 1.534, occasionally approaching 1.544 in relatively coarsely crystallized material. Ordinary quartz has ω 1.5442, ϵ 1.5533 for Na light. Some measured values for the indices of refraction of chalcedony are given in Table 28. The indices of refraction decrease further on dehydration.

The divergence in optical properties from those of ordinary quartz is caused principally by the fact that the measurements relate to more or less disoriented aggregates of fibers.[6] Strain in an interlocking aggregate also may be a factor. The low values for the indices of refraction derive in part from the presence of pores or tubules that contain water or that are empty. The optical properties of chalcedony also have been interpreted[7] in terms of form-birefringence, in which the fibers have been assumed to be embedded in opal, but the presence of opal has not been proved and is very doubtful. A chalcedonic type of GeO_2 (the polymorph isostructural with quartz) has been synthesized hydrothermally as fan-shaped aggregates of fibers; this material has lower apparent indices of refraction and lower birefringence than crystals of GeO_2 (crystals: ω 1.697, ϵ 1.724; fibers: ω 1.633, ϵ 1.653, fiber elongation mostly positive, some negative).[8] The anomalous optical properties, in particular the seemingly biaxial character, have led to the description of various types of fibrous silica as species separate from quartz. It also has been suggested that most of these substances are variants of a single biaxial substance distinct from quartz; orthorhombic symmetry is usually assigned. X-ray studies[2] of the so-called quartzin, lutecin, and pseudochalcedonite, however, have shown that this material and all other chalcedony so far investigated give a powder diffraction pattern identical with that of quartz. The so-called lussatite (lussatin), has proved on x-ray study to be a fibrous variety of cristobalite (which see).

Haloes or fringes of coarsely fibrous or feathery quartz, so-called pressure shadows, occur around porphyroblasts in schistose metamorphic rocks such as pyrite crystals in slate.[16]

Physical Properties. Chalcedony fractures readily across the banding and parallel to the fibers to give an uneven or splintery surface with a rather waxy luster. Fracture also may take parallel to the layering. The material takes a good polish with a slightly waxy to subvitreous luster. Chalcedony is sub-translucent to translucent, turbid, often varying somewhat in successive layers. The color of chalcedony in the narrow sense is always rather pale, in gray, grayish blue, milky blue, pale blue, grayish green, greenish blue, yellowish, or pale yellowish brown tones; also grayish white to slightly milky and almost colorless. Chalcedony occurs less commonly in bright or dark shades of blue-green or yellow; also faint pink or reddish. A bluish cast of color in reflected light and a pale reddish brown tint in thick sections in transmitted light are often observed and are caused by Tyndall scattering.[17] This effect also is shown by flint, blue quartz, and opal. The blue color of some chalcedony may disappear on heating.[18] The color generally is uniform or exhibits only slight differences in successive layers. As the banding becomes more distinct, particularly with the appearance of brown and reddish brown colors, usually interlayered with milky or white material of reduced translucency, chalcedony passes indefinably into the subvariety called agate. Other subvarieties derive from pigmentation by foreign materials. Brown and reddish brown types pigmented by dispersed iron oxides include the uniformly colored subvarieties called sard and carnelian and the banded agates in part. The green subvariety chrysoprase is colored by nickel.

Chalcedony is more or less porous, containing isolated pores or more generally minute tubular or thread-like openings parallel to the fibers. The porosity varies in different layers or color bands and is at a minimum in white or milky material. Some types adhere to the tongue, or rapidly imbibe water sometimes with a faint hissing sound or with the development of bubbles on a wet surface. The porosity of chalcedony and of its subvarieties, including agate, permits them to be dyed by artificial means.

The hardness is slightly variable and lower than that of coarsely crystallized quartz, usually about $6\frac{1}{2}$. The specific gravity also varies, depending on admixture, porosity, and content of water, usually 2.57–2.64.

Chemistry. Chemically, chalcedony contains about 90–99 per cent SiO_2, the higher values obtaining in light-colored or milky material, with the rest H_2O and varying amounts of Fe_2O_3, Al_2O_3, etc., as impurities. The brown and reddish brown types are relatively high in Fe_2O_3. Some of the few reported analyses of the fine-grained varieties of silica are given in Table 29. The water in chalcedony is not essential and is held by capillary and adsorptive forces. Infrared absorption study[19] indicates that (OH) also is present. The (OH) may be held in substitution for O in (SiO_4) tetrahedra located in a strained and disordered region between adjacent, interlocking fibers. Hydroxyl also may be present in the interior structure of the fibers, especially since this substitution has been identified in coarsely crystallized quartz.[20] A large part of the water is lost by desiccation at room temperature or by heating to about 100° and is believed to be held in open, interconnecting tubules or interstices, chiefly of microscopic or submicroscopic dimensions, between the fibers. Molecular water also may be entrapped in closed pores and is then not lost until much higher temperatures are reached, in the range 350–800°, with water bonded as (OH) also yielded in the higher range. Loss of water is accompanied by a decrease in refractive index and in specific gravity. The decrease in the ω index produced by dehydration at low temperatures usually is less than about 0.005.

Alkaline solutions readily dissolve large amounts of silica from chalcedony and its subvarieties. This has led to the hypothesis,[21] first suggested in 1833, that chalcedony contains opal as an interstitial material, between the fibers of quartz, that is leached out. Opal is easily and completely soluble in alkaline solutions. The bands usually are differentially attacked and rendered more distinct. Bands with a relatively low index of refraction are more rapidly attacked. The attack in hot water or steam has been found to be less selective. The high chemical reactivity of chalcedonic silica with alkalies contributes to the deterioration of portland cement concrete-aggregate made with chert or flint.[22] Direct evidence of the presence of opal in chalcedony as a general condition, however, is lacking. X-ray study[2] of chalcedony shows the presence

TABLE 29. ANALYSES OF CHALCEDONY, CHERT, FLINT, ETC.

	1	2	3	4	5	6	7
SiO_2	93.74	90.3	[99.37]	96.75	72.46	98.19	99.47
Al_2O_3	...	3.1	0.08	0.25	2.52	Nil	0.17
Fe_2O_3	3.23	1.73	0.05	0.50	17.88	0.02	0.12
CaO	...	0.94	...	...	0.14	0.30	0.09
MgO	...	1.28	0.03	...	...	0.21	0.05
K_2O	...	...	0.004	...	1.08	0.03	0.07
Na_2O	...	0.70	0.07	...	0.06	0.08	0.15
H_2O^-	0.55	...	...	...	2.66	0.33	...
H_2O^+	0.90	1.95	0.39	2.50	3.04	1.11	...
Ign.	...	...	...	...	...	...	0.12
Rem.	...	...	0.003	...	...	...	...
Total	98.42	100.00	100.00	100.00	99.84	100.54	100.24

1. Chalcedony, Waldshut, Baden.[24] 2. Chalcedony, reddish. Marienbad, Bohemia,[25] 3. Carnelian. Gobi desert, Mongolia.[26] Rem. is C. 4. Chalcedony. Mt. Olympus, Asia.[27] 5. Jasper. Victor mine, Cripple Creek, Colo.[28] 6. Flint nodule. England.[29] Rem. is CO_2 0.24, P_2O_5 0.03, Cl tr. 7. Novaculite. Arkansas.[30]

	8	9	10	11	12	13	14
SiO_2	98.17	95.50	93.54	87.30	95.81	93.54	98.93
Al_2O_3	...	0.10	2.26	5.70	1.27	2.26	0.14
	0.83						
Fe_2O_3		1.95	0.48	1.30	2.20	0.48	0.06
CaO	0.05	...	0.09	0.30	...	0.66	0.04
MgO	0.01	Tr.	0.66	1.20	...	0.23	0.02
K_2O	...	...	...	...	...	0.51	Tr.
Na_2O	...	Tr.	...	...	...	0.37	Tr.
Ign.	0.78	1.43	0.93	1.90	0.28	0.72	0.17
Rem.	...	1.02	0.79	1.40	1.43	1.00	0.56
Total	99.84	100.00	99.86	99.80	100.14	99.86	99.92

8. Chert. Belleville, Mo.[31] 9. Chert. Ireland.[32] Rem is $CaCO_3$ 0.87, FeO 0.15. 10. Chert. Mt. Diablo, Calif.[33] Rem. is FeO 0.79. 11. Chert. Devonshire, England.[34] Rem. is FeO 1.40. 12. Chert. Jonesboro, Mo.[35] Rem. is $CaCO_3$. 13. Chert. Franciscan formation, California.[36] Rem. is MnO 0.79, H_2O^- 0.21. 14. Chert. Flint Ridge, Ohio.[37] Rem. is TiO_2 0.005, FeO 0.08, MnO 0.01, CO_2 0.02, C 0.18, H_2O^- 0.27.

only of quartz, without diffraction evidence of the presence of significant amounts of either cristobalite, which is the chief constituent of opal, or amorphous silica. Electron microscope photographs[23] show a porous structure but no interstitial material. The effect of alkalies is primarily a consequence of high porosity in a microcrystalline aggregate, affording a large internal surface area, which may be intensified, in one view of the internal structure, by the greater reactivity of a strained, dislocated zone of silica that constitutes the structural transition between adjacent fibers. It should be noted, however, that opal and chalcedony often occur associated in nature, or are admixed on a gross scale, and that chalcedony occurs as a product of the phase conversion or recrystallization of cristobalite in nature (see further under *Lussatite* and *Opal*).

Chalcedony and other microfibrous or microgranular types of silica do not produce a sharp or measurable heat effect when heated through the high-low inversion point. Differential stresses set up in a mosaic of intergrown fibers and grains have the effect of spreading the heat over a considerable range of temperature. Grinding the material to a particle size approaching that of the individual grains, usually down to a few microns diameter or below, produces measurable effects.

Synthesis. Spherulitic chalcedony has been obtained experimentally by heating "amorphous silicic acid" with potassium tungstate with water in a closed tube at 80° for 144 days.[38] Chalcedony also has been synthesized[39] by the action of sodium carbonate solution on obsidian at 320–360°; by heating silica glass in water at 400°; and by heating water from the hot springs of Plombières, France, which contains alkali silicates in solution, in a glass tube sealed within a steel bomb at about 400° for 2 days. Chalcedony with a typical microstructure has been synthesized[48] by heating silica glass tubing, and also cristobalite, in slightly alkaline water solutions at 400° and 340 atm. pressure. The devitrification is zonal, the glass being converted successively to cristobalite, keatite, and chalcedony or more coarsely fibrous quartz. No change is effected in acid solutions. A fibrous type of GeO_2 analogous to chalcedony has been synthesized hydrothermally.[8]

Occurrence. Chalcedonic silica occurs in many types of geological environments, and in general is deposited at relatively low temperatures and pressures. Ordinary chalcedony of light color and without marked banding is common as a late hydrothermal deposit or alteration product in acidic to basic igneous rocks, tuffs, and breccias. It also is formed in hypogene hydrothermal veins of near-surface types, and in both the meteoric circulation and the zone of weathering. Chalcedony is especially common as crusts, vein-fillings, and cavity-fillings in altered basic igneous flow rocks and may then be associated with zeolites, carbonates, and chloritic or celadonitic alteration products. The fine-grained varieties of quartz, because of their chemical and physical stability, also occur in detrital deposits and residual deposits, and many of the important localities for chalcedony and its subvarieties are of this nature.

Among the many well-known localities for chalcedony may be mentioned Tresztyan, Transylvania, where light to dark blue material occurs as veinlets and geodes, in part as cubical pseudomorphs after fluorite, in rhyolitic rocks; France, at many places in the Paris Basin in gypsum and marls and as residual concretionary masses in the soil; also Puy-de-Dôme, Plateau Central, in basalt. Fine specimens are abundant at the zeolite localities in the Faroe Islands and at numerous localities in the Deccan traps of India. Chalcedony is abundant in the agate localities of Uruguay and Rio Grande do Sul, Brazil; in Cuba, especially near Madruga, Havana Province; and in the Canal Zone and elsewhere in Panama, chiefly in volcanic ash. Numerous occurrences are found in lavas along the shores of the Bay of Fundy, Nova Scotia. In the United States, fine chalcedony pseudomorphs after coral and sponges are found in Tampa Bay, Hillsborough County, Florida. The pseudomorphs are usually hollow and are coated on the inside by thin botryoidal crusts of chalcedony. Chalcedony, together with agate and other varieties of microcrystalline quartz, is found in great abundance in association with acidic igneous rocks and volcanic ash in central Oregon, particularly in Crook County; also as water-worn pebbles in the rivers and on the Pacific beaches, and similarly on the Olympic Peninsula, Washington. Chalcedony occurs abundantly at localities in Sweetwater County, Wyoming; in the gravels of the Yellowstone River, Montana; in a large area in Custer County, South Dakota. Chalcedonic silica colored shades of red or orange by disseminated cinnabar or spotted by inclusions of this mineral occurs in Nye County, Nevada; Jefferson County, Oregon; near Morton, Lewis County, Washington; at the New Almaden mine, Santa Clara County, and in San Bernardino County, California; and numerous other places; usually dull in luster and virtually opaque, but sometimes clear.

Silicified wood, often chalcedonic in microstructure but occasionally fine granular, brightly colored, and varying into types usually called jasper, is extremely common. Well-known localities in the United States[40] include the Petrified Forest National Park near Holbrook, Arizona; the Yellowstone National Park, Wyoming; Kittitas County, Washington; near Eden in Sweetwater County, Wyoming. (See further for localities mentioned beyond under *Agate, Moss Agate, Carnelian*, etc.) In a good specimen of petrified wood not only the external form but also the general structure of the trunk (bark, wood, and pith), as well as the growth rings, medullary rays, and the cellular structure itself are preserved, so that the species may often be determined by microscopic examination. The petrifying material may be wholly chalcedonic (or opaline) silica or wholly or in part microgranular quartz, sometimes with the interior of the cells a single anhedron of quartz.[41] The wood may become silicified after partial decomposition, with attendant deformation of form and structure, or after transport so that the outer parts have been removed by abrasion. Internal structures such as worm borings may be preserved. Hollow casts have been found of tree trunks destroyed when engulfed in hot lava or volcanic ash, the cavities lined with chalcedony or with crusts of colorless quartz or amethyst. Silicified wood is commonly associated with volcanic ash,[40] a substance which alters readily with the liberation of silica.

Iris Agate, Rainbow Agate. Chalcedony and agate are sometimes found that show spectral colors or an iris effect. The colors originate in a periodic structure in the chalcedony that acts as a diffraction grating. The structure appears in cut sections as uniformly spaced bands that range up to at least 17,000 an inch. The bands are perpendicular to the fiber axis, which is [11$\bar{2}$0] or another horizontal direction, and extend across adjacent fibers over wide areas. The spacing of the bands may vary within a single specimen. The alternate bands are of slightly different index of refraction. The diffraction effects are best seen in sections a millimeter or so in thickness that are cut perpendicular to the bands (parallel to the fibers).

Iris agate has been known since at least the eighteenth century (*Iris chalcedonius*; *Regenbogenchalcedon*). The origin of the color play, discussed by Marx in 1827, was first explained in 1843 by Brewster,[42] who observed three orders of reflection from an agate with 17,000 bands per inch and was able to resolve the D spectral line. The nature of the periodic structure recognized by Brewster and later investigators is, however, problematic. It may be noted that between crossed Nicols in transmitted light the sections show regularly alternating light and dark bands at right angles to the fiber elongation; i.e., the fibers go in and out of extinction along their length. This indicates that the fibers are twisted,[5] the dark bands marking the position where the optic axis of the fiber is parallel to the line of sight. When adjacent fibers are all in twist phase, a periodic planar variation in index of refraction is produced perpendicular to the fiber elongation, and this may be the cause of the diffraction effects. When adjacent fibers are slightly out of phase the moiré pattern or zigzag extinction bands commonly seen in chalcedony with twisted fibers is produced.

Historical. Colorless quartz and virtually all its varieties, including amethyst and the fine-grained massive types, were known to and used by the ancient civilizations of Asia Minor and the Aegean region. Most evidence of this familiarity has come from extensive collections[43] of amulets, engraved seals and finger stones, beads, and ornamental and utilitarian objects of various kinds obtained in archeological exploration, and in part from early written accounts.

The names by which these types of quartz were called in ancient times are in many instances quite uncertain. The descriptions of the earliest writers, including Theophrastus[44] and Pliny,[45] are often vague or ambiguous for lack of criteria other than color or resistance to fire that would adequately characterize the materials, and the nomenclature and systematic relations of these substances remained confused in later works by Agricola,[46] Boetius de Boodt,[47] and other writers on gem materials up through the seventeenth century. In some instances it is apparent that the meaning of the name changed with time. A clearer understanding of the chemical and physical properties of these substances with an attendant clarification of the nomenclature was reached toward the end of the eighteenth century in the works of Haüy, Werner, and others.

The name chalcedony has a long and complicated history that has not been fully traced. The present application as a broad term to include sard, agate, flint, and related substances with a fibrous microstructure is recent in origin. A narrower reference to pale-colored or grayish types of material without marked banding extends back to about the sixteenth century, although uncommon before the eighteenth century, and with some overlap of earlier meanings. The names chalcedony and calcedony, *calcedoine* in Old French, are adapted from the Latin *calcedonius*, used in the Vulgate version of the Scriptures to render the Greek χαλκηδών in reference to a stone forming the third foundation of the New Jerusalem or Heavenly City (Rev. xxi: 19). The *charchedonia* or *carcedonius* of Pliny, brought from North Africa by way of Carthage, has been ascribed to chalcedony, with charchedonia sometimes used as equivalent to chalcedony, but the material appears from the description of Pliny to have been some other mineral.

The name chalcedony itself ultimately derives from Chalcedon or Calchedon, an ancient maritime city of Bithynia on the Sea of Marmora. The name of the town is erroneously given as Carthage in some editions of Theophrastus's *On Stones*. Mines were worked in ancient times in the neighborhood of Chalcedon, and there was a trade in ornamental stones brought thence from Cappadocia and other places. The name chalcedony may once have been applied to an ornamental material brought from the region of the Chalcedonians, in the sense of turquoise, schmirgel, and other names that refer to a trade source, but there is no evidence that Chalcedon was a source or a locality for the mineral now called chalcedony.

The banded and variegated subvarieties of chalcedony were valued in antiquity, and many types received specific names, with *onychion* (onyx, pt.) perhaps employed as a general term in reference to them, and the uniformly colored types also were separately named. The pale-colored types of chalcedony without marked banding, which occur in association with these substances, were less valued. Chalcedony in this sense is believed to have been included by Pliny under the name *cerachates* (from the Latin *cera, wax*), stated to be an abundant material, and *leucachates* (Greek *leucos, white*). The older term *iaspis* doubtless also included ordinary chalcedony of pale yellowish, greenish, and grayish tints.

Agricola used chalcedony in reference to a banded stone. The identity of the *murrhina* or *murrha* used for cups and vessels in Roman times is doubtful. It has been attributed to ordinary chalcedony, but may have been a white or colored type of porcelain or perhaps a collective term for varieties of chalcedony, fluorite, and other materials used for the purpose.

References

1. On the microstructure of chalcedony, flint, etc., see Michel-Lévy and Munier-Chalmas: *Bull. soc. min.*, **15**, 161 (1892); Wallerant: *Bull. soc. min.*, **20**, 52 (1897); Lacroix: *Min. de France*, **3**, 120 (1901); Timofejeff: *Trav. Soc. Imp. Nat. St. Petersburg, Sect. Geol. Min.*, **35**, 157 (1911); Hein: *Jb. Min.*, Beil. Bd. **25**, 182, 1908; Leitmeier: *Cbl. Min.*, 1908, 632: Cayeux: *Les roches séd. de France, Mém. la carte géol. France*, Paris, 1929; Tarr: *Univ. Missouri Stud.* **1**, no. 2, 1926; Correns and Nagelschmidt: *Zs. Kr.*, **85**, 199 (1933); Lengyel: *Föld. Közl.*, **66**, 129, 278 (1936); Storz: *Die sekundäre authigene Kieselsaüre*, Berlin (*Mon. Geol. Paleo.*, Ser. II, **5**), 2 Pts., 1931; Wetzel: *Cbl. Min.*, 1913, 356; Laird: *Trans. Roy. Canadian Inst.*, **20**, 231 (1935); Neuwirth: *Min. Mitt.*, **3**, 32 (1952); Slavík: Vest. král. Česke spol. nauk, no. 16, 1942, and no. 16, 1946; Braitsch: *Heidelberg. Beitr. Min. Pet.*, **5**, 331 (1957); Pelto: *Am. J. Sc.*, **254**, 32 (1956).

2. Raman and Jayaraman: *Proc. Indian Ac. Sci.*, **38**, 199 (1953).

3. On x-ray studies of the identity, orientation, etc., of chalcedony and other fibrous types of silica see: Washburn and Navias: *Proc. Nat. Ac. Sci.*, **8**, 1 (1922); Correns and Nagelschmidt (1933); Blattman and Hägele: *Zbl. Min.*, 313, 1937; Kolazkowska: *Arch. min. soc. sci. Warsaw*, **12**, 82 (1936); Laves: *Naturwiss.*, **27**, 705 (1939); Midgley: *Geol. Mag.*, **88**, 179 (1951); Jensen and Andersen: Royal Vet. and Agricult. Coll., Copenhagen, Yearbook 1956; Braitsch (1957); Jayaraman: *Proc. Indian Ac. Sci.*, **38**, 441 (1953); Raman and Jayaraman: *Proc. Indian Ac. Sci.*, **41**, 1 (1955); Neuwirth (1952); Novák: *Bull. soc. min.*, **70**, 238 (1948); Ichinose: *Mem. Coll. Sci. Kyoto Univ.*, **18A**, 315 (1935).

4. Lacroix (1901), p. 127; Michel-Lévy and Munier-Chalmas (1892).

5. Bernauer: *Jb. Min.*, Beil. Bd. **55**, 92 (1927); Braitsch (1957); Michel-Lévy and Munier-Chalmas (1892); Lacroix (1901), p. 124.

6. See treatment by Braitsch (1957).

7. Donnay: *Ann., soc. géol. Belg.* **59**, B289 (1936); Laves (1939).

8. White, Shaw, and Corwin: *Am. Min.*, **43**, 580 (1958).

9. Neuwirth (1952).

10. Braitsch (1957).

11. Wetzel: *Cbl. Min.*, 356, 1913.

12. Wülfing: *Sitzber. Heidelberg Ak. Wiss., Math.-nat. Kl.*, 20, 1911.

13. Wallerant (1897).

14. Correns and Nagelschmidt (1933).

15. Jones: *Am. Min.*, **37**, 578 (1952).

16. Pabst: *Am. Min.*, **16**, 55 (1931).

17. Weymouth and Williamson: *Min. Mag.*, **29**, 573 (1951); Folk and Weaver: *Am. J. Sci.*, **250**, 498 (1952); Pelto (1956).

18. Aubert de la Rüe: *Bull. soc. min.*, **55**, 165 (1932).

19. Pelto (1956).

20. Brunner, Wondratschek, and Laves: *Naturwiss.*, **24**, 664 (1959).

21. Doelter: *Handbuch der Mineralchemie*, **2**, Pt. 1, 173, 1914, with earlier literature; Fuchs: *J. Chem. u. Phys.*, **7**, 10 (1833) and *Ann. Phys.*, **31**, 577 (1833); Correns and Nagelschmidt (1933); Pelto (1956); Oakley: *Sci. Progress*, **34**, 277 (1939).

22. Mather: *Proc. Highway Res. Board,* **30,** 218 (1951); Pelto (1956); McConnell and Irwin: *Am. Min.,* **30,** 78 (1945).
23. Folk and Weaver (1952); Pittman: *Soc. Econ. Paleo. and Min. Spec. Publ.* **7,** 125, 1959.
24. Graeff: *Zs. Kr.,* **15,** 377 (1889).
25. Kersten: *Jb. Min.,* 656, 1845.
26. Heintz: *Ann. Phys.,* **60,** 519 (1843).
27. Klaproth: *Beitr.,* **4,** 325 (1807).
28. Hillebrand: *U.S. Geol. Surv.* 16*th Ann. Rpt.,* Pt. 2, 127, 1895.
29. Weymouth and Williamson (1951).
30. Griswold: *Rpt. Arkansas Geol. Surv.* **4,** 1890.
31. Schneider: *U.S. Geol. Surv. Bull.* **228,** 297, 1904.
32. Hardman: *Sci. Trans. Roy. Dublin Soc.,* **1,** 85 (1878).
33. Melville: *Bull. Geol. Soc. Amer.,* **2,** 411 (1921).
34. Hinde and Fox: *Quart. J. Geol. Soc.,* **51,** 629 (1895).
35. Hovey: *Missouri Geol. Surv.,* **7,** 727 (1895).
36. Davis: *Univ. Calif. Publ., Bull. Dept. Geol. Sci.* **11,** 268, 1918.
37. Stout and Schoenlaub: *Bull. Geol. Surv. Ohio* 46, ser. 4, 82, 1945.
38. Leitmeier: *Jb. Min.,* Beil. Bd. **27,** 244 (1909).
39. Koenigsberger and Müller: *Cbl. Min.,* 1906, 339; Nacken: *Natur u. Volk (Ber. Senckenberg. Naturfor. Ges.),* **78,** 2 (1948); Daubrée: *Études synth. de géol. exper.,* Paris, 1879, p. 174.
40. Murata: *Am. J. Sci.,* **238,** 586 (1940).
41. Barksdale: *Am. Min.,* **24,** 699 (1939); Polkunov: *Chem. Abs.,* **52,** 16, 994 (1958).
42. Brewster: *Phil. Mag.,* **22,** 213 (1843) and *Ann. Phys.,* **61,** 134 (1844); see also Marx: *Archiv gesamte Naturlehre (Kastner's Archiv),* **12,** 220 (1827); Jones: *Am. Min.,* **37,** 578 (1952); Raman and Jayaraman (1953).
43. Walters: *Cat. of the Engraved Gems and Cameos, Greek, Etruscan and Roman,* British Museum, London, 1926; Maffei: *Gemme antiche,* Rome, 1708, 4 vols.; Agostini: *Gemme antiche,* Rome, 1702, and *Gemme antiche figurate,* Rome, 1686; Blümner: *Tech. und Terminologie der Gewerbe und Künste die Griechen und Römern,* Leipzig, 1875–87.
44. Theophrastus: *On Stones,* 4th century B.C. (English transl. and commentary, Caley and Richards, Columbus, Ohio, 1956).
45. Pliny: *Natural History,* 37th Book, A.D. 77 (English transl. by Holland, London, 1601, modernized by Ball: *A Roman Book on Precious Stones,* Los Angeles, 1950).
46. Agricola: *De natura fossilium,* 1546 (English transl. by M. C. and J. A. Bandy, *Geol. Soc. Amer. Spec. Paper* **63,** 1955); also Agricola, *Ausgewählte Werke,* 3 vols., Berlin, 1956, with commentary.
47. de Boodt: *Gemmarum et lapidum historia,* Jena, 1609.
48. White and Corwin: *Am. Min.,* **46,** 112 (1961).

SARD AND CARNELIAN

The name sard is applied to the uniformly colored light to dark brown semi-precious, translucent types of chalcedony, varying in color to chestnut brown, orange-brown, and reddish brown tints. Carnelian is a uniformly colored red to reddish brown or flesh-red chalcedony. The most valued kind is a deep blood red in transmitted light and a blackish red in reflected light. All gradations occur to pale red, yellowish red, and brownish red colors, and carnelian passes indefinably into sard. Both carnelian and sard may grade into more or less distinctly banded types. Carnelian differs from jasper of similar color in being translucent and, usually, in possessing a fibrous rather than a granular microstructure.

The name sard is the older of the two, and in ancient times included, under the name *sardion*, the reddish type later distinguished under the name carnelian (see *Historical* beyond). The reddish carnelian is pigmented by colloidally dispersed hematite. Both carnelian and sard occur in the same way as ordinary chalcedony and agate and are often associated with these more common types. Some reddish material sold as carnelian is obtained by the heat treatment of greenish, yellowish, or brown chalcedony, including sard. Sard is easily imitated by dying chalcedony. Sard and carnelian of fine color and translucency are rare. India has been an important commercial source[3] of sard and carnelian since at least the fourth century B.C. Ratnapur was a famous old source, with cutting centers at Broach and Cambay. Prolonged exposure to the sun and baking in clay pots were employed to improve the color of the material. Good specimens have been found in many other localities. In the United States countless occurrences of these subvarieties of chalcedony are known, but they rarely afford material of any significant gem value. Among the more abundant localities may be mentioned Specimen Ridge and Amethyst Mountain in Yellowstone National Park, Wyoming, and for a long distance along the Yellowstone River in Montana; near Chehalis in Lewis County, Washington; near Colorado Springs, El Paso County, Colorado; in a wide area around Fairburn, Custer County, South Dakota; on Isle Royale in Lake Superior, off Michigan; on Flint Ridge in Licking County, Ohio. Very numerous localities are known in central Oregon. Carnelian occurs as a replacement of bone in Dinosaur National Monument in Moffat County, Colorado, and Uintah County, Utah; chalcedony occurs similarly at Crawford, Darris County, Nebraska.

Historical. Sard, the *sardion* of Theophrastus, is one of the few mineral names whose original meaning is known and still obtains after the passage of several thousand years. It is the oldest name known definitely to apply to a type of silica, aside from *crystallos*. The name comes from Sardis, capitol of the ancient kingdom of Lydia, and probably derives from the substance having been handled in trade at that place rather than as a locality where it occurred. Sardis is located on the Hermus River about 70 miles inland from the Aegean coast of Turkey. Herodotus stated that the Lydians were the first to introduce gold and silver coinage, and Theophrastus indicates that touchstone or Lydian-stones used for assaying (probably water-smoothed pieces of slate)[1] came from the nearby Tmolus Mountains. The original *sardion* included both the brown and the reddish types of chalcedony. The reddish type was later distinguished by the gem merchants of the Middle Ages under the name *carneolus* (carnelian), from the Latin *carneus*, *fleshy*, in allusion to the color. The name also is given as cornelian and is said to have been derived from the medieval Latin name *cornus* or *cornum* for a species of dogwood, the cornelian cherry (*Cornus mas*, Linn.), with a reddish berry. Both types of sard were employed in Grecian and Roman times more than any other semi-precious stone, chiefly for engraved seals and ring stones. Herodotus in the fifth century B.C. stated that each Babylonian had his seal, and Xenophon the Greek stated (400 B.C.) that rings set with an intaglio were worn by many of his fellow soldiers.[2]

References

1. Theophrastus: *On Stones*, 4th century B.C. (English transl. and commentary, Caley and Richards, Columbus, Ohio, 1956).
2. Pliny: *Natural History*, 37th Book, A.D. 77 (English transl. by Holland, London, 1601, modernized by Ball: *A Roman Book on Precious Stones*, Los Angeles, 1950).
3. Ritter: *Die Erdkunde*, Berlin, 6, 1836, pp. 603–607; Fulljames: *Trans. Bombay Geog. Soc.*, 2, 74 (1838); Ball: *Econ. Geol.*, 26, 681 (1931).

MOSS AGATE

This name is given to chalcedony, usually gray, bluish, or milky in color and translucent to subtransparent, that contains branching inclusions with a fancied resemblance to plant life, including trees, shrubs, moss, algae, and seaweed. A resemblance to trees is often apparent, leading to the description as dendritic structure and corresponding to the original *dendrachates* of Pliny, but the name moss agate is now almost universally employed. A moss agate is shown in Fig. 92. In other types included under the name, but sometimes designated by special terms, the inclusions have a lace-like, feathery, ribbon-like, flowery, or plume-like structure, and in some types the inclusions appear as aggregates of tubes or filaments or are streaky. Sagenitic agate contains acicular inclusions. Occasionally the inclusions are so densely aggregated as to make the material virtually opaque. The inclusions may be essentially planar, or have this appearance in cut and polished sections, but they often have a three-dimensional form in the enclosing chalcedony. Moss agates are generally cut as thin, flat plates with an oval or a round outline. The included material appears as a finely divided pigment impregnating the chalcedony. It usually is black in color and is attributed to a manganese oxide, or is brownish red, red, orange-red, and yellowish red, then doubtless due to an iron oxide; less commonly green, gray, or yellowish white. Black and reddish dendrites may occur together, or the color may vary gradationally. Black dendrites may turn red on strong heating.

Under magnification the filaments of the inclusions are seen to be tubular, and to vary erratically in diameter along their length; sometimes there is a lighter-colored and relatively translucent zone or halo in the chalcedony immediately adjacent to the filaments. The inclusions have formed[1] by the

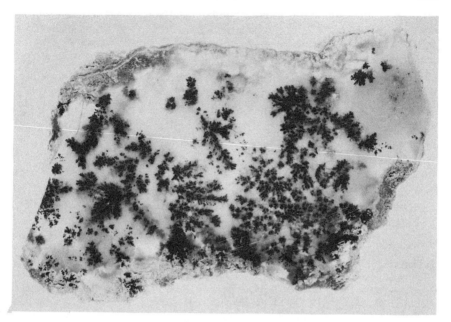

Fig. 92. Moss agate. Collier County, Kansas. U.S. Nat. Mus. coll.

precipitation of material that diffused into a mass of gelatinous silica before it crystallized into chalcedony (see further under *Agate*). In some chalcedony geodes the inclusions may be seen extending into the cavity from the walls. The inclusions are believed to be similar in origin to the "chemical gardens" or filamentous and dendritic growths formed by introducing pieces of soluble metallic salts into silica gel or other viscous, colloidal media. Although the inclusions are inorganic in origin, the close resemblance to moss or other organic forms has occasionally led to the belief that they are actual remnants of plant life.[2] Dendritic and other types of inclusions also occur in opal, forming moss opal.

Moss agate is very abundant at some localities for chalcedony, although mostly of poor quality, and the material is widespread. A locality that provided material in ancient times is near Mocha in Yemen at the entrance to the Red Sea (whence the common name Mocha-stone; Germ., *Mokkastein*). Also produced commercially on the Kathiawar Peninsula and in the area of the Deccan basalt flows, India, as well as from the agate localities in Uruguay and Brazil. A number of important localities for moss agate are known in the United States. Found as waterworn pebbles and nodules in the gravels of the Yellowstone River, Montana, from roughly Billings to the North Dakota boundary. In Wyoming, massive moss agate showing a translucent white or bluish white matrix occurs at Hartville, Platte County, and pebbles and nodules of moss agate are found abundantly in surface gravels in Fremont County in the area of the Sweetwater River. The Sweetwater moss agate and chalcedony often show a greenish yellow uranyl fluorescence in ultra-violet radiation. Also in the gravels of the Rio Grande River for many miles along the course above and below Laredo, usually reddish brown or yellow-brown in color and densely aggregated. A number of important occurrences are known in Oregon, notably in Crook County in a large area around Prine-ville, where an orange-red, reddish brown, and black plume type is found, and on the Priday Ranch near Madras. Fine moss agate also has been ob-tained in Horse Canyon near Cache Peak, Kern County, California. A beautiful black and red type of plume agate is found on the Woodward Ranch in the Alpine area, Brewster County, Texas. Moss agate and plume agate of a reddish to golden color occur in the neighborhood of Villa Ahumada in Chihuahua, Mexico.

References

1. Gergens: *Jb. Min.*, 801, 1858; Liesegang: *Die Achate*, Dresden and Leipzig, 1915; Brown: *Smithsonian Inst. Ann. Rpt. for 1956*, 329, 1957.
2. Razoumovsky: *Jb. Min.*, 627, 1836 (abstr.); Brown (1957); Lillie and Johnston: *Biol. Bull.*, **23**, 135 (1917) and **26**, 225 (1919); Leduc: *Mechanism of Life*, London, 1911, Ch. 10.

AGATE

Agate is a common and important subvariety of chalcedony with a distinct banding in which successive layers differ in color and in degree of translucency. All gradations, however, occur between agate and ordinary chalcedony. Agate generally occurs as a cavity-filling. The individual bands or layers are continuous, generally of uniform thickness and color, and are concentric to the external surface of the nodule or mass as determined by the shape of the cavity in which it was formed. Individual layers may range up to a centimeter or more in thickness in their gross appearance, but under the microscope in sections cut perpendicular to the banding these layers may be seen to be composed of many thinner layers of essentially the same color and trans-lucency. The gross layers often are only paper thin. The layers may show sharp salient and re-entrant angles, resembling in section the plan of a bastion or fortress, and are then called fortification agate. Other descriptive prefixes also are employed according to the nature of the pattern, such as landscape agate, brecciated agate, ruin agate, and star agate. Nodules containing con-centric spherical layers, appearing as concentric circles when sectioned, are called eye agate. Agates may form in cavities containing irregular or stalac-titic projections and may then show bands concentric to both the walls and to the projections as intersected in section. In some agates the layers may be curved and concentric in the outer part but internally strike across the stone in a series of plane, parallel layers (ribbon agate) representing two stages of cavity-filling. Occasionally two sets of parallel bands may be obtained tilted at an angle. The nodules do not always consist entirely of agate. The central part often consists of coarsely crystallized quartz, either colorless or amethystine, rarely smoky, as more or less prismatic crystals pointing toward the center and terminated by rhombohedral faces if there is a central cavity. Crystals of other minerals, such as calcite, siderite, goethite, or zeolites, may be observed in the central cavity. Dickite has been observed.[11] The central parts of agate geodes also may be filled by white or milky opaline silica, sometimes almost hyaline, or showing a play of colors, and without marked banding. Some agates or, more properly, geodes show only a thin wall of chalcedonic silica, often encrusted with quartz or amethyst, with a large central cavity. Almost all agate nodules show one or more narrow canals or channelways (*Einflusskanälen, Spritzlöcher*),[1] extending out from the central part to the surface where they may appear as funnel-shaped depressions. The layers of the agate bend sharply outward along the course of the canal. Polished agate sections are shown in Figs. 93–95.

Agates ordinarily range from small nodules the size of a nut up to about a foot in size, but examples, often irregular in shape, weighing hundreds of pounds have been found. The surface of the nodules may be rough to fairly smooth, often with tiny pore-like depressions, or with small warty projections or small-botryoidal. Agates from the amygdaloidal cavities of altered basic

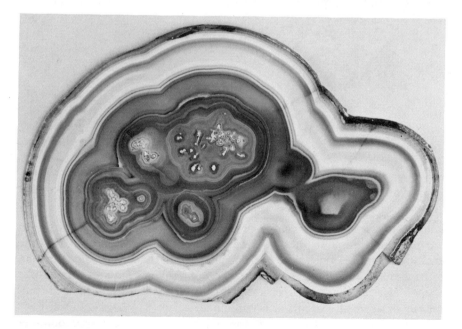

FIG. 93. Agate. Brazil. U.S. Nat. Mus. coll.

FIG. 94. Agate with horizontal banding. In rhyolite. Oregon. U.S. Nat. Mus. coll.

igneous rocks may have a more or less thick skin of dull green earthy to compact material, and when in place the agate may be separated from the enclosing rock by a thick zone of such material. In shape, agate nodules tend to be more or less elliptical or flattened on one side, or have an almond-like shape; also irregular, more or less elongate, rarely spherical. Banded chalcedony or agate also occurs as a vein-filling.

Color. The natural colors of the different layers are those found in ordinary chalcedony and include white, milky white, gray, bluish white, grayish green, yellowish brown, brown, reddish brown, and, rarely, green, blue, lavender, or black. Natural agates used for ornamental purposes (and some mineral specimens) are generally artificially colored or are heated to convert the brown colors to reddish brown and red. Polished agates often are soaked in oil to increase the luster and the translucency. The porosity of different layers varies considerably, and is least in the natural white or milky

FIG. 95. Agate, closely banded, with interior filled with coarse quartz crystals. Germany. U.S. Nat. Mus. coll.

layers, which have a relatively coarse fibrous microstructure. Special names often are applied according to the color of the natural or dyed material. Onyx in the strict sense consists of milky white or white bands alternating with black or deep brownish black bands; also with brown bands in sardonyx and red in carnelian onyx. Onyx is employed in the carving of cameos, the figure being carved in relief in the white band with the dark band as background, and of intaglios, the figure being incised in the dark band to reveal the white material. Plane-parallel ribbon agates are employed for the purpose. In jasp-agate bands of translucent chalcedonic material alternate with opaque layers of red-brown jasper.

Most commercial agate for ornamental purposes is artificially colored.[2] A wide variety of colors, including blue and green of various shades and citron yellow, can be produced, either directly by dyes or other pigmenting substances, by the reaction of reagents successively diffused into the agate, or by the thermal decomposition of introduced substances. The artificial coloring is done after the agate has been cut and shaped. Black agate is made by immersing carefully cleaned material in a hot solution of sugar for a period up to weeks in length and then, after washing, carbonizing the sugar by immersing in concentrated sulfuric acid and heating. A brownish black to brown color can be produced by heating stones impregnated with a dilute solution of sugar. A black color also can be produced by careful heating of material soaked in a solution of cobalt nitrate and ammonium sulfocyanate. Blue agate, sometimes called Swiss lapis or false lapis lazuli, is obtained by impregnating the stone with potassium ferrocyanide and then warming in a solution of ferrous sulfate to form a precipitate of Berlin blue. The depth of color and tint can be varied with the concentration and pH of the solutions. Green and bluish green colors can be produced by impregnating the stone with a chromium salt, such as chromium alum or potassium dichromate, and then heating. An apple green color can be obtained with nickel salts. Red and reddish brown colors can be obtained by heating natural brown stones [containing hydrous ferric oxide (goethite) as pigment] or by first impregnating the stone with a solution of an iron salt. Yellowish brown to greenish yellow tints may be obtained by heating well-dried agate that has been treated with concentrated hydrochloric acid. The depth of penetration of the color varies with the porosity of the material and the duration of treatment, but usually only a thin surface layer that can be removed or tested by grinding is affected. Agates colored by organic dyes may bleach or change color on exposure.

Occurrence. Agate is very common and is widely distributed. It occurs in the same general way as ordinary chalcedony but with greater dependence on cavity-filling as a mechanism of deposition. The great bulk of commercial agate is obtained from a belt of altered basic lavas, chiefly amygdaloidal melaphyres, about 400 miles in length and extending from the region of Salto on the Uruguay River, Uruguay, to Porto Alegre in the province of Rio

Grande do Sul, Brazil. The agates occur together with other quartz varieties, including amethyst, in cavities in the rock or either loose in a residual ferruginous clay or as detrital cobbles. Agates and other subvarieties of chalcedony are obtained in commercial amounts at numerous places on the Deccan Plateau of India. Important localities are at Ratnapura on the lower Narbuda River and in the country north of Rajkot on the Kathiawar Peninsula. Commercial production also has been obtained in Madagascar, in northern Bohemia, in older times from the region of Idar and Oberstein in Germany, and from numerous minor localities. Found in Armenia, Georgia, Azerbaidzhan, and other regions in the USSR.[3]

Among the many well-known localities in the United States[4] may be mentioned the Keweenaw Peninsula and Isle Royale, Michigan, where agate, generally known as Lake Superior agate, occurs in the basic volcanic rocks and as waterworn masses on the beaches and in the streams; also in the adjacent parts of Minnesota, and widely in glacial drift and in river gravels to the south of this general region, as at Muscatine, Iowa. Numerous old localities in the eastern United States are mentioned by Robinson (1825) and by Cleaveland (1816),[5] as in the diabase sills of Franklin and Hampshire counties, Massachusetts, especially near Deerfield. Ocasionally found in good specimens in the diabase (trap rock) quarries around Paterson and at nearby places in New Jersey. Agate and related materials occur in a wide area in Shannon, Custer, and Pennington counties, South Dakota, notably in the neighborhood of Fairburn, and on Amethyst Mountain and elsewhere in Yellowstone Park, Wyoming. At numerous localities in Oregon, together with moss agate, especially on the Priday Ranch, near Madras, Jefferson County. Fine-banded agates, in part with reddish, lavender, or purple tints, are found around Villa Ahumada and southward in Chihuahua, Mexico. Agate of good quality is found in Canada on Michipicoten Island in Lake Superior and along the shores of the Bay of Fundy, Nova Scotia.

Origin. The agate geodes found in basic lavas and other igneous flow rocks or welded tuffs are believed to have formed by the deposition of silica in gas cavities. Small cavities may be spheroidal, but larger ones, unstable under flowage of the enclosing lava, take on a flattened and elongate shape (amygdules). In a highly viscous material expansion of the gas cavity by rupture may require less energy than spherical expansion, giving angular cavities which may be symmetrically developed. The chalcedony nodules ("thunder eggs") from the rhyolitic lavas and welded tuffs of Oregon and other localities[6] often show in section a five-pointed star or central cavity resulting from a shear pattern with pyritohedral symmetry. Other nodules show a roughly cubical cavity with each face comprising a flat four-sided pyramid pointing toward the center (Fig. 94); irregularities on one face are matched by a counterpart on an opposite face, indicating that all the faces were originally together and were forced apart by the gas pressure. These chalcedony nodules generally contain or border on a spherulitic aggregate of intergrown feldspar and cristobalite, the crystallization of which is believed to have initiated the release of gases dissolved in the rhyolitic glass. The chalcedony filling the cavities is believed to have been initially a gelatinous mass which subsequently crystallized in microcrystalline form. The banding of agates is essentially a rhythmic deposition from the enclosed siliceous solution or silica sol on the walls of the cavity, but diffusion in the gel, leading to the formation of Liesegang bands, and other processes have been suggested. The peripheral canals may represent the expulsion of fluid during the progressive syneresis of the gel, but the origins of this feature and of some others of agate are problematic.[7] The silica apparently was introduced hydrothermally while the enclosing rock was still hot and was not derived by the chemical alteration of the rock immediately adjacent to the geodes, although such peripheral alterations are common. The altered zones in basic rocks contain chiefly celadonite, saponite, chloritic and serpentinous material, and carbonates, and in the rhyolitic rocks chiefly montmorillonite. Silica gel with agate-like banding can be obtained experimentally[8] by the diffusion and rhythmic precipitation of metallic salts therein.

Historical. The name agate comes from the occurrence in the river Achates, the modern Drillo, in southwestern Sicily. The ancient term *achates*, however, is not wholly synonymous with agate. Some types that we now know by this name were classed as varieties of *onychion* (or onyx, which then included chalcedony, alabaster, and other materials), while others irregularly veined or variegated, classed under *achates*, do not correspond to present usage. The old usage still survives in names used for certain irregularly marked or colored but not banded types of chalcedony, such as moss agate. Many types of *achates* were distinguished and separately named by Pliny and others chiefly as varieties, according to the color, the type of banding, fancied resemblances to various objects (such as eye agates), and other characters. In 1747 the Swedish mineralogist Wallerius remarked that to enumerate all the varieties of agate is impossible and unnecessary (but efforts continue). The visualization of scenes in these marked or variegated agates was as popular then as now.[9] Agricola commented that moss agates do not contain as many images of birds as of beasts of burden and men, perhaps a sign of the times. Pliny mentioned that Pyrrhus, King of Epirus, owned an *achates* in which could be seen "the nine Muses and Apollo with his lyre truly pictured, not by art nor man's hands but drawn by nature," and Wallerius cited an agate depicting the passage of the children of Israel across the Red Sea. Some eighteenth century authors classified the *achates figurate* into *achates anthropomorphos*, *achates zoomorphos*, and *achates phytomorphos*—the last becoming our moss agate.

One of the most famous and largest specimens of worked agate is a nearly flat dish 75 cm. (29$\frac{1}{4}$ in.) in diameter, cut from a single mass, preserved in the

Treasure Chamber of the Kunsthistorisches Museum in Vienna.[10] The dish was fashioned at Trier in the fourth century A.D. and became one of the heirlooms of the House of Habsburg. The commercial trade in agate has centered for centuries at Idar and Oberstein, on the Nahe river, Germany. The trade originated there in the Middle Ages, possibly in Roman times, utilizing material from the immediate neighborhood; but since about 1827, when wandering agate polishers from Idar discovered the deposits of Uruguay and Brazil, the raw material has been imported from South American and other foreign sources. The practice of artificially coloring agates, known to the Romans, was developed commercially at Idar-Oberstein, beginning in the very early 1800's. Wallerius (1747) described how to stain agates brown with silver solutions.

Agates and moss agate hold a peculiar fascination and since earliest times have been among the most widely employed semi-precious natural materials. Sardonyx was ranked by Pliny as the fifth most valuable gem stone in his time, coming after diamond, pearl, emerald, and opal, and followed by the red carbunculi (spinel, ruby, and garnet) with gold in tenth place and silver, sapphire, and topaz well down in the list. The popularity of agate has varied widely, reaching peaks at the time of the Crusades, when much antique material was brought back to Europe, and in the middle nineteenth century with the importation of abundant material from Uruguay and Brazil. Agate and related materials also were extensively employed in Renaissance times, especially in Italy, as mosaic and ornamental stones. The collecting and polishing of agates as a hobby by individuals has reached a high level in the western United States during the past three or four decades, stimulated by the discovery of outstanding localities in Oregon, California, and other western states.

References

1. See Liesegang: *Die Achate*, Dresden and Leipzig, 1915; Pilipenko: *Bull. soc. nat. Moscou, Sect. Geol.*, 12, 279, 296 (1934); and ref. 7.
2. See Bauer: *Edelsteinkunde*, Leipzig, 1932 (third ed. by Schlossmacher, p. 694); Hoffman: *Jb. Min.*, Abt. A, 77, 238 (1942); Nikiforov: *Chem. Abs.*, 34, 2226 (1940).
3. Vanyushin: *Mem. soc. russe min.*, 67, 141 (1938); see also Fersman: *Precious and Colored Stones of Russia*, Petrograd, 1922 (*Mon. Russ. Ac. Sci.* 3, 1, 280).
4. Many U.S. localities for chalcedony, agate, etc., of interest for specimen material are mentioned by Kunz: *Gems and Precious Stones of North America*, New York, 1890, 1892; Dake: *Northwest Gem Trails*, Portland, Oregon, 1950; Sinkankas: *Gemstones of North America*, New York, 1959. See also topographical state mineralogies such as Murdoch and Webb: *Minerals of California, Division of Mines Bull.* 173, 1956; Northrop: *Minerals of New Mexico*, Albuquerque, N.M., 1959; and Gordon: *Mineralogy of Pennsylvania, Acad. Nat. Sci. Philadelphia Spec. Publ.* 1, 1922.
5. Robinson: *Catalogue of American Minerals and Their Localities*, Boston, 1825; Cleaveland: *Treatise on Mineralogy and Geology*, Boston, 1816, second ed., 1822.
6. Ross: *Am. Min.*, 26, 727 (1941).
7. On the origin of agate see Fischer: *Jb. Min.*, Abh. 86, 367, 1954; Leitmeier in Doelter: *Handbuch der Mineralchemie*, 2, Pt. 1, 186, 1914; Pilipenko (1934); Knöll: *Kolloid-Zs.*,

101, 296 (1942); Liesegang (1915) and *Cbl. Min.*, 1919, 184; Heinz and Linck: *Chem. der Erde*, **4**, 501, 526 (1930); Ingerson: *Am. Min.*, **38**, 1057 (1953); Nacken: *Natur u. Volk*, **78**, 2 (1948); Kaspar: *Veda Prirodni, Prague*, **23**, 97 (1944).

8. Liesegang (1915); Fillinger, McConnell, and Oline: *J. Chem. Educ.*, **22**, 440 (1945); Copisarow: *Nature*, **149**, 413 (1942).

9. Examples described by Weidhaas: *Rocks and Min.*, **11**, 131 (1936); Fersman: *Unterhaltende Mineralogie*, Leningrad, 1931, p. 57; Sowerby: *British Mineralogy*, **3**, 39 (1819); Ball: *A Roman Book on Precious Stones*, Los Angeles, 1950, p. 207.

10. Fillitz: *Katalog der Weltlichen und der Geistlichen Schatzkammer*, Guide no. 2, Kunsthistor. Museum, Vienna, 1954 (abridged English ed., 1956).

11. Allen: *Am. Min.*, **21**, 457 (1936).

CHRYSOPRASE; PLASMA; PRASE

Chrysoprase is an apple green, translucent subvariety of chalcedony with usually a fibrous but sometimes a microgranular structure. The color in general is bright but not deep and grades into pale green to yellowish or grayish green. Small spotted or irregular areas of a white or yellowish color may be present. The color may pale on exposure to sunlight or by slight heating. The material is pigmented by disseminated particles of a hydrated nickel silicate. Chrysoprase is relatively brittle, and large pieces are more or less cracked. It typically occurs as a secondary mineral in veinlets and crevices in serpentine. Well-known localities include Frankenstein, Kosemütz, and Gläsendorf in Silesia; the nickel deposit at Riddle, Douglas County, Oregon; and localities in serpentine near Porterville, Exeter, and Lindsay, Tulare County, California. A considerable commercial production has come from the California localities, including vein masses up to a foot thick suitable for slicing for mosaics. Green or greenish blue chalcedony containing chrysocolla and somewhat resembling chrysoprase or turquoise has been found in copper mines in the Globe district, Gila County, Arizona. Chrysoprase is easily imitated by dyeing chalcedony with solutions of chromium or nickel salts.

There is no evidence that chrysoprase was among the various green stones used by the ancients for ornamental purposes. *Chrysoprasius* is mentioned by Pliny and by Agricola, but in reference to plasma, prase, beryl, and other green materials. The name chrysoprase comes from χρῡσό-ς, *gold*, and πράσν, *leek*.

Plasma is a microgranular or microfibrous variety of quartz colored various shades of green, including dark leek green, oil green, and dull apple green, by disseminated particles of silicates variously attributed to celadonite, chlorite, microfibrous amphibole, and other species. Plasma is virtually opaque with a conchoidal to smooth fracture. It has been employed for mosaic and carving purposes and for intaglios. Plasma has been obtained commercially from India, Madagascar, Egypt, and other places abroad; at Durkee, Oregon, and in Clarke County, Washington.

Prase is more translucent than plasma and has a leek green color (Greek *prasios, leek green*). The color usually is caused by dense aggregates of minute fibers of hornblende; also by chlorite. Prase includes both massive granular material and euhedral crystals. Material classed as prase has been found in Vermont; in Bucks, Delaware, and Chester counties, Pennsylvania; and at many other localities in the United States and abroad.

HELIOTROPE; BLOODSTONE

These substances are a type of chalcedony or plasma, usually with a uniform green or grayish green color, and subtranslucent to opaque, that contains red spots of iron oxide or red jasper resembling spots of blood. They are also found with yellowish or brownish spots, and this variety is less highly valued. Bloodstone has been obtained commercially on the Kathiawar Peninsula and elsewhere in India, and is known from Egypt, China, and other places. Specimens have been obtained in Chatham County, Georgia, and at other places in the United States. Point agate and polka-dot agate are similar but have smaller and more numerous spots or flecks of color, usually set in a translucent matrix. Agate of this type occurs at Pony Butte, Jefferson County, Oregon, and at a few other localities in the United States. The name heliotrope derives from the Greek ἥλιος, *sun*, and -τροπος, *turning*, and refers to the red appearance obtained when translucent green chalcedony with red spots is immersed in water in sunlight.

FLINT; CHERT; NOVACULITE

Flint. There is no sharp mineralogical distinction between flint and chert. The term flint is used principally for the siliceous nodules found chiefly in chalk and marly limestone, such as in the Upper Cretaceous chalk beds of western Europe and other regions. Massive bedded deposits are generally called chert. Flint is a tough, very compact microcrystalline material. The color of freshly broken nodules typically is light to dark gray; also bluish gray, brown, brownish gray to brownish black, smoky black, and black. The blackish tints have been attributed to carbonaceous matter.[1] They may be optical effects, however, caused by a condition of the surface in which light is absorbed with very little reflection, analogous to the black color of finely divided metals or of a coplanar surface of fine steel needle points directed toward the light.[2] In transmitted light, flint shows a brownish color similar to that in chalcedony (which see). In shape, flint nodules generally tend to be elliptical or discoidal, with the plane of flattening parallel to the bedding, or are irregular with knobby, nodular, or tuberose projections. The size usually is in the range from a few inches up to a foot or so, but nodules several feet in diameter have been found; flint occurs also as anastamosing aggregates or veinlets.

The surface of weathered nodules and worked objects of flint may be chalky white in appearance,[3] sometimes stained yellowish brown or brown by iron oxide. The surface layer or patina is porous, and gives the x-ray diffraction pattern of quartz. The white color is caused by the scattering of light by the fine particles and voids. The material becomes identical in appearance with the core when the interstices are filled with a liquid with an index of refraction near those of quartz. The patination can be produced experimentally in alkaline solutions, which penetrate and dissolve the quartz along grain boundaries and cracks. The presence of admixed opaline or amorphous silica has been claimed in chert and flint, as in ordinary chalcedony and agate, on the basis of leaching experiments in alkaline solutions (see under *Chalcedony*). Although opaline silica occurs in bedded sedimentary deposits formed by the accumulation of the siliceous tests of organisms such as radiolaria, or of sponge spicules, there is no direct evidence of interstitial opaline silica as a general condition in flint and chert. The depth of the natural patina is a function of the fiber size and texture of the flint, the composition and concentration of the solvent responsible for the attack, the duration of the solvent action, etc.

Flint generally breaks with a broad conchoidal fracture, giving clean and smooth surfaces and sharp cutting edges, but all gradations occur to the typical splintery or irregular fracture of chert. The experience of professional flint knappers is that flint nodules have a grain that permits easy flaking in certain directions. Freshly mined nodules are said to be relatively easy to crush. The luster of broken surfaces is dull to faintly waxy. Very thin splinters and edges are translucent; larger pieces are subtranslucent to virtually opaque. Under the microscope, flint commonly shows spherulitic, fan-shaped, and crudely rectangular tapering or irregular areas of fibrous chalcedony with the usual negative optical elongation, at least in part representing sections through replaced organic remains, together with a fine-grained matrix of crenulated and interlocking units that range down in size to submicroscopic dimensions. These units, which may resemble quartz grains, apparently are subparallel bundles of fibers and show undulose extinction. The indices of refraction of the fibers and of the aggregates, so far as can be determined, are comparable to those seen in chalcedony and are less than those of coarsely crystallized quartz. The density of flint also is comparable to that of chalcedony and less than that of ordinary quartz. Banding on a microscopic scale with a parallel alignment of the fibers, such as seen in agate, is not a feature of flint or of chert, which in general has a random arrangement in the matrix. Flint, like chert, often contains organic remains, chiefly sponge spicules but also the silicified tests of foraminifera, fragments of lamellibranchs, brachiopods, etc. These areas are generally made up of relatively coarsely fibrous chalcedony with a parallel or radial arrangement. Rhombohedral outlines of silicified calcite crystals may be observed, together with unreplaced calcite or dolomite or fragments of calcitic fossil material.

Chemically, flint and chert consist essentially of silica, usually in the range 97–99 per cent SiO_2 after physically admixed calcite or other material is deducted, with the remainder water or ignition loss in the range from a few tenths of a per cent up to 1 per cent or so, and usually small amounts of Al_2O_3, Fe_2O_3, and alkalies. The CaO, MgO, and CO_2 often reported are generally present as unreplaced calcite or dolomite. The FeO reported in some analyses probably is present in these carbonates, and also in glauconite, pyrite, etc. Typical chemical analyses are cited in Table 29. The flint nodules in chalk and related rocks have formed by replacement, probably in most instances during the diagenesis of the sediment.

Chert. The color of chert usually is grayish white, gray or pale bluish gray, occasionally dark gray to black, greenish, yellowish, reddish, or brownish. The color of flint generally is darker than that of chert. With increasing admixture of iron oxide, chert becomes shades of yellow, brown, or red in color and grades into jaspery chert and jasper. Ferruginous cherts of this nature are developed on a very large scale in some regions, and are often associated with pillow lavas and greenstones derived from spilitic rocks in areas of volcanic activity. Irregular color mottling is often observed in chert, and to a certain extent in flint, and a rude banding parallel to the bedding may be observed. Chert has a dull chalky to vitreous luster and is translucent on thin edges and virtually opaque in large masses. Surfaces of chert as seen under the electron microscope may show a spongy texture or exhibit minute pores or holes.

Under the microscope, chert shows the same general features as flint but tends toward a coarser grain size. The principal and usually the only constituent in chert and flint is chalcedonic (fibrous) silica. Randomly interlocking grains of microcrystalline granular quartz similar to that constituting novaculite also may be observed. Some chert and flint of extremely small grain size appear isotropic under the microscope because of aggregate polarization. Microscopic isotropic globules of opaline silica either isolated or in chains or in dense aggregates occur in some chert. Unreplaced remains of calcitic fossils, together with disseminated glauconite, pyrite, chlorite,[4] and clayey material, are often observed. The best burrstones or millstones are obtained from massive forms of chert with a coarse cellular structure, produced by the dissolving out of calcareous fossils; a superior type is obtained from the fresh-water Oligocene cherts of the Paris basin, France.

Chert occurs as bedded deposits, often of great lateral extent, and as discontinuous lenses; the beds usually range from a few inches up to 10 ft. or so in thickness, but sometimes considerably more, and are enclosed in or are interstratified with chalk, limestone, or dolomite. Bedded chert and flint nodules are found in sedimentary rocks of all ages from Proterozoic to Tertiary. The calcareous Paleozoic rocks of North America, particularly in the midcontinent region, are rich in chert; also the Franciscan formation of Jurassic age in California; the Jurassic and Cretaceous chalk and limestone of western Europe; etc. The bedded chert deposits in general appear to have formed[5] by replacement of limestone or chalk during the diagenesis of the rock, although direct inorganic chemical precipitation from sea water and the recrystallization of the opaline silica of rocks consisting of the remains of siliceous organisms such as radiolaria, diatoms, and sponge spicules are additional factors.

Novaculite. Novaculite is a white rock of uniform grain size and without lamination that is wholly composed of microgranular quartz. Under the optical microscope or electron microscope novaculite shows a randomly oriented aggregate of sharply defined polyhedral blocks or grains of quartz with smooth, slightly curved surfaces, somewhat resembling the air cells in foam. The rock may be a product of the low-grade metamorphism of chert beds. The well-known novaculite of Arkansas is of Devonian-Mississipian age.

Historical. The origins of the names flint and chert are not known. Flint, also written flynt and flynte in Old English, may be of local English origin and has been traced back to ca. A.D. 1000. The name possibly is cognate with the Greek *plinthos, a tile* or *brick*. In early writings, and particularly in poetic and literary usage, the name often is employed for hard stone in general. The mineralogical nature of the material first appears in seventeenth century English works. Nicols (1652) said[6] flint or silex was grayish, firmly compacted, without pores and black within, and equated it with the German *kieselstein*. Webster (1661) spoke[7] of ". . . Saxum corneum, or Flint, Hornstein, which appeareth most hard, of the color of Horn, whose parts are so continuate, that one cannot discern one from another . . . ." Under the name *flinta* the Swedish mineralogist Wallerius (1747) gave[8] a clear description of flint nodules, which he noted were commonly found loose on the ground or in chalk. The name flint is not found in most seventeenth and eighteenth century German mineralogical works, where flint fell within the meaning of *hornstein* and, more specifically, of *feuerstein*. The *ostacius* of Pliny, a hard material used for the points of weapons, may have been flint.

Flint, because of the sharp cutting edge to which it breaks, was invaluable to early man as a weapon and tool. It has been used since early times to strike sparks with steel. Flint was long employed as an ignition agent in guns. A modification of the original wheel-lock gun (invented in 1517) that used flint instead of pyrite as a source of sparks came into general use in Europe about the middle of the sixteenth century. The flint lock, a further modification that was highly developed in England, reached its widest use about the beginning of the nineteenth century. The preparation of gun flints was a specialized art;[13] a good workman could finish 1000 in 3 days. An important local source of flint since earliest times has been at Brandon in Suffolk, England. Flint nodules from the chalk of western Europe have long been used as an ingredient of ceramic bodies, after calcining to cristobalite, and flint once was an ingredient in the manufacture of certain glass (flint glass).

Chert, perhaps originally chirt, is believed to be a local English term that was taken into geological use. It may be of onomatopoeic origin. The name chert may be of more recent origin than flint and, unlike flint, is not found in literary usage. It was well established in meaning in 1679, when a reference appeared[9] to "beds of Chirts," and in 1729 Martyn[10] stated that "Chert, this is a kind of Flint . . . called so, when it is found in thin Strata." The *Derbyshire Miners' Glossary* (1824)[11] lists chirt and states that it is petrosilex, found in mountain [Carboniferous] limestone, and when ground fine is used with flint and clay in making Staffordshire pottery ware. Sowerby (1817) said, "Hornstone or chert is found in many parts of Great Britain . . . ." J. D. Dana in 1844 stated,[12] "Chert is a term often applied to hornstone, and to any impure flinty rock, including the jaspers." Later usage has been quite varied. An occurrence in bedded form would seem to be the essential point of difference from the nodular flint.

The name hornstone (hornstein) probably originated with the Saxon miners in the Middle Ages. In its early meaning it included various fine-grained or dense types of quartz, more or less translucent, with a resemblance in their splintery fracture and general appearance to horn. Hornstein was mentioned in various sixteenth century German mineralogical works and long had general use but has since been largely supplanted, especially in English, by other terms of narrower meaning. The English flint, as a mineral or rock name, corresponds to the French *silex* and the German *hornstein* in part; as a material for striking sparks, it is the French *pierre à feu, pierre à fusil*, or *silex pyromaque*, and the German *feuerstein*. Chert, used chiefly in English-speaking countries, corresponds to the French *silex* and *silex de la craie* and is included in the German *hornstein*.

The name novaculite is derived from the Latin *novacula, razor* or *sharp knife*, and alludes to the use of this material for whetstones and razor hones.

References

1. Tarr: *Univ. Missouri Stud.*, **1**, no. 2, 1926; Heintz: *Ann. Phys.*, **60**, 519 (1843); Wöhler and Kraatz-Koschlau: *Min. Mitt.*, **18**, 464 (1899).
2. Wood: *Physical Optics*, New York, third ed., 1934, p. 106.
3. Moir: *Sci. Progress*, **20**, 249 (1925); Gehrcke: *Phys. Zs.*, **31**, 970 (1930); Heinz: *Chem. der Erde*, **4**, 501 (1930); Schmalz: *Proc. Prehistoric Soc.*, **26**, 44 (1960); Hurst and Kelly: Science, **134**, 251 (1961); Cayeux: *Bull. soc. min.*, **53**, 60 (1930).
4. Sudo: *Japan. J. Geol. Geog.*, **23**, 109 (1953).
5. On the origin of chert and flint and on the source of the silica see reviews in Pettijohn: *Sedimentary Rocks*, second ed., New York, 1957; Symposium on Silica in Sediments, *Soc. Econ. Paleo. and Min. Spec. Publ.* **7**, 1959.

6. Nicols: *A Lapidary* . . ., London, 1652, p. 84.
7. Webster: *Metallographia* . . ., London, 1661, p. 354.
8. Wallerius: *Mineralogia, eller Mineral-Riket.*, Stockholm, 1747, p. 80.
9. See under "Flint" in Oxford Dictionary, Oxford, **2**, 1893.
10. Martyn: *Phil. Trans.*, **36**, 30 (1729).
11. Mander: *Derbyshire Miners' Glossary*, Bakewell, 1824, p. 15.
12. Dana: *System of Mineralogy*, New York, second ed., 1844, p. 411 (also in later editions, to sixth).
13. Ure: *Dictionary of Arts, Manufact., and Mines*, New York, **1**, 1853.

JASPER

Jasper is a type of massive, fine-grained, or dense quartz that contains relatively large amounts of admixed material, chiefly iron oxide, up to 20 per cent or more. It typically has a dark tile-red to dark brownish red color, also brown, brownish yellow, or ocher-yellow; less commonly grayish green, green, dull blue, brownish black, or black. Jasper often is variegated, in spotted, cloudy, and especially banded types. The bands generally are broader than in agate and are usually planar rather than concentric. The coloring material in reddish types of jasper is finely divided hematite, and in the brown and yellow types, goethite. Clay is also a coloring substance, affording white, yellowish, or gray material, often with a porcelaneous appearance, and chlorite. The fracture of jasper generally is smooth and even, grading to flat-conchoidal. The material is opaque or virtually so, and the luster is dull.

With decrease in the amount of foreign pigmenting substance jasper grades into translucent fine-grained material which may be variously termed chert, hornstone, novaculite, etc. The color of this material usually ranges through gray-black, gray, pale yellowish, brownish, and reddish brown shades to nearly white. The microstructure generally is fine granular, especially in metamorphic types, and may have been produced by the recrystallization of originally fibrous or spherulitic material.[1] More or less large amounts of fibrous silica, often as spherulitic or confused aggregates, or as small bundles which simulate anhedral quartz grains, also may be present in jasper, and in this respect jasper grades into chalcedony and its subvarieties. Banding on a microscopic scale with the fiber length perpendicular to the bands, a feature typical of chalcedony, is not characteristic of jasper, although a coarse color layering often is present. The name jasper is based on gross features rather than microstructure, primarily on the color, the near opacity, and the dull luster, and an even or smooth rather than splintery fracture, in hand with a relatively large content of admixed foreign material, chiefly iron oxide. Unlike agate and other types of chalcedony, most of which occur on a small scale or as local deposits, jasper occurs in part as very extensive beds or formations of sedimentary or metamorphic origin, affording large blocks of material, and to some extent this has been a further factor in the application of the name.

Jasper may be traversed by veinlets of light-colored chalcedony or occur as a breccia cemented by other varieties of quartz. It may contain irregular areas or spherules of colorless quartz crystals, or exhibit concentrically banded and variously pigmented radial growths of fibrous silica. Jaspilite is a metamorphic rock composed of alternating layers of jasper with black or reddish fine-grained hematite that occurs in the Precambrian iron formation of the Lake Superior region, in India, and in other localities. Itabirite is a finely laminated or schistose quartz-hematite rock containing relatively coarse hematite crystals. The term jasponyx has been applied to thin-banded jasper containing alternate light or dark bands; band jasper and ribbon jasper are similar. Lydian-stone, also called touchstone or basanite, is a velvety black jasper that has been used for thousands of years in testing the color of the streak of gold alloys as a measure of the gold content. Fine-grained black igneous rocks and slate also were used for the purpose. A skilled person can by this method obtain under favorable circumstances a precision of about 1 part in 100 in estimating the gold content of a gold-silver alloy.[2]

Occurrence. Jasper is a very common and widely dispersed material. In part it occurs like chalcedony, or with it, as a filling of cavities and veins and as crusts. It occurs as nodules or veinlets in iron ores or in altered igneous rocks and as a silicification along the contact of igneous rocks. Waterworn pebbles and cobbles of jasper are common in detrital deposits. Jasper also occurs as bedded deposits on a large scale. Variegated red to brown and yellow cloudy jasper is a common petrifying material of wood. Green jasper and banded green and red jasper suitable for ornamental purposes are found abundantly at localities in the sourthern Ural Mountains and in the Altai Mountains, Russia.[3] Pebbles of brown and of red jasper are common in the stony areas of the north African deserts, and similarly in other parts of the world. Jasper also has been obtained commercially in Sicily, northern Bohemia,[4] Germany, India, etc.

Very numerous localities for jasper of ornamental quality are known in the United States. A bed of yellow, red, and white banded jasper occurs at Collyer, Trego County, Kansas, and banded jasper occurs at Brackettville, Kinney County, and elsewhere in Texas. A red orbicular type of jasper occurs near Cave Creek, Maricopa County, Arizona. Chert or jasper containing red, brown, orange, and yellow spherules, often banded, and ranging up to an inch or so in size, occurs at Morgan Hill, Santa Clara County, and at various localities in the San Francisco Bay area, California. A red jasper occurs near Burlington, Chittenden County, Vermont, and jasper occurs with agate at Diamond Hill, Cumberland County, Rhode Island. Numerous minor localities are known in New York, Pennsylvania, Maryland, and other eastern states. Conglomerates containing pebbles of jasper and other colored varieties of fine-grained quartz may become cemented into a hard, compact rock and when cut and polished make attractive ornamental stones, such as the Sioux Falls jasper of Minnehaha County, South Dakota. Jasper has been extensively employed since ancient times as an ornamental material, chiefly in mosaics and for carving vases, dishes, and other objects.

Historical. The name jasper as applied to the opaque, dull types of fine-grained quartz with a red, reddish brown, yellow, or green color is a modern usage, and appears in this sense in some texts of the eighteenth and early nineteenth centuries. The name itself derives from the *iaspis* of the ancients, but this term was then applied chiefly to translucent to semi-transparent stones that included all bright-colored chalcedony, especially that with a bluish green or green color but also with yellowish, smoky, blue, and rose tints. Agricola also used jasper in this sense, in the sixteenth century, but he included material that was opaque and found in very large masses and that corresponds to the modern jasper. Jasper and agate were used in a more or less generic sense by some eighteenth century writers, the opaque or impure varieties of fine-grained quartz being classed under jasper, and the pellucid types, such as chalcedony and opal, as agate. The opacity or near opacity of the material has continued to be one of its defining characteristics, but sharp boundaries do not exist in any regard between this and other types of fine-grained quartz. It has been suggested that the *haimatitis* and *xanthe* of Theophrastus corresponded to our red and yellow jasper, respectively, and that the *haematitis* of Pliny (his *haematites* being iron oxide) also was red jasper. In any case the material known today as jasper was extensively employed in antiquity.

The word jasper, with the Latin *iaspis*, is of oriental origin corresponding to the Persian *iashm* and *jashp* and the Assyrian *ashpu*.

References

1. Dunn: *Econ. Geol.*, **48**, 58, 729 (1953); Spencer and Percival: *Econ. Geol.*, **47**, 365 (1952).
2. Gowland: *J. Inst. Metals*, **4**, 11 (1910); Steuer: *Min. Mitt.*, **21**, 357 (1902); Rose: *Metallurgy of Gold*, London, 1915, p. 554.
3. Fersman: *Precious and Colored Stones of Russia*, Petrograd, 1922 (*Mon. Russ. Ac. Sci.*, **1**, no. 3, 280).
4. Slavík: *Věst. král. České spol. nauk*, no. 16, 1942, and no. 16, 1946.

INCLUSIONS IN QUARTZ

The inclusions in quartz are of two main types: liquid, comprising solution entrapped either during the initial growth of the crystal or at a later period in (healed) cracks; and solid, comprising crystals of other species either mechanically enclosed during the growth of the quartz crystal or formed later by exsolution.

First then on this occasion I remember, that a very ingenious and qualify'd Lady who had accompany'd her Husband in an Embassy to a great Monarch, assur'd me, that she brought thence among several Rich Presents and other Rarities (some whereof she shew'd me) a piece of Christal, in the midd'st of which there was a drop of Water, which by its motion might be very easily observ'd . . . . (Robert Boyle, *Essay about the Origine and Virtue of Gems*, London, 1672)

LIQUID INCLUSIONS

Liquid inclusions range in size from microscopic dimensions up to sizable cavities that may contain a milliliter of more of liquid. Pores or cavities that range from microscopic dimensions up to a size barely visible to the unaided eye are virtually always present in quartz. Sometimes they are so abundant as to give the quartz a turbid or white appearance. They may be distributed uniformly but more commonly tend to be localized along irregular or wavy zones. Irregular planes of liquid inclusions may cross adjoining quartz grains in an aggregate without break or offset, as in metamorphic and igneous rocks[1] and in fine-grained vein quartz. The planes may have a high degree of preferred orientation, and it is believed that they represent shear fractures penetrated by intergranular liquid and then healed by deposition of silica. Whitish, irregular, or curved planes or zones of bubbles occur in transparent euhedral crystals of quartz, especially toward the base of the crystal, and are referred to as veils in the commercial grading of quartz. The liquid inclusions in euhedral crystals also may be distributed in planar arrangement parallel to the external crystal faces. The small pores and cavities usually have a slightly irregular or rounded shape, but may be bounded by crystal faces (negative crystals). Larger cavities tend to be irregular and more or less flattened, often parallel to the rhombohedral faces, and larger openings may be joined by small irregular channels. Although liquid inclusions are found in all types of quartz, they probably are more common and larger in size in the quartz of pegmatites and of quartz veins.

The inclusions ordinarily are composed of two phases, a liquid and a gas. Others contain three phases, a liquid plus a solid (usually present as a single crystal moving freely in the liquid phase) and a gas bubble, or two immiscible liquids plus a gas bubble. A crystal also may be present in the latter case, giving a four-phase inclusion.

The chemical composition of the inclusions is very poorly known in detail. A few analyses have been made directly on the liquid and gaseous components, and additional analyses have been reported of bulk samples of quartz in which the minor constituents that were found probably came from liquid inclusions. The range of composition[2] in the inclusions is from nearly pure water to more or less concentrated or saturated water solutions containing chiefly Na, K, Ca, Cl, SO_4, and CO_2 and passing with increasing content of CO_2 through two-liquid inclusions, containing both a water solution and liquid CO_2, to inclusions containing chiefly liquid CO_2. A gas phase is present in all instances at room temperature. The characteristic minor constituents are NH_3, H_2S, and N. An analysis of liquid inclusions in Swiss smoky quartz gave[3]: H_2O 83.4, CO_2 9.5, CO_3 1.8, SO_3 0.5, Cl 1.6, Na 2.0, K 0.7 Li 0.2, Ca 0.3 per cent. The average content of dissolved material has been estimated in one study[7] as from 10 to 20 per cent by weight and in another study[8] as in the range 20–30 per cent. Analyses of quartz in bulk generally show H_2O, CO_2, and N in determinable amounts, and these substances doubtless came from inclusions; the small loss of weight commonly found by igniting quartz presumably is in part at least of the same origin. The $H_2O(-)$ and gases found in igneous rocks may come in part from liquid inclusions. Gas analyses of granite[6] show CO_2, CO, N, H, S, Cl, F, and H_2O. Hydrocarbons also have been reported in quartz.[22] The liquid inclusions found in the limpid quartz crystals from Herkimer County, New York, often contain a movable black speck of solid hydrocarbon. Two types of hydrocarbon have been found[5] in a Herkimer crystal, a yellow type melting at 70–80° and a brown type melting at 200–220°. A 4500 gram sample of smoky quartz from the Tiefen glacier, Uri, Switzerland, yielded 0.5–0.6 gram of ammonium carbonate.[4]

Microscopic crystals are sometimes observed suspended in the liquid phase. These are almost always cubical in shape and optically isotropic, and are believed to be an alkali halide, probably NaCl in most instances, although KCl has been reported. They dissolve when the inclusion is heated. Inclusions of anhydrite have been figured.[20]

The two-liquid inclusions show a zonal structure: an outer zone of water solution, an intermediate zone of liquid CO_2, and a central bubble essentially composed of gaseous CO_2. The phenomena obtained on heating depend on the relative volumes of these phases,[9] the boundary between the liquid and gaseous CO_2 phases in some circumstances disappearing at temperatures below the critical point of CO_2 (31°). Certain specimens of massive smoky quartz from the pegmatite at Branchville, Connecticut, are remarkable for the great abundance of three-phase inclusions containing liquid and gaseous CO_2. When struck with a hammer the material explodes with a noise, and when heated it decrepitates with such violence that bits may fly through the air to a considerable distance. Analysis of the gas from these inclusions gave:[10] CO_2 98.33, N 1.67 per cent, plus traces of H_2S, SO_2, and NH_3; small amounts of Na and Cl, presumably present as NaCl, since small cubical crystals were observed in some of the cavities, and a hydrocarbon also were detected. The gas collected amounted in different specimens to from 0.97 to 1.65 times the volume of the quartz. The quartz has a relatively low specific gravity and shows a well-developed cleavage on $\{10\bar{1}1\}$.

A find of colorless to smoky quartz crystals containing exceptionally large inclusions of liquid CO_2 was made in 1882 in a quartz vein containing emerald, hiddenite, rutile, and muscovite in Alexander County, North Carolina.[11]

About a half-ton of terminated crystals was found loose in clay at the bottom of an opening about 3 ft. wide and 7 ft. deep. Most of the crystals ranged up to about 6 in. in length and weighed a pound or two, but one weighed 25 lb. The crystals contained large numbers of irregular or flattened cavities, mostly about $\frac{1}{4}$–1 in. in size but ranging up to $2\frac{1}{2}$ by $\frac{1}{4}$ in. The external rhombohedral and prism faces on the crystals were depressed and cavernous, and the liquid cavities were arranged roughly parallel to these surfaces. Some crystals were mere skeletons with sufficient internal partitions to retain the liquid. Crystals were noted visually to contain two liquids and a movable bubble that disappeared on slight warming. When originally found the crystals were stored in a log cabin, but the temperature dropped below freezing during the night and almost all of them exploded with sharp reports. Some of the crystals remained as coherent masses of fragments cemented by ice. Unbroken crystals taken into the sunlight were heard to make a noise likened to boiling and audible for a distance of several feet. Experimental freezing of CO_2 inclusions shows supercooling of the water.

Cavities containing liquid and gas bubbles in quartz and other minerals were described by Davy (1822)[12] and Brewster (1826 and later),[13] and they were studied in detail from a genetic point of view by Sorby (1858).[14] A review of the extensive literature is available.[16] The inclusions are useful in geologic thermometry,[15] under the assumption that the cavities were just filled with liquid at the temperature and pressure of formation. The degree of filling or size of the bubble at room temperature is determined by these conditions and by the composition of the solution. If the inclusions are heated the liquid will expand and ultimately fill the cavity: this is the temperature of formation if the original pressure did not significantly exceed the vapor pressure of the solution. If the pressure was greater, then on reheating the bubble would disappear at a temperature lower than that of formation. Corrections to be added to the observed temperature of filling to give the temperature of formation[17] can be calculated from the known PVT relations of water and from the pressure estimated from geologic evidence to have obtained at the time of formation. A source of error is in the departure of the chemical composition of the inclusions from pure water. Leakage from the inclusions and the sometimes uncertain distinction between primary and secondary inclusions[18] are added factors. The temperature of filling of the cavities can be measured optically. The beginning of decrepitation during heating also can be employed.[19] Measurements of the temperature of formation of quartz crystals from different localities[20] on the basis of fluid inclusions have given values in the range from 82° to 530°. The direction of flow of mineralizing solutions also can be investigated by study of the inclusions.[21]

References

1. Cf. Tuttle: *J. Geol.*, **57**, 331 (1949); Dale: *U.S. Geol. Surv. Bull.* **738**, 17, 1923.
2. Koenigsberger: *Min. Mitt.*, **19**, 149 (1900) and *Jb. Min.*, Beil. Bd. **14**, 533 (1901);

230 QUARTZ

Newhouse: *Econ. Geol.*, **27**, 421 (1932); Farber: *Danmarks Geol. Unters.*, ser. 2, no. 67, 1941; Zirkel: *Jb. Min.* 801, 1870, and 802, 1890; Vogelsang: *Ann. Phys.*, **137**, 56, 69, 265 (1869); Harrington: *Am. J. Sci.*, **19**, 345 (1905); Hartley: *J. Chem. Soc. London*, **31**, 241 (1877) and **30**, 237 (1876); Johnson: *Sitzber. bayer. Ak. Wiss.*, *Math. nat. Kl.*, 321, 1920; Holden: *Am. Min.*, **10**, 203 (1925); Travers: *Proc. Roy. Soc. London*, **64**, 30 (1898); Warburg and Tegetmeier: *Ann. Phys.*, **35**, 455 (1888) and **41**, 18 (1890); Salm-Horstmar: *Jb.*, *Min.*, 54, 1853; Karpinsky: *Gornyi J.*, **2**, 96 (1880); Nacken: *Cbl. Min.*, 12, 35, 1921; Lindgren: *U.S. Geol. Surv.*, *17th Ann. Rpt.*, Pt. 2, 130, 1895; Laemmlein and Klevtsov: *Mem. soc. russe min.*, **84**, 47 (1955); Inshin: *Chem. Abs.*, **53**, 21,457 (1959); Roedder: *Econ. Geol.*, **53**, 235 (1958).

3. Koenigsberger and Müller: *Zbl. Min.*, 76, 1906.
4. Forster: *Ann. Phys.*, **143**, 177 (1871); Wöhler and Kraatz-Koschlau: *Min. Mitt.*, **18**, 304 (1899).
5. Keith and Tuttle: *Am. J. Sci.*, Bowen Vol., 244, 1952; see also Dunn and Fisher: *Am. J. Sci.*, **252**, 489 (1954).
6. Shepherd: *Am. J. Sci.*, **35A**, 326 (1938); Tilden: *Trans. Roy. Soc. London*, **60**, 453 (1897); Chamberlain: *The Gases in Rocks*, Carnegie Inst., Washington, No. 106, 1908.
7. Farber (1941).
8. Roedder (1958).
9. Hawes: *Am. J. Sci.*, **21**, 203 (1881).
10. Wright: *Am. J. Sci.*, **21**, 209 (1881).
11. Hidden: *Trans. New York Ac. Sci.*, **1**, 131 (1882); Taber: *J. Geol.*, **58**, 37 (1950).
12. Davy: *Phil. Trans.*, 367, 1822.
13. Brewster: *Trans. Roy. Soc. Edinburgh*, **10**, 1 (1826), **16**, 7 (1845), **20**, 547 (1853), and **23**, 39 (1862); *Trans. Geol. Soc. London*, **3**, 455 (1835).
14. Sorby: *Quart. J. Geol. Soc. London*, **14**, Pt. 1, 453 (1858); *Monthly Microscop. J.*, **1**, 220 (1869).
15. F. G. Smith: *Historical Development of Inclusion Thermometry*, Toronto, 1953; Ingerson: *Am. Min.*, **32**, 375 (1947) and *Econ. Geol.*, **50**, 341 (1955); Cameron, Rowe, and Weis: *Am. Min.*, **38**, 218 (1953); Correns: *Geol. Rundschau*, **42**, 19 (1953); Deicha: *Les lacunes des cristaux leurs inclusions fluides*, Paris, 1955.
16. Smith (1953).
17. Kennedy: *Am. J. Sci.*, **248**, 540 (1950) and *Econ. Geol.*, **45**, 533 (1950).
18. Laemmlein: *Zs. Kr.*, **71**, 237 (1929); Cameron *et al.*, (1953).
19. Scott: *Econ. Geol.*, **43**, 637 (1948); Smith and Peach: *Econ. Geol.*, **44**, 449 (1949); Peach: *Am. Min.*, **34**, 413 (1949) and *J. Geol.*, **59**, 32 (1951); Koizumi and Hosomi: *J. Geol. Soc. Japan*, **59**, 478 (1953).
20. See Ingerson (1947); Cameron *et al.* (1953); Konta: *Rozpravy Česko. ak.*, Cl. II, **60**, no. 13 (1950).
21. Smith: *Econ. Geol.*, **45**, 62 (1950) and **49**, 530 (1954).
22. Sjögren: *Geol. För. Förh.*, **27**, 113 (1905); Reese: *J. Am. Chem. Soc.*, **20**, 795 (1898); Tschermak: *Min. Mitt.*, **22**, 197 (1903); Holden: *Am. Min.*, **10**, 203 (1921); Brewster (ref. 13).

ENHYDROS

This name is given to thin-walled chalcedony geodes that contain entrapped water, usually with a large air space. The content of water is apparent either visually, through the translucent shell, or by its motion when the geode is tilted. Most examples soon dry out after they have been taken from the enclosing rock. Because of the relatively low temperatures and pressures at which these geodes form, the large and rather variable air space, which may amount to half or more of the volume, as seen in museum specimens, must be largely or entirely caused by partial evaporation of the liquid. The best examples of enhydros are known from the agate-producing regions of Uruguay and Brazil. Among other localities, they are known from Monte Tondo near Vicenza, Italy, from New South Wales, and in the United States from near Yachats in Lincoln County, Oregon, and from Pescadero Beach, San Mateo County, California. Although most examples are only a few inches in size, specimens are known containing several hundred milliters of water, and dry or broken chalcedony geodes several feet in diameter have been found. The original *enhydros* or *enhygros* of ancient writers has been interpreted[1] as a reference to quartz crystals containing fluid inclusions, but the descriptions clearly indicate a type of geode. Agricola spoke of enhydros as round, smooth, and white, with a liquid inside that sways back and forth when moved, and that may even drip water.

Hollow polyhedral casts are often observed that have formed by the deposition of a layer of quartz on the sides of a cavity bounded by flat, intersecting surfaces of other crystals, such as platy crystals of calcite. The casts may be drusy inwards. One example[2] from a pegmatite in Swaziland, Africa, contained virtually pure water with a gas composed of N 75, O 22, and CO_2 3 per cent under 14 atm. pressure. Another example from Victoria, Australia, contained Na, Mg, and Ca sulfate and chloride in solution.[3] Similar occurrences of hollow quartz casts containing liquid are known from Iredell County and from Rutherfordton, Rutherford County, North Carolina. Empty or solid casts are known from a number of other localities.

References

1. Moore: *Ancient Mineralogy*, New York, 1859, p. 190.
2. Mountain: *Trans. Roy. Soc. South Africa*, **29**, 1 (1942).
3. Liversidge: *Rec. Australian Mus.*, **2**, 1892.

Crystal is subject to numerous defects, sometimes presenting a rough, solder-like substance, or else clouded by spots upon it; while occasionally it contains some hidden humor within, or is traversed by hard and brittle inclusions . . . while, in other instances, it contains filaments that look like flaws . . . . (Pliny, *Natural History*, Book 37, Ch. 11, A.D. 77)

SOLID INCLUSIONS

Quartz frequently encloses solid foreign material of various kinds.[1] This section describes those instances in which macroscopic inclusions are present either in single crystals of quartz or in relatively coarse-grained massive varieties. The fine-grained and microcrystalline varieties of quartz are often admixed with finely dispersed foreign material and are described elsewhere.

The best-known and most highly prized type of inclusion in quartz doubtless is rutile. The rutile occurs as long thin needles or hairs, sometimes straight but more often gently curved, with a reddish brown to yellow color. The needles or hairs range up to 10 in. or more in length. They may be randomly arranged or, more commonly, occur as rather open divergent sprays or subparallel groups. They may be so abundant as to render the quartz nearly opaque, giving the whole a reddish brown color, but usually are widely dispersed. Fine specimens may be sawn or rounded off and polished to display the inclusions (Figs. 96, 97, and 98). The surface of the inclusions may have a silvery luster, caused by an air gap at the boundary with the quartz, and bubbles or particles of foreign material may be seen adhering to the rutile needles. The finest specimens of rutilated quartz have been obtained from localities in Minas Gerais and other states in Brazil, and from Madagascar, but good examples although usually of smaller size are of worldwide occurrence. Specimens have been found of quartz crystals containing inclusions of platy hematite crystals from which radiate in parallel sets at about 60° rutile needles and hairs oriented on {0001} of the hematite. Exceptional specimens were found in the period 1830–1850 as glacial boulders of transparent colorless to smoky quartz containing hair-like and needle-like rutile crystals in the vicinity of Hanover, New Hampshire, and also at places in northern Vermont.

During the eighteenth century polished specimens of rutilated quartz, also called Venus' hair-stone, enjoyed a high vogue in France and England and sold at extremely high prices. The material has been known since ancient times, and apparently constitutes the *chrysothrix* or golden hair of the Orphic poem *Lithica*, probably written in the fourth century A.D., and also the *Veneris crines* of Agricola (1546), described as containing red hair-like inclusions. Small specimens are sometimes cut as cabochon stones. Most examples are found in colorless and transparent quartz, but identical inclusions also occur in smoky quartz. A broken disc-like mass of transparent smoky quartz about 14 in. in diameter and 6 in. thick containing a great number of glistening golden brown rutile needles was found in England in 1942 in a shipment of commercial quartz from Brazil. Amethyst less commonly contains rutile enclosures and then generally of small size. Amethyst more typically contains small acicular enclosures of goethite,[2] which are easily mistaken for rutile, and also hematite. Rutile is sometimes present in quartz as twinned, reticulated or net-like aggregates, and is then called sagenite (from σαγήνη, a net). The twinned net-works more often are found implanted on the surface of the crystals, or only partly enclosed, and are not oriented to the quartz. Extended tree-like twinned intergrowths also may be observed. The rutile of rutilated quartz has been mechanically enclosed during the growth of the quartz crystal. Specimens of acicular and hair-like rutile crystals that are not embedded in quartz crystals are surprisingly rare. Rutile also occurs as oriented microscopic needle-like inclusions in rose quartz and blue quartz, where it has formed by exsolution.

FIG. 96. Rutile needles enclosed in quartz crystal. Minas Gerais, Brazil. Polished section, approximately 6 × 9 in. Harvard coll.

FIG. 97. Rutile inclusions in quartz. Switzerland. U.S. Nat. Mus. coll.

Tourmaline frequently occurs as acicular inclusions in both colorless and smoky quartz, particularly in crystals from pegmatites. When the inclusions are very small and numerous, the quartz may appear black or bluish black.[3] The color of the tourmaline inclusions also includes dark green, pale green, bluish green, and pink shades. The identity of the inclusions is not always apparent at sight, and there is the possibility of confusion with greenish acicular inclusions of actinolite and other minerals. The tourmaline inclusions range in size from minute needles or asbestiform aggregates up to coarse prismatic crystals which may protrude from the surface of the quartz crystal. Fine examples of tourmaline inclusions are found in the pegmatites of New England and in Minas Gerais, Brazil. Zoned smoky quartz crystals containing hair-like crystals and wispy aggregates of tourmaline are found in a pegmatite in southwestern Jefferson County, Montana. The crystals, sometimes black and opaque in the outer parts because of the tourmaline, often show cappings or sceptre-like overgrowths of amethyst and make handsome specimens. Tourmaline crystals and radial groups of small crystals also occur embedded in massive granular quartz at numerous localities.

Acicular or fibrous crystals of an amphibole variously called actinolite, hornblende, or byssolite but usually without specific identification are widespread as inclusions both in single crystals of quartz and in the granular quartz of veins and of metamorphic rocks. Some of the finest examples are found in the transparent quartz crystals of the Alpine veins of Switzerland, but scarcely inferior specimens are widespread. In the United States good examples have been found at various places in North Carolina and in Upper Providence township, Delaware County, Pennsylvania. The inclusions usually are green in color and under magnification may exhibit the prism outline of amphibole. The fine fibrous and asbestiform types tend toward a light green or grayish green color and may be so abundant as to render the quartz crystal virtually opaque. A common feature of these inclusions is a tendency to be confined to particular stages of growth of the quartz or to one side of the crystal. Amphibole and other types of inclusions in quartz crystals may be dissolved out, leaving hollow molds. Translucent to milk-white granular quartz containing black needles of hornblende was formerly obtained for

FIG. 98. Quartz crystal, showing phantoms and inclusions of rutile. Minas Gerais, Brazil. Crystal approximately 6 in. in height. Harvard coll.

ornamental purposes on Calumet Hill near Cumberland, Rhode Island, and similar material occurs widely.

When asbestiform fibers are densely aggregated, sufficiently thin, and parallelly aligned, the quartz may reflect a milky wave of light, especially when cut as cabochons with the length of the stone perpendicular to the elongation of the fibers. These stones, called quartz cat's-eye[4] or occidental cat's-eye in distinction from the true oriental or chrysoberyl cat's-eye, which they somewhat resemble, are translucent with a rather greasy luster and a yellow, brownish, or, more commonly, a grayish green or green color. The finest examples are obtained from India and Ceylon. Gem material has been obtained from Crystal Park, Yuma County, Arizona. The fibers may be dissolved out by natural or artificial treatment, leaving hollow tubes which still afford the cat's-eye effect. The so-called tiger-eye and hawk-eye ornamental stones from South Africa are a type of quartz with a finely fibrous structure produced by the partial or complete silicification of cross-fiber veins of crocidolite. The color of tiger-eye is yellow to yellow-brown and brown, and polished stones exhibit a fine golden luster. A polished surface cut parallel to the fiber length usually exhibits a series of yellow to brown bands, the fibers in each being not quite parallel, so that tilting the stone alternates the cat's-eye effect from band to band. Tiger-eye often is cut as cabochons and is widely used in small ornamental objects of all kinds. The crocidolite in its unaltered state is blue, and when infiltrated by quartz along the fibers affords the blue material called hawks-eye. Oxidation of the crocidolite either before or after silicification produces the golden brown colors. The color also can be modified artificially. In some tiger-eye the crocidolite has been wholly destroyed, leaving hollow tubes that may be partly filled with hydrated iron oxide. A chatoyant fibrous quartz formed by the replacement of goethite has been obtained on the Cuyuna iron range, Minnesota.

Ill-defined green moss-like inclusions found in quartz crystals from Brazil and other localities probably are a very finely fibrous amphibole. Good examples were found in Ashe County, North Carolina. Chlorite occasionally occurs as inclusions in quartz crystals, more commonly as thin coatings on the surface or partly embedded in the surface of the crystal. Examples are found in the Alpine veins of Switzerland and from Dauphiné, France. Goethite occurs as short needles and prismatic enclosures in quartz crystals, especially in the amethystine variety, comprising part of the material called "cupid's darts" (*flêches d'amour*, Fr.). A well-known locality is on Wolf Island in Lake Onega, Russia, and similar specimens have been found in the United States in North Carolina and at Florissant, Colorado. Good examples are cut as cabochon or heart-shaped stones.

Stibnite has been found[5] as black needles with a metallic luster in quartz crystals from veins in Nevada, Japan, and other places. A variety of other minerals has been occasionally noted as enclosures in quartz crystals, including arsenopyrite, pyrrhotite, sphalerite, stephanite, pyrite,[6] hematite (Switzerland), anatase, brookite (Arkansas and Switzerland), pyrochlore, zircon, anhydrite,[7] native copper (Lake Superior), epidote (notably Chaffee County, Colorado), hiddenite, ilmenite, magnetite, siderite, calcite, dolomite (Herkimer County, New York), topaz, sphene, and mica. The quartz may be cracked around such inclusions, because of differential strains set up on cooling, or a smoky color may be produced as a halo around radioactive inclusions. Granular quartz sometimes contains abundant disseminated crystals of dumortierite, as at Lincoln Hill and near Oreana in Pershing County, Nevada, and at Riampotsy, Madagascar;[8] the color usually ranges from rose-pink to purplish and lavender blue, yielding attractive polished slabs and cabochons.

Granular vein quartz containing grains or veinlets of native gold is sometimes cut and polished for brooches or other ornamental purposes. This material was popular in the United States in the latter nineteenth century and was supplied principally from California.[9]

The term aventurine (or avanturine) is applied to translucent granular to compact varieties of quartz enclosing glistening spangles of mica, variously silvery, reddish brown, white, or green in color, giving the aggregate a speckled or somewhat metallic sheen. Reddish brown aventurine quartz with a coppery reflection is more highly valued, and resembles the aventurine varieties of feldspar containing exsolved plates of hematite (sunstone). Other types of aventurine quartz are white and reddish white with a silvery appearance, black with white flecks, reddish yellow, etc., and some types contain disseminated plates or crystals of chlorite (?), hematite, goethite, or other species. Green aventurine containing flakes of a bright green chromium-containing mica occurs in Madras, India, and in China, where it is valued as highly as some types of jade; a similar material occurs in Rutland County, Vermont. Aventurine quartz is frequently used for vases, bowls, and other ornamental objects, and small uniform pieces with a bright uninterrupted sheen are often set as ringstones, brooches, and the like. Extensive deposits as beds in mica-schist are found at several places in the Urals, and a fine reddish white aventurine occurs near Kolivan in the Altai Mountains, Russia. This material has been worked into vases and bowls, up to 3 or 4 ft. in dimensions, and fine examples are preserved in museums in Russia and Europe. Aventurine quartz was known to the ancient Greeks as *sandastros* and was obtained from India. Aventurine quartz is simulated poorly by disseminating metallic copper in soft glass. An inclusion of quartz in diamond has been described.[11]

Euhedral quartz crystals frequently contain finely disseminated particles of clayey or ferruginous matter which, if supplied intermittently during the growth of the crystal, may outline successive shells or stages of growth (phantom quartz, ghost quartz). As many as a dozen concentric zones may be noted in a single crystal. Frequently the zones are distinct only on one side of the crystal, the foreign material being supplied unidirectionally by a current or by gravitational settling. Finely divided particles of chlorite and other species and local concentrations of liquid inclusions also may show this distribution.[10]

References

1. Early observations in Scheuchzer: *Helveticus, sive itinera per Helvetiae Alpinus regiones*, **1**, 233, 1708; Kenngott: *Sitzber. Ak. Wiss. Wien, Math. nat. Kl.*, **9**, 402 (1852); Blum, Leonhard, Seybert, and Söchting: *Die Einschlüsse von Mineralien in kristallisierten Mineralien. Natuurkund.*, Verh. Hollandsche Maatsch. Wetensch., Haarlem, 1854; Tschermak: *Min. Mitt.*, **22**, 197 (1903); Zodac: *Rocks and Min.*, **12**, 44 (1937).
2. Holden: *Am. Min.*, **10**, 203 (1925).
3. Collins: *Min. Mag.*, **1**, 115 (1877); McCarthy: *Am. Min.*, **13**, 531 (1928).
4. Wibel: *Jb. Min.*, 367, 1873; Fischer: *Min. Mitt.*, 117, 1873; Peacock: *Am. Min.*, **13**, 241 (1928); Thiesmeyer: *Am. Min.*, **22**, 701 (1937).
5. Stearns: *Am. Min.*, **20**, 59 (1935); Tschermak (1903).
6. Brech: *Nature*, **135**, 917 (1935).
7. Beaugey: *Bull. soc. min.*, **12**, 396 (1889); Kenngott: *Jb. Min.*, 301, 1859.
8. Michel-Lévy, Emberger, and Sandréa: *Bull. soc. min.*, **82**, 77 (1959).
9. Kunz: *Gems and Precious Stones of North America*, New York, 1890, p. 117.
10. Grigoriev: *C. R. ac. sci. URSS*, **44**, 198 (1944); Johnston and Butler: *Bull. Geol. Soc. Amer.* **57**, 602, 1946.
11. Colony: *Am. J. Sci.*, **5**, 400 (1923).

BLUE NEEDLES

The remarkable so-called blue needles, whose presence in colorless quartz for optical or piezoelectric applications constitutes a serious defect, also may be mentioned in this section, although these apparent inclusions actually are needle-like voids.[1] They are not dissolved-out inclusions of a foreign mineral, and their origin is unknown. The blue needles are difficult to detect, although of common occurrence, and are best sought by examining the quartz crystal in the beam of an arc-light or other intense source of light while the crystal is immersed in a tank of liquid approximating the quartz in index of refraction. The needles appear as V-shaped clusters. The V always points towards the base of euhedral crystals, never to the terminated end, and is arranged in the quartz with the plane of the V parallel to an *r* or *z* rhombohedral face, with the sides of the V parallel to the *rz* edges. The needles range up to several centimeters in length and are wholly enclosed in the quartz. In a basal section, the V's at different levels in the crystal overlap optically and cause a reticulated appearance. The bluish color is a light-scattering effect; the inclusions sometimes appear white or silvery in color. Under magnification small projections may be seen extending at right angles to the length of the needles. The needle-like voids may be the directions of intersection of parting or cleavage surfaces parallel to the *r* and *z* rhombohedra, perhaps caused by strain on cooling. In sawn sections the needle-like openings are directions of relatively rapid attack by solvents.

Reference

1. Gordon: *Am. Min.*, **30**, 284 (1945).

PSEUDOMORPHS

Quartz, including its microcrystalline varieties, has been found as pseudomorphs after many different minerals. Much of the descriptive literature, now very old, can be found in the summary work by Blum, *Die Pseudomorphosen des Mineralreichs* (1843; appendices in 1847, 1852, 1863, and 1879), and in the works on chemical geology by Roth (1879)[1] and Bischoff (1863–1866)[2]. A modern review is available.[3] Blum made collections of specimens to illustrate his works, and several of these collections are preserved in Europe and in the United States (as at Yale University). The study of the pseudomorphs formed by quartz and other minerals played a leading role in the early development of mineral chemistry and, in particular, of mineral paragenesis. Breithaupt's work of 1849, *Die Paragenese der Mineralien*, and the studies by Haidinger (1831, 1844),[4] which included an effort to classify pseudomorphs on the basis of opposite electrochemical affinities, are of interest in this connection.

Pseudomorphs of quartz after calcite are especially common. They range from solid substitution pseudomorphs without volume change after euhedral crystals, or after rock masses (limestone), to hollow incrustation pseudomorphs defined by a thin shell of superdeposited quartz. The so-called gash or hackly quartz pseudomorphs[5] found in many low-temperature hydrothermal veins are after thin and often interlaced crystals of calcite flattened on {0001}. These pseudomorphs, which appear as open gashes (molds) in massive granular or chalcedonic vein quartz, were formed by the deposition of quartz upon crystals of calcite which were then or later removed in solution. Similar pseudomorphs form after crystals of barite. Hackly box-work pseudomorphs may form by the penetration of quartz along cleavage or parting planes of calcite.[6]

Quartz pseudomorphs after fluorite[7] and barite also are very common, particularly in hydrothermal vein occurrences. Substitution pseudomorphs of quartz have been found after siderite, rhodochrosite, smithsonite, dolomite, barytocalcite, cerussite, aragonite, gypsum, anhydrite, celestite, glauberite, thenardite, anglesite, datolite, beryl, andalusite, laumontite, apophyllite, stilbite, analcite, heulandite, spodumene, feldspar, mica, various amphiboles and pyroxenes, wernerite, hemimorphite, wollastonite, apatite, pyromorphite, scheelite, corundum, garnet, mellite,[16] and other species. Incrustation pseudomorphs of quartz and its varieties have been found after pyrite, marcasite, bournonite, pyrrhotite, galena, hematite, wolframite, huebnerite, wulfenite, and other minerals.

Hollow quartz molds after anhydrite and glauberite are common at West Paterson, Great Notch, Upper Montclair, and other zeolite localities in the diabase sills of northern New Jersey.[8] The anhydrite molds are rectangular, often with thin internal partitions representing penetration along cleavage planes; the glauberite crystals are prismatic with a rhomboidal cross-section. Quartz casts after glauberite are found, and similar molds with prehnite as the incrusting mineral are common. Remarkable hollow incrustation pseudomorphs after euhedral crystals of an unidentified mineral, ranging over a foot in size, have been found in cavities in granite pegmatite at Greenwood, Maine.[9] Another pseudomorph from this locality consisted of a euhedral pollucite crystal about 10 in. along an a-axis, a cube modified by trapezohedral faces, completely replaced by quartz. Casts of quartz formed by deposition in a regular cavity left by the dissolution of an earlier-formed mineral are of common occurrence, as of quartz after halite cubes in some sedimentary rocks. Quartz pseudomorphs after cubical crystals of halite (?) have been found in fossil wood in Oregon.[15] Plane-sided but asymmetrical quartz bodies, often hollow or with a radiated internal structure, may form by the infiltration of silica into the hollow interstices between interlocking crystals of other species, such as calcite.[10] Open molds or impressions of other minerals often are found on or in crystals or masses of quartz, including anhydrite, actinolite, chlorite, calcite, fluorite, and barite, and the identity of the original mineral, which was dissolved away, often is not apparent. Fine examples have been found in the Alpine veins of Switzerland. Quartz crystals with incisions left by the dissolution of embedded albite crystals have been found in pegmatites.

Inversion pseudomorphs (paramorphs) of quartz after tridymite and cristobalite are well known, and all the reported occurrences of high-quartz are represented by paramorphs in low-quartz. The microcrystalline varieties of quartz very commonly are found as the petrifying material of wood. Quartz also occurs as the petrifying material of shell and bone, but may then be a substitution after calcite or apatite as the original petrifying material. Veinlets of fibrous quartz have been described[11] from a number of localities as pseudomorphs variously after fibrous amphibole, chrysotile, or chlorite; they are also reported after fibrous gypsum and calcite,[12] and as pseudomorphs after crocidolite (see *Inclusions in Quartz*).

Relatively few minerals have been found as true substitution pseudomorphs after crystals of quartz, and these but rarely. Here may be mentioned the well-known pseudomorphs of dense talc after quartz crystals from Göpfersgrün and Wunsiedel, Bavaria, and a few other localities;[13] also pseudomorphs of pectolite after quartz crystals from a zeolite locality at West Paterson, New Jersey.[14] Incrustation pseudomorphs of various minerals after quartz crystals are more common, as of hematite, pyrite, cassiterite, goethite, and calcite. The quartz of the graphic intergrowth with microcline in pegmatite (graphic granite) sometimes is dissolved away, leaving hollow molds in the feldspar.

References

1. Roth: *Allgem. u. chem. Geol. Berlin*, **1**, 1879; appendix in 1890.
2. Bischoff: *Lehrb. chem. phys. Geol. Bonn*, second ed., 3 vols., 1863–1866; also an English transl.

3. Frondel: *Bull. Am. Mus. Nat. Hist.* **57**, 389, 1935.
4. Haidinger: *Trans. Roy. Soc. Edinburgh*, **11**, Pt. 1, 73 (1831); *Ann. Phys.*, **52**, 161, 306 (1844).
5. Morgan: *Econ. Geol.*, **20**, 203 (1925); Sekanina: *Min. Abs.*, **7**, 500 (1940); Chirvinsky: *Bull. intern. ac. sci. Bohême*, **23**, 231 (1923); Lacroix: *C. R.*, **210**, 353 (1940); Atlee: *Am. J. Sci.*, **35**, 139 (1838).
6. Whitlock: *New York State Mus. Bull.* **251**, 134, 1924.
7. Murdoch: *Am. Min.*, **21**, 18 (1936).
8. Schaller: *U.S. Geol. Surv. Bull.* **832**, 1932.
9. Palache and Lewis: *Am. Min.*, **10**, 405 (1925).
10. Hidden: *School Mines Quart.*, **7**, 334 (1885); Vaux: *Proc. Ac. Nat. Sci. Philadelphia*, **78**, 17 (1926).
11. Richards: *Am. Min.*, **10**, 429 (1925); Thiesmeyer: *Am. Min.*, **22**, 701 (1937).
12. von Lasaulx: *Jb. Min.*, 1874, 165; Tschermak: *Sitzber. Ak. Wien*, **46**, 468 (1862) and *Zs. deutsch. geol. Ges.*, **17**, 68 (1865); Fischer: *Min. Mitt.*, 117, 1873: Weinschenk: *Zs. Kr.*, **14**, 305 (1888).
13. Roth (1879), **1**, 110, 292; Sandberger: *Sitzber. Ak. Wiss. München*, 12, 1872; Weinschenk (1888).
14. Glenn: *Am. Min.*, **2**, 43 (1917).
15. Staples: *Am. J. Sci.*, **248**, 124 (1950).
16. Déverin: *Schweiz. min. pet. Mitt.*, **17**, 530 (1937).

OCCURRENCE

Silicon constitutes 27.7 per cent by weight of the Earth's crust and, after oxygen, is here the second most abundant element. The major part of the silicon is combined in silicates, chiefly as feldspars, which comprise the principal minerals of the crust, and as pyroxenes and amphiboles. The remaining silicon is present as the oxide, SiO_2. It has been estimated[1] that this forms about 12 per cent of the entire lithosphere. It is chiefly contained as quartz in igneous rocks. The other polymorphs of silica are relatively rare, tridymite and the opaline types of cristobalite being the more abundant.

In detail, quartz occurs in a great variety of ways in nature, but only a few generalities can be given here. It is an important rock-forming mineral and occurs as a characteristic constituent of many igneous, metamorphic, and sedimentary rocks. Quartz is the principal constituent of sandstone and quartzite and occurs as unconsolidated sands and gravels. It commonly occurs in massive, granular form as veins or as dike-like bodies, and it is a characteristic gangue mineral in metalliferous deposits of hydrothermal origin. Most of the specimens of crystallographic interest are from cavities in vein deposits of various types. Quartz occurs abundantly in granite pegmatites, often as solid masses of large size in the central core. It is deposited, sometimes on a large scale, in granular or microcrystalline form by meteoritic or ascending hydrothermal waters as a replacement of pre-existing rocks, especially adjacent to veins or other channelways. Quartz occurs as a secondary deposit in cavities of sandstone, limestone, and other sedimentary rocks, making geodes of crystals or of chalcedony, and similarly lining cavities or crevices in basalt and other igneous rocks. The secondary enlargement[21] of the quartz grains of sandstones, by the deposition of quartz in crystallographic continuity with the original grain (the rounded outline of which may still be seen in thin section), is an important factor in the cementation of such rocks. Secondary quartz has been found in halite beds,[23] perhaps formed by the decomposition of clay by HCl released through hydrolysis of $MgCl_2$. Secondary quartz also occurs in coal[24] and as concretionary masses or isolated crystals in gypsum beds.[14] The free silica in meteorites is chiefly tridymite, but quartz has been observed.[15] The microcrystalline varieties of quartz typically are low-temperature hydrothermal deposits or are formed by surface or near-surface waters at essentially ordinary conditions of temperature and pressure. They are also formed during the diagenesis or the low-grade metamorphism of sediments, as from opaline silica of sedimentary, biochemical origin. The occurrences of the microcrystalline varieties of quartz, including chalcedony, agate, flint, and other types, have been described under the separate descriptions in the section on *Fine-Grained Varieties*.

Quartz crystals of a size and perfection suitable for technical purposes or of outstanding size and beauty as mineral specimens are found in relatively few places. A few of the localities for rock-crystal of high quality have already been cited (see *Euhedral and Coarse-Grained Varieties*).

Madagascar is a traditional source for fine rock-crystal.[25] Commercial exports, to France, began in the seventeenth century, but material was earlier supplied to India, Europe, and probably China by Arab settlers established on the northern end of the island in the ninth and tenth centuries. Most of this quartz was used for ornamental purposes. Production of clear quartz in the period 1909–1921 was 33 tons. The quartz crystals occur in cavernous veins in quartzite and also, to a much smaller extent, as crystals, chiefly smoky, in pegmatites. Alluvial and eluvial deposits were early worked. Madagascar rock-crystal, at least as seen on the market, is of unusually large average size, and clear crystals weighing hundreds of pounds apparently were common.

Colorless quartz of a quality suitable for optical and piezoelectric applications is now obtained principally from the states of Minas Gerais, Goiaz, and Bahia in Brazil.[2] The quartz occurs in deeply weathered veins, pipes, and pockets in siliceous Precambrian sedimentary rocks, with associated residual and alluvial deposits. The veins consist almost wholly of coarsely crystallized milky or gray quartz, with crystal lined cavities or with a comb structure extending from the walls. Crystals weighing from a few hundred pounds up to a ton or more are common, but most of the production has come from small crystals, up to a pound or so in weight, or from the clear terminations of larger crystals. The yield of usable quartz is only a small fraction of 1 per cent of the vein material. The veins have formed by the filling of fractures and cavities with silica transported by aqueous solutions at low to moderate temperatures. Associated minerals are few and rare. The Brazilian deposits have been known for well over a century, with a continuing production of quartz for optical purposes; they were worked intensively for quartz of piezoelectric grade during World War II. About 33 million pounds of commercial quartz crystals were exported from Brazil during 1910–1945.

A small production of quartz crystals of industrial grade has been obtained from Africa, Switzerland, Australia, Colombia, Japan, Burma, Russia,[16] and other places. In the United States, a very small production of industrial quartz together with much fine specimen material has been obtained from numerous small deposits in the Ouachita Mountains of western Arkansas.[3] The quartz occurs as hydrothermal veins and sheeted zones, apparently formed at relatively low temperatures and pressures, in sandstone, shale, and chert of Paleozoic age. Calcite, adularia, clay minerals, chlorite, sulfides, and brookite occur in very subordinate amounts. The great bulk of the crystals are under 200–300 grams in weight, but crystals have been found weighing up to 600 lb. Brazil twinning is very prevalent. Most of the transparent Arkansas quartz is water clear and brilliant, lacking the very faint cast of smoky color found in much Brazilian quartz; sometimes it is more or less smoky, or has a yellowish brown color, or occurs with chlorite phantoms. Very minor occurrences of industrial quartz have been reported in Virginia, North Carolina,[4] California,[5] and Idaho.[6]

Veins of the Alpine type, of which the classic occurrences are in Switzerland[19] but which are found in Russia[28] and numerous places elsewhere in the world, are important repositories for well-crystallized quartz. These veins, which contain actinolite, chlorite, epidote, muscovite, adularia, sphene, hematite, magnetite, rutile, fluorite, monazite, etc., as characteristic minerals, often well crystallized in cavernous openings, are in general found in metamorphic rocks. They owe their mineralization to solutions connected with the metamorphism or to igneous activity in so far as this is associated with the metamorphism. The chemical components of the vein minerals may be largely derived by the alteration of the rocks immediately adjacent to the fissures. A monographic description has been published[29] of the quartz from the Swiss occurrences. Veins of massive quartz also occur in metamorphic rocks.[20]

A few additional localities for colorless quartz that have afforded notable specimen material may be mentioned. Magnificent groups of drusy colorless quartz crystals have been found in the quartz veins at La Gardette near Bourg d'Oisans, Dauphiné, France,[7] and drusy crusts of crystals up to 5 ft. or more across, together with fine clusters of crystals, have been obtained from the veins in Arkansas. Fine drusy specimens were formerly obtained at the Ellenville lead mine, Ulster County, New York, and outstanding specimens have been found occasionally at many other places. Handsome groups of crystals, together with material of industrial grade, have been obtained from a deposit at Lyndoch, Ontario, Canada. Brilliant doubly terminated colorless crystals occur very abundantly in Herkimer County, New York, as at Little Falls, Salisbury, Middleville and Saratoga Springs[18] in cavities in the Little Falls dolomite and loose in the soil; also at Diamond Isle and Diamond Point, Lake George, Warren County, New York. The Herkimer "diamonds" generally range from almost microscopic dimensions up to an inch or so in size, but occasionally doubly terminated crystals, usually much flawed and having rough surfaces, have been found up to 18 lb. in weight. The crystals are short prismatic to almost equant in shape. Similar crystals occur in sandstone and clay slate in Com. Marmaros in northeastern Hungary bordering on Galicia. Beautiful clear crystals occur in cavities in the marble of Carrara, Italy.[26] Fine crystals, sometimes complexly developed, occur in cavities in granite at Baveno, Italy,[27] and near Striegau, Silesia. Drusy crystals sometimes containing disseminated hematite occur in the iron mines of Cumberland, England, and at Antwerp, Jefferson County, and Fowler, St. Lawrence County, New York. Large well-developed crystals of colorless quartz are found in cavities in pegmatites, but most of the material from this type of occurrence is more or less smoky in color. The more important or interesting localities for smoky quartz, and also for amethyst, rose quartz, and other colored varieties, have been mentioned in another section (see *Euhedral and Coarse-Grained Varieties*). A lengthy list of older localities that have afforded material of crystallographic interest has been published.[8]

Quartz crystals often occur of enormous size.[2] In some areas in Brazil single-crystals weighing up to 5 tons are abundant, and crystals weighing up to 25 tons are not rare. What is probably the largest quartz crystal known was found at Manchão Felipe near Itaporé, Goiaz. It measured about 20 ft. in length and 5 ft. across a prism face, and was estimated to weigh over 44 tons. At this locality, seven well-formed quartz crystals were once exposed, each with rhombohedral faces a meter or more in length. A group of quartz crystals in a deposit in Minas Gerais was estimated to weigh over 120 tons, and yielded about 2 tons of clear quartz. A pale smoky crystal from the Ariranha mine, north of Teófilo Otoni, Minas Gerais, preserved in the museum at Belo Horizonte, is 7 ft. in length and 11 ft. in circumference, and weighs more than 5 tons.[9] A milky white crystal 11½ ft. long and 5¼ ft. in diameter and estimated to weigh 14 tons was found in a mine near Betpak-dala in the Balkhash steppe, Siberia.[11] A crystal from the Neroika region in the subpolar Urals weighed about 3700 lb.[2] Another large crystal from pegmatite in Volhynia, Russia,[22] was almost 9 ft. long and 5 ft. across and weighed 9–10 tons. These giant crystals generally are milky or turbid and much flawed. Very large crystals of smoky quartz (which see) also have been found.

Concretionary geodes[13] of quartz crystals are of common occurrence. A famous locality is in the region of Keokuk, Iowa, where geodes up to several feet in diameter occur in great numbers in Mississipian limestone or weathered out in the soil; they are found similarly in Kentucky, Illinois, and Missouri. The outermost layer of the geode usually is chalcedony. The Keokuk geodes sometimes contain crystals of calcite, dolomite, galena, pyrite, sphalerite, millerite, goethite, and other species resting upon the quartz crystals; also containing bitumen, clay or limonite.

References

1. Clarke: *U.S. Geol. Surv. Bull.* **770**, 33, 1924, and *U.S. Geol. Surv. Bull.* **228**, 20, 1904.
2. Johnston and Butler: *Bull. Geol. Soc. Amer.*, **57**, 602, 1946; Campbell: *Econ. Geol.*, **41**, 773 (1946); Stoiber, Tolman, and Butler: *Am. Min.*, **30**, 245 (1945); Kerr and Erichsen: *Am. Min.*, **27**, 487 (1942); Knouse: *Trans. Am. Inst. Min. Met. Eng.*, **173**, 173 (1947); Walls: *Trans. Opt. Soc. London*, **21**, 157 (1920).
3. Engel: *U.S. Geol. Surv. Bull. 973-E*, 173, 1952; Miser: *Econ. Geol.*, **38**, 91 (1943).
4. Mertie: *U.S. Geol. Surv. Bull. 1072-D*, 1959.
5. Durrell: *Calif. J. Mines and Geol.*, **40**, 423 (1945).
6. Herdlick: *U.S. Bur. Mines Rept. Investig. 4209*, 1948.
7. Lacroix: *Min. de France*, **3**, 1901, p. 95.
8. Hintze: *Handbuch der Mineralogie*, **1**, Berlin, Lfg. 9, 1905, pp. 1353–1440.
9. Kerr and Erichsen (1942).
10. Campbell (1946).
11. Komarov: *Min. Abs.*, **11**, 459 (1952).
12. Laemmlein: *Min. Abs.*, **7**, 44 (1938).
13. Bassler: *Proc. U.S. Nat. Mus.*, **35**, 133 (1908); Van Tuyl: *Am. J. Sci.*, **42**, 34 (1916).
14. Brownell: *Univ. Toronto Stud., Geol. Ser.*, **47**, 7, 1942; Lacroix (1901), p. 109; Tarr and Lonsdale: *Am. Min.*, **14**, 50 (1929).
15. Merrill: *Am. Min.*, **9**, 112 (1924).
16. For some localities see Fersman: *Precious and Colored Stones of Russia, Mon. Russ. Ac. Sci.*, **1**, no. 3, 1922; Gordienka: *Geram. Abs.*, **10**, 598 (1932); Ivanov: *Pamir Quartz.* Moscow, 1940; Vertushiov: *C. R. ac. sci. URSS*, **51**, 55 (1946); Ermakov:. *C. R. ac. sci. URSS*, **48**, 46 (1945); Laemmlein and Osadchev: *C. R. ac. sci. URSS*, **50**, 441 (1945).
17. Henderson: *New Zealand J. Sci.*, Sect. B, **25**, 162 (1944).
18. Rowley: *Rocks and Min.*, **26**, 528 (1951).
19. Niggli, Koenigsberger, and Parker: *Die Mineralien der Schweizeralpen*, Basel, 1940, 2 vols.
20. Chapman: *Am. Min.*, **35**, 693 (1950).
21. Sorby: *Proc. Geol. Soc. London*, **36**, 62 (1880); Irving and Van Hise: *U.S. Geol. Surv. Bull.* **8**, 8, 1884; Irving: *Am. J. Sci.*, **25**, 401 (1883); Pettijohn: *Sedimentary Rocks*, New York, second ed., 1957.
22. Osadchev: *Mem. soc. russe min.*, **75**, 238 (1946).
23. Yorzhemski: *Dokl. Ak. Nauk USSR*, **66**, 915 (1949).
24. Hoehne: *Glückauf*, **85**, 661 (1949) and *Chem. Erde*, **18**, 235 (1956); Leskevich: *Dokl. Ak. Nauk SSSR*, **124**, 575 (1959).
26. Aloisi: *Atti soc. toscana sci. nat.*, **25**, 3 (1909).
27. Gallitelli: *Per. min. Rome*, **6**, 105 (1935).
28. Grigoriev: *Cursillos y Conferencias (Madrid)*, **7**, 63 (1960) (Symposium, Int. Min. Assoc., Zurich meeting, 1959).
29. Friedlaender: *Beitr. Geol. Schweiz, Geotech. Ser.*, Lfg. 29, 1951.

HISTORY OF THE NAMES CRYSTAL AND QUARTZ

The name quartz and the term crystal both are very old, and their meanings have changed with time. Crystal was originally applied as a name for what we now recognize as colorless euhedral crystals of quartz, and this usage continued for over 20 centuries. The Roman natural historian Pliny, in Book 37 of his *Natural History*, remarked:

> *Crystallus*, however, is formed in an opposite way, namely from cold, for from a liquid it is congealed by extreme cold in the same way as is ice. This is proved by the fact that *crystallus* is only found where the snow of winter is frozen hard; so that we can confidently say that it is really ice and nothing else. For this reason the Greeks have given it their name for ice, that is *cristallos* . . . . Europe produces excellent *crystallus*, to be specific as to the locality, on the crests of the Alps . . . . Strange as it seems, *crystallus* grows in six-angled forms: nor can a sound reason be given for this, particularly as the faces are not exactly of the same shape; but the sides between each edge are so absolutely smooth and even that no lapidary in the world could make the faces as flat and polished.

The term *cristallos* in reference to rock crystal also was used by Theophrastus in the third century B.C. Water-clear and flawless quartz crystals were called by the Greeks *aconteta, without flaw*.

That rock crystal was ice was believed by many authorities through the Middle Ages and seems to have been common opinion as late as the seventeenth century, although Steno (1669), Robert Boyle (1672), and others still earlier had refuted it. The same notion was expressed independently of the Greeks in China, Japan, and other countries. It has been noted[1] that various primitive peoples, doubtless because of the resemblence of white quartz to packed snow, and of limpid quartz to ice, together with the cool touch, have also drawn this analogy. Thus, certain Eskimo of Alaska have believed that quartz is the center of ice masses frozen so hard that they become stone, and the Ojibway Indians called white flint by a name meaning ice stone. Claudian, a Roman poet of about A.D. 400, wrote several verses on the chalcedony *enhydros*, indicating that the wintery cold was not sufficient to convert all the water into stone.

The name crystal (*crystallos, crystallus*) was applied to quartz, more specifically to the transparent, crystallized or euhedral type, until the seventeenth and eighteenth centuries, when it began to be applied in a generic sense to all bodies with polyhedral external form. The name quartz, already in local use for some centuries, then slowly came into more or less general application in place of crystal as a specific designation for silica. The extension of the meaning of the ancient term *crystallus* doubtless was primarily a consequence of the developing interest in the seventeenth and eighteenth centuries in the growth, properties, and internal structure of crystals, as in the work of Steno (1669) and Hooke (1665) and in the optical studies of Bartholinus (1669) and Huygens (1690). The generality of the new meaning is indicated by the treatise by M. A. Cappeler,[2] *Prodromus crystallographiae*, published in Lucerne in 1723.

The name quartz or quartzum was first applied only to the white, granular-massive types of this mineral, chiefly vein quartz. The name is not common in sixteenth and seventeenth century works, where it ordinarily was used together with *crystallus* in its original sense. In the eighteenth century quartz came into more general use for the massive material, and the euhedral, transparent types, the original *crystallus*, were commonly designated as *Crystallus Montanus* and *Bergkrystall* in the sense of a specific kind of crystal. The name *crystallus* was still used by some authors, but then usually in such context as to make its narrow meaning clear. Wallerius in his *Mineralogia* of 1747 made separate categories of *Silex* or *Kieselsten* (including here agate, carnelian, chalcedony, flint, opal, etc.), *Petrosilex* or *Hälleflinta* (types of jasper and certain rocks), *Quartz* (various types of granular and colored quartz, including drusy quartz under the name *Quartzum Crystallisatum*), and placed amethyst and euhedral colorless quartz (*Berg-Crystall*) in the general category of *Crystals and Gems*, together with ruby, sapphire, topaz, diamond, beryl, etc. The separate use of *quartzum* and *Crystallus Montanus*, etc., also is found in works by Woltersdorff (1748), Browne (1756), Cronstedt (1758), Vogel (1762), and others of this period.

In some works, such as those of Schmiedel (1753), de Bomare (1762), and J. G. Lehman (1766), descriptions equivalent to crystallized quartz, in the modern sense, are employed, and the idea seemed generally prevalent that quartz could crystallize, but the uniting of both the granular massive material and the euhedral transparent crystals under the single name quartz did not become general until almost the nineteenth century. Schmiedel's work, *Erz Stüffen und Berg Arten mit Farben genaü abgebildet* (Nürnberg, 1753), mentions "quarz adern," "undurchsichtigen Quarz, der theils milchfarb ist," "Crystall-quartzen," and "Quarz . . . überall mit kleinen Crystall-Nestern besezt, die mehrentheils sechseckig und oben zugespitzet sind . . ." and provides labeled illustrations to make clear the meanings. *Quarz* and the adjectivial form *quarzige*, also written *querzige*, appear in the *Beschreibung Allerfurnemisten Mineralischen Erz . . .* of Lazarus Ercker, a textbook of assaying published in 1629. Ercker spoke of quartz as white and massive and as distinct from hornstone, and he noted that both types occur in gold veins. The name *quartzum* also appears in works by Schwenckfelt (1600), Worms (1655), Webster (1661), and other seventeenth century writers. The *Principia mineralogiae* of Scopoli (1772) classified the siliceous earths as follows: I. Silex Achatinus: opacus, Jaspis; pellucidas, Achates. II. Silex Quartzosus: amorphus, Quartzum rude; crystallinus ignobilis, Crystallus. III. Silex Arenarius. Sage in his *Elemens de Mineralogie Docimastique* of 1775 classed *amethystus* and *quartzum granulare* together with *Crystallus Montanus* and *Crystallus Madagascum* in the general category of Quartzum, and the name quartz was used essentially in its present signification by de l'Isle (1783), Haüy (1801, 1822), and Werner (1817).

The name rock-crystal still continues in general use, as does the name crystal in its ancient sense (but only in reference to glass objects that simulate in clarity and value the similar objects carved from rock-crystal, or *crystallus*, in earlier times). The use of crystal for glass dates back to at least the sixteenth century. Limpid crystals of quartz early became accepted as a standard of clearness or transparency, leading to familar terms such as "crystal clear." The notions of clarity and of purity tend to become associated and are often used as near synonyms. This has unfortunately encouraged to some extent the use of crystal-clear but unanalyzed quartz and Iceland spar (calcite) crystals as standards for the measurement of the physical properties of these substances. The chemical composition and properties of these substances can and do vary independently of the color or lack of it.

The origin and meaning of the name quartz itself and its early history are obscure. The name is not mentioned in most works on minerals or on mining written in the sixteenth century or earlier. The earliest known mention[3] of quartz is in a fourteenth century German verse written in Saxony or Bohemia, which describes the effort of a miner to obtain financial assistance from a farmer. The name also appears in a manual of ore deposits, *Eyn Nützlich Bergbüchlein*, ascribed to von Kalbe (Calbus) and first published in Augsburg probably in 1505 and republished in 1518 and later. The name is of Saxon origin, and it has been suggested[4] to have come from the miner's term *Querklufterz* (cross-vein ore), applied to small, quartz-rich cross-veins often enriched in metallic constituents that transect the main vein mass. This term may have been contracted through *querertz* and *quertz to Quarz* in German and quartz in English. Agricola mentioned *quertz* and *quartzum* in his *Bermannus sive de re metallica* (1530), where the following passage appears (in translation):[5]

The following is another kind of mineral, that appears to be almost transparent. Our German miners call it quartz. This quartz is sometimes almost white, sometimes somewhat yellowish, and sometimes blends into gray . . . . The fourth kind is by far the hardest. One is frequently reminded of horn by its color, and thereby it bears its German name hornstone. Its particles frequently are so intergrown that one cannot distinguish them at first glance.

It is clear that originally the name quartz had a restricted meaning and did not include rock-crystal, which, for instance, in Agricola's *De natura fossilium* (1546) is throughout called by its ancient name *crystallus*. The original quartz also was distinct from hornstone, a fine-grained, translucent type of silica, often yellowish or bluish in color, that included our chalcedony, doubtless together with very fine-grained granular vein quartz and flinty material. In a concordance of Latin and German terms appended to some editions of *De natura fossilium* Agricola specifically equates *hornstein* and *feuerstein* (flint) with the *silex* of the Latins. The terms *querkluft, querklufterz, quergang*, etc. (also often with *querch* in place of *quer*), appear frequently in Agricola's writings. M. J. Mathesius, a friend of Agricola's in Joachimsthal, in his *Sarepta* (Leipzig, 1570), a work[6] on mining and metallurgy with many historical and theological comments, indicates that the name quartz derives from the term *quaderz, bad ore*—presumably gangue material. A psalm for miners given in this work contains the lines

> Gott der du schaffst Kies, Glanz und Quertz,
> Verwandel solchs bey uns in Erz.

The name also has been suggested[7] to derive from the Old Slavic *tvrudu, hard*, with transliteration of the initial letters *tv, tw, zy*u, *kw, q*u into the West Slavic *kwardy*, Czech *tvrdy*, Polish *twardy*, and ultimately, through the Slavic miners of Bohemia, into the Germanic *quarz*. Another suggested[3] derivation is from the Old High German word for gravel, *crioz*, which by the well-known replacement of the *c* (or *k*) by *qu* passed into *grioz* (*griess*) and the variant *quarz*. This derivation recalls that of *silex*, applied in Roman times to any hard pebble or stone found in fields and, anticipating flint, to denote hard-heartedness. Silex, from which silicon and silica derive, is usually credited in a mineralogical context to Pliny, but was earlier used by Cicero and doubtless others.

The earliest known mention of quartz in English is in the second edition of Jorden's *A Discourse of Naturall Bathes and Minerall Waters*, published in London in 1632. Here it is said, "There are also certain stones which we call fluores, which doe naturally shoot in divers forms: as Christall six squares Sparr, which the Dutch [Germans] call Sput or Quertz, shoots into points like diamonds: as we see in those Cornish or Bristoll stones." The 1631 edition does not contain this statement. The reference in part is to cubes of fluorite; the Bristol crystals, however, are quartz. Another early mention of quartz in English is in Webster's *Metallographia, or, An History of Metals* (London, 1661), where it is said that native gold occurs in " . . . Mines or Pits, partly alone, partly its little sparks do as it were cleave to a certain white kind of stone, which in the German Tongue is called quartz . . . ." The English work on gem stones by T. Nicols, *A Lapidary: or, the history of pretious stones, with cautions for the undeceiving of all those that deal with pretious stones* (London, 1652), does not mention quartz but describes amethyst and other varieties and states, "The crystall is a well known diaphanous gem, like unto most pure water congealed into a transparent perfectly perspicuous body of six sides, which in its extremety doth seem to intend them all to one point . . . ." Nicols also mentions the "diamond of Bristoll."

Quartz seems to have been an unfamiliar term even a century later in England. Pryce's *Mineralogia Cornubiensis* (London, 1778), a work on mining in Cornwall, included the name quartz in a glossary (pyrite and other minerals stated as being left out as too well known) with a description as a hard, opaque, and sometimes semi-transparent crystalline stony mass. Pryce also stated that the name is of German origin, and clearly placed it as a massive material: " . . . the first is a coarse quartz, which is the most impure, and covets no particular form; the second is Crystal, which forms hexagonal columns, cuspides and pyramids . . . " and, it is added, "Quartz is undoubtedly the most debased kind of Crystal." Dutens (1779)[8] said, "Der Crystall bildet sich im Quarz, der gleichsam seine Mutter ist . . . "

Linnaeus (*Systema naturae*, 1735) believed that minerals exhibited only a few fundamental crystal forms, which were impressed by a form-giving principle inherent in certain salts. Quartz was thought to have the form of saltpeter, and Linnaeus employed the name *Nitrum quartzosum*, together with *crystallum*, for the mineral. J. D. Dana used *Hyalus rhombohedrus* as an alternative name for quartz in the first two editions of his *System* (1837, 1844). His binomial Latin system of nomenclature, with classes, orders, and genera, was abandoned in the third edition (1850).

The varietal nomenclature of quartz runs to hundreds of trivial names and contains many etymological and other problems. The derivation of the very few varietal names retained in this work has been discussed in other sections. The remaining names, mostly disused or of uncertain application or local meaning, are almost wholly devoid of any scientific usefulness. Summary lists have been published. One writer of 2000 years ago, his patience tried after relating the meaning of a very long list of trivial names applied to gem materials, mostly varieties of silica, complained of the "infinity of names for precious stones all devised by the shrewd Greeks" and lapidaries, and added that it "should be well understood, that one and the same stone changes its name according to the various spots, marks and warts that characterize it: according to the many lines that cross them, the various veins in them, and the different colors of the veins."

References

1. Ball: *Smithsonian Instit., Bureau Am. Ethnology, Bull.* **128**, 30, 1941; *A Roman Book on Precious Stones*, Los Angeles, 1950, p. 221.
2. Weber: *Schweiz. min. pet. Mitt.*, **3**, 113 (1923).
3. Taube: *Sci. Monthly*, **58**, 454 (1944).
4. Tomkeieff: *Min. Mag.*, **26**, 172 (1942).
5. Agricola: *Ausgewählte Werke*, Bermannus, Berlin, 1955, **2**, p. 164.
6. Mathesius: *Sarepta, darinn von allerley Bergwerck und Metallen*, Leipzig, 1570, p. 169, line 9, and p. 1043 top. This reference found through Taube, ref. 3. On the origin of the names quartz and amethyst see also Keferstein: *Mineralogia Polyglotta*, Halle, 1849, pp. 70 and 79, and Yener: *Maden Tetkik ve Arama*, **8**, 27 (1943).
7. Klemm: *Der Aufschluss*, **4**, 125 (1953).
8. Dutens: *Abh. von den Edelstein*, Nürnberg, 1779, p. 73.

HIGH-QUARTZ

HIGH-QUARTZ. Beta-quartz. High-temperature quartz. (In some early usage[1] high-quartz was called alpha-quartz and low-quartz was called beta-quartz.)

High-quartz is stable at 1 atm. from about 573° to 870°. At higher temperatures it converts to tridymite, but the conversion, of the reconstructive type, is extremely sluggish, and high-quartz can be heated to its melting point. The melting point[27] may be near 1460°. On cooling, high-quartz undergoes a displacive or high-low type of inversion to low-quartz (ordinary quartz) at about 573°. This inversion varies considerably in temperature at 1 atm., as a function of compositional variation (see further under *Quartz, High-Low Inversion*).

The high-low inversion was recognized[2] by Le Châtelier in 1889, and the first extended effort to identify high-quartz in nature was made by Wright and Larsen[3] in 1909.

STRUCTURE

Hexagonal; hexagonal trapezohedral—6 2 2. Space group $C6_2 2$ or $C6_4 2$ (enantiomorphous). Unit cell contents Si_3O_6.

UNIT CELL DIMENSIONS

a_0 in Å	c_0	Ref.
4.999	5.457 (at 575°)	5
5.02	5.48	6

CELL DIMENSIONS AT DIFFERENT TEMPERATURES[32]

Temp.:	579°	590°	610°	665°	730°
c_0	5.4464 kX	5.4473	5.4470	5.4464	5.4459
$d_{10\bar{1}0}$	4.3181 kX	4.3196	4.3196	4.3196	4.3195

The atomic positions (in $C6_2 2$) are:

Si: (c) $\frac{1}{2}, \frac{1}{2}, \frac{1}{3};$ $\frac{1}{2}, 0, 0;$ $0, \frac{1}{2}, \frac{2}{3}$

O: (j) $u, \bar{u}, \frac{5}{6};$ $\bar{u}, u, \frac{5}{6};$ $u, 2u, \frac{1}{2};$ $\bar{u}, 2\bar{u}, \frac{1}{2};$

$2u, u, \frac{1}{6};$ $2\bar{u}, \bar{u}, \frac{1}{6};$ with $u = 0.197$

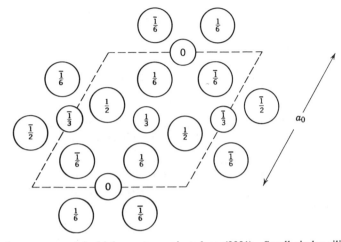

F<small>IG</small>. 99. Atomic arrangement in high-quartz, projected on (0001). Small circles silicon, large circles oxygen.

The crystal structure of high-quartz, like that of all other crystalline forms of silica, is based on a three-dimensional linkage of (SiO_4) tetrahedra by sharing of the oxygens (Fig. 99). The structure of low-quartz is a slight distortion of that of high-quartz. Both substances are enantiomorphous; a crystal of a given hand inverts into a crystal of the same hand. In the conversion from the high to the low forms the Si atoms shift in position in such a way as to destroy one set of two-fold axes and to convert the six-fold into three-fold axes. No Si-O-Si bonds are broken.

MORPHOLOGY

The crystal habit of high-quartz is holohedral in character, and the general forms of the crystal class, the right and left hexagonal trapezohedra, are not often observed. The assigned symmetry, 6 2 2, is based on x-ray studies[7] and on etching tests[8] made on material while heated into the high-quartz region. The axial ratio of low-quartz, approximately 1.100 at room temperature, decreases with increasing temperature to the inversion, where it drops from about 1.094 to 1.092 and then remains essentially constant in the high-quartz region. The interfacial angles of high-quartz thus are very close to those of low-quartz, allowance being made for the morphological expression of the different point-symmetries. The hexagonal pyramid $\{10\bar{1}1\}$, for example, is close in angle and identical in appearance to a low-quartz crystal bounded by equally developed faces of the two positive and negative rhombohedra, $\{10\bar{1}1\}$ and $\{01\bar{1}1\}$. An angle table for high-quartz, based on the ratio of the unit cell dimensions first cited, is given in Table 30.

High-quartz typically has a dipyramidal habit: usually $\{10\bar{1}1\}$ alone, sometimes with small modifying faces of $\{10\bar{1}0\}$; also combinations of $\{10\bar{1}1\}$ with $\{20\bar{2}1\}$ or steeper pyramids to $\{50\bar{5}1\}$; as $\{10\bar{1}1\}$ modified by obtuse pyramids. The crystals are almost always much rounded and rough on the surface, in addition to being more or less cracked paramorphs composed of low-quartz, and do not permit accurate morphological measurement. Not all the reported forms have been identified with certainty. High-quartz is also found as deeply embayed crystals, and as more or less skeletonized growths.[9] In sections the

TABLE 30. ANGLE TABLE FOR HIGH-QUARTZ

$a:cc = 1:1.092; \quad p_0:r_0 = 1.261:1$

Forms	ϕ	ρ	M
$m\ 10\bar{1}0$	30° 00′	90° 00′	60° 00′
$\pi\ 10\bar{1}2$	30 00	32 14	74 32
$r\ 10\bar{1}1$	30 00	51 35	66 56
$j\ 30\bar{3}2$	30 00	62 8	63 46
$l\ 20\bar{2}1$	30 00	68 22	62 18
$e\ 50\bar{5}1$	30 00	80 59	60 24
$\xi\ 11\bar{2}2$	0 00	47 30	90 00

Less common and rare: $70\bar{7}5$, $10.0.\overline{10}.7$, $13.0.\overline{13}.9$, $30\bar{3}1$, $70\bar{7}2$, $40\bar{4}1$, $10.0.\overline{11}.2$, $80\bar{8}1$, $11.0.\overline{11}.1$, $13.0.\overline{13}.1$, and $40.0.\overline{40}.1$.

crystals may show concentric growth zones.[9] The high-quartz phenocrysts of rhyolite often contain large inclusions of glass, which may in turn contain a spherical gas bubble.[29] At one occurrence as phenocrysts in porphyry, right- and left-handed crystals have been found in about equal abundance. The habit of high-quartz varies in different types of occurrence:[25] the form $\{10\bar{1}1\}$ is typical of phenocrysts in acid eruptive rocks, and the crystals often are simple bipyramids; $\{10\bar{1}1\}$ modified by $\{30\bar{3}2\}$ and $\{10\bar{1}0\}$ is common on crystals found in cavities in acid eruptive rocks; and a steep pyramidal habit with $\{20\bar{2}1\}$ modified by $\{30\bar{3}1\}$, $\{40\bar{4}1\}$, and $\{50\bar{5}1\}$ is typical of crystals from cavities in basic eruptive rocks. The trapezohedron $\{21\bar{3}2\}$ has been observed but is very rare.

TWINNING

Twins are very common in high-quartz, comprising the majority of the crystals at some occurrences. The twins are of the contact type with inclined axes and distinct re-entrant angles. In virtually all instances the twinned crystals are of the simple bipyramidal habit $\{10\bar{1}1\}$ and occur as phenocrysts in volcanic rocks. The twins are not distorted, as in the Japan Law of low-quartz.

(1) Esterel Law. Twin plane $\{10\bar{1}1\}$; angle cc 76°26'. The most common twin law in high-quartz. Often found as a group of two crystals of equal size, with composition plane parallel to the twin plane; also with composition plane perpendicular to the twin plane. Also as more complex groups: as triplets of three crystals of about equal size, one crystal twinned on two alternate upper faces with the other two crystals, or with two parallel faces of one crystal each twinned with another crystal; as a large single crystal with smaller crystals in twinned position on two or more faces; etc. Twins on the Esterel Law are also found combined with twins on the Verespatak and other laws. The Esterel Law is the analogue of the Reichenstein-Griesernthal Law in low-quartz. Twinning was first recognized[10] in high-quartz in crystals twinned on this law from a quartz diorite porphyry at Esterel, near Cannes, in the south of France. Also recognized in quartz phenocrysts in volcanic rocks at Four-la-Brouque, Auvergne, France; Verespatak, Hungary; the Ural Mountains, Russia; Cornwall; Dôgo, Oki, Japan;[11] Saubach, Vogtland, and Pöbelknochen, Saxony, Germany;[12] Samshvildo, Georgia, and Kafan, Armenia, USSR. Of about 1500 specimens from Cornwall,[13] 600 were twinned on the Esterel Law, 300 on the Verespatak Law, and 35 on other laws.

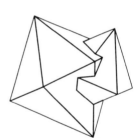

Twin plane (30$\bar{3}$2)

FIG. 100

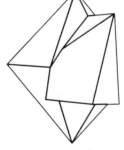

Twin plane (20$\bar{2}$1)

FIG. 101

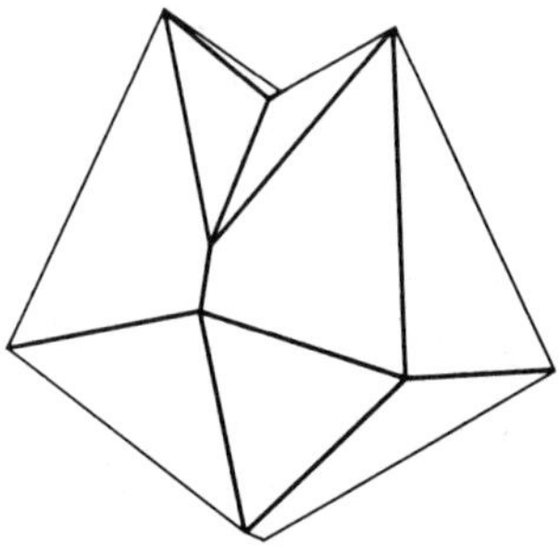

Twin plane (21$\bar{3}$1)

FIG. 102

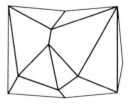

Twin plane (21$\bar{3}$3)

FIG. 103

Twin plane (21$\bar{3}$3) vertical

FIG. 104

(2) Sardinian Law. Twin plane $\{10\bar{1}2\}$; angle cc 115°10′ (64°50′). The composition plane bisects the acute angle between the c-axes, and is at right angles to the twin plane. Known[14] from Esterel, France; Saubach and Pöbelknochen, Germany. Extremely rare. Also found as a twin law in low-quartz (Sella's Law).

(3) Belowda Law. Twin plane $\{30\bar{3}2\}$; angle cc 55°24′. (See Fig. 100.) Rare. Known[13] from quartz porphyry at Wheal Coates and the Belowda Beacon china-clay pit, Cornwall, and from Esterel.

(4) Cornish Law. Twin plane $\{20\bar{2}1\}$; angle cc 42°58′. (See Fig. 101.) Rare. Found[13] at Belowda Beacon and Wheal Coates, Cornwall.

(5) Verespatak Law. Twin plane $\{11\bar{2}2\}$; angle cc 84°34′. The second most common twin law in high-quartz. Composition plane parallel to twin plane. Sometimes as a grouping of two individuals of about equal size. More commonly, asymmetrically developed, with a small crystal astride the edge of a large crystal, or as triplets similar to those of Esterel twins; also as more complicated groupings. Combinations of Verespatak and Esterel twins are common. Known from Verespatak, Hungary,[15] in rhyolite; Esterel, France;[16] Cornwall;[13] the Ural Mountains, Russia; at Samshvildo and Kafan, Russia;[17] at Saubach, Germany.[12] The Verespatak Law is the analogue of the Japan Law in low-quartz.

(6) Breithaupt's Law. Twin plane $\{11\bar{2}1\}$; angle cc 48°54′. Very rare. Known[18] from a porphyry at Four-la-Brouque, Auvergne, France.

(7) Wheal Coates Law. Twin plane $\{21\bar{3}1\}$; angle cc 33°8′. (See Fig. 102.) Very rare. Known only from Cornwall.[13]

(8) Pierre-Levée Law. Twin plane $\{21\bar{3}3\}$; angle cc 83°30′. (See Figs. 103 and 104.) Very rare. Known from Pierre-Levée and Four-la-Brouque, France, and Cornwall.[13]

(9) Samshvildo Law. Described[17] as having $\{10\bar{1}1\}$ and $[\bar{2}113]$ of the twinned crystals coincident, with a turn to the right or left of 39°28′ about the perpendicular to $\{10\bar{1}1\}$; angle cc 30°40′. From Samshvildo, Georgia, Russia.

All the described twin laws can be further specialized by specification of the hand of the twinned crystals, but no such observations have been made. Twins with parallel axes are not known. Brazil twinning can be introduced into high-quartz by heating low-quartz twinned on the Brazil or Combined Law. Dauphiné twinning, however, and the rotatory component of Combined twins disappear when low-quartz is heated over the inversion point because this twin operation then becomes an operation to identity. Dauphiné twinning, on the other hand, is in general produced when high-quartz is cooled below the inversion point (see further under *Quartz, Twinning*). It has been suggested[33] that the twins of high-quartz in igneous rocks have formed by the swimming together and the mutual orientation while in contact of isolated crystals suspended in the magma.

PHYSICAL AND OPTICAL PROPERTIES

The low-quartz paramorphs are white to grayish or milky white in color and translucent to opaque. Usually cracked or fissured. Observations[34] on the fracture of quartz while heated at 650–700° indicate that high-quartz has a good cleavage on $\{10\bar{1}1\}$, better than the $\{10\bar{1}1\}$ cleavage in low-quartz, with probably a less good cleavage on $\{10\bar{1}0\}$. Natural low-quartz paramorphs sometimes show a parting on $\{10\bar{1}0\}$ that is suggestive of a cleavage in the original high-quartz. The density of low-quartz decreases with increasing temperature, more rapidly as the inversion point is approached, where it abruptly decreases and then very slowly increases in the high-quartz region. The density at 600° is about 2.53 and at 1100° about 2.54. The indices of refraction of low-quartz decrease with increasing temperature, at a constantly accelerated rate as the inversion to high-quartz is approached. At the inversion point, the indices decrease abruptly, and then increase with increasing temperature but at a diminishing rate in the high-quartz region.[19] At 580° (Na): ϵ 1.5405, ω 1.5329; at 765°: ϵ 1.5431, ω 1.5356 (absolute values). See also Table 16, page 130. The birefringence of high-quartz has been measured[20] directly as a function of temperature to over 800°. The rotatory power[21] of high-quartz increases with increasing temperature, but much more slowly than that of low-quartz. Piezoelectric.[26]

COMPOSITION

Silicon dioxide, SiO_2. Chemical analyses are lacking. High-quartz probably undergoes the same general type of compositional variation as low-quartz but to a somewhat greater extent. Analyses of quartz paramorphs coupled with a microscopic study to investigate the possibility of exsolution accompanying the inversion would be of interest. High-quartz synthesized hydrothermally at 890° in the presence of much Al and Li has been found[22] to invert (at about 556° on heating) to low-quartz with relatively large cell dimensions, indicating the presence of Al and Li in solid solution. Quartz phenocrysts from rhyolite, perhaps representing high-quartz, show[23] a relatively low inversion temperature corresponding to a relatively high content of Al. High-quartz has been synthesized at 1 atm. at temperatures above 870° by heating "silicic acid" in the presence of certain elements, such as Ca, which catalyze or stabilize the phase;[30] also by the entrance of minor amounts of Mn^2 and Al into solid solution[31] at 1200°. High-quartz commonly appears in the ternary system Li_2O-Al_2O_3-SiO_2 at temperatures over its stability range, and probably contains Al and Li in solid solution.[35]

Solid solutions isostructural with high-quartz that can be quenched to room temperature have been found in the system[36] $MgO\text{-}Al_2O_3\text{-}SiO_2$ and the system[37] $Li_2O\text{-}Al_2O_3\text{-}SiO_2$. High-quartz or a solid solution related thereto that does not invert on cooling to room temperature also has been obtained[38] by heating montmorillonite at 1000°.

OCCURRENCE

Found in nature as a product of direct crystallization, chiefly as phenocrysts in acidic volcanic rocks, such as rhyolite, liparite, and dacite, and in cavities in such rocks. Also said to occur in pegmatite,[28] granitic igneous rocks, etc., but without conclusive evidence. Quartz phenocrysts of bipyramidal habit have been observed in granite.[24]

The identification of high-quartz in natural occurrences is in general based on arguments that the rock crystallized at temperatures and pressures in the stability field of high-quartz, and on the crystallographic characters of the material itself. The latter evidence is of two kinds: the absence of morphological or other features diagnostic of primary low-quartz, and the presence of the dipyramidal habit and of the twin laws apparently characteristic of high-quartz. The latter evidence is not rigorous, in the absence of features revealing the trapezohedral symmetry (see further under *Quartz, High-Low Inversion*). Both tridymite and cristobalite can crystallize metastably in the stability field of high-quartz.

References

1. Mügge: *Jb. Min.*, Festbd., 1907, p. 181. See discussion in Sosman: *Properties of Silica*, New York, 1927, p. 43.
2. Le Châtelier: *C. R.*, **108**, 1046 (1889).
3. Wright and Larsen: *Am. J. Sci.*, **27**, 421 (1909).
4. Bragg and Gibbs; *Proc. Roy. Soc. London*, **109A**, 405 (1925); Wyckoff: *Am. J. Sci.*, **11**, 101 (1926).
5. Bragg and Gibbs (1925).
6. Wyckoff (1926).
7. Rinne and Gross cited in Mügge: *Cbl. Min.*, 609, 1921; also ref. 4.
8. Nacken: *Jb. Min.*, I, 71 (1916); see also earlier inconclusive work by Friedel: *Bull. soc. min.*, **25**, 112 (1902) and Mügge (1907).
9. Laemmlein: *Zs. Kr.*, **75**, 109 (1930) and *Inst. Petrog. Ac. Sci. URSS*, no. 3, 71, 1933.
10. Drugman: *Min. Mag.*, **16**, 112 (1911).
11. Tomita: *J. Geol. Soc. Tokyo*, **35**, 419 (1928).
12. Bindrich: *Cbl. Min.*, 203, 1925A.
13. Drugman: *Min. Mag.*, **21**, 366 (1928).
14. Drugman: *Min. Mag.*, **19**, 295 (1922).
15. Balogh: *Múzeumi Füzetek, Kolozsvár*, **2**, no. 3 (1913).
16. Drugman: *Zs. Kr.*, **53**, 271 (1913).
17. Laemmlein: *Trav. lab. crist. ac. sci. URSS*, no. 2, 123, 1940.
18. Drugman: *Bull. soc. min.*, **53**, 95 (1930) and **51**, 187 (1928).
19. Rinne and Kolb: *Jb. Min.*, II, 138 (1910) give data as a function of wavelength for

temperatures over the range −45° to 765° for a crystal from Galicia. See also Sosman (1927), who has recalculated the data to absolute values.

20. Wright and Larsen: *Am. J. Sci.*, **27**, 421 (1909); Mallard and Le Châtelier: *Bull. soc. min.*, **13**, 123 (1890).
21. Le Châtelier: *Bull. soc. min.*, **13**, 119 (1890).
22. Frondel and Hurlbut: *J. Chem. Phys.*, **23**, 1215 (1955).
23. Tuttle and Keith: *Geol. Mag.*, **91**, 61 (1954); Tuttle: *Am. Min.*, **34**, 723 (1949); Tuttle: *J. Geol.*, **60**, 107 (1952).
24. Johannsen: *Descriptive Petrography*, Chicago, 1958, **2**, p. 131.
25. Kalb and Klotsch: *Zbl. Min.*, 1941A, 66; Kalb: *Verh. Naturalist. Ver. Rheinland* **98A**, 173 (1939).
26. Osterberg and Cookson: *J. Franklin Inst.*, **220**, 361 (1935); elastic constants—Kammer, Pardue, and Frissel: *J. Appl. Phys.*, **19**, 265 (1948).
27. Martinez: *C. R.*, **223**, 612 (1946); Cusack: *Proc. Roy. Irish Ac.*, **4**, 399 (1897).
28. Cf. Quensel: *Geol. För. Förh.*, **64**, 283 (1942).
29. Laemmlein and Kliya: *Dokl. Ak. Sci. USSR*, **94**, 233 (1954).
30. Schulman, Claffy, and Ginther: *Am. Min.*, **34**, 68 (1949); Birks and Schulman: *Am. Min.*, **35**, 1035 (1950); Bailey: *Am. Min.*, **34**, 601 (1949).
31. Claffy and Ginther: *Am. Min.*, **44**, 987 (1959).
32. Jay; *Proc. Roy. Soc. London*, **142A**, 237 (1933).
33. Laemmlein: *C. R. ac. sci. URSS*, 1930A, 709; see also Frondel: *Am. Min.*, **25**, 69 (1940).
34. Bloss: *Am. J. Sci.*, **255**, 214 (1957).
35. Roy and Osborne: *J. Am. Chem. Soc.*, **71**, 2086 (1949).
36. Schreyer and Schairer: *Ann. Rept. Director Geophys. Lab.*, *Yearbook Carnegie Inst.*, *Washington*, no. 59, 97, (1960).
37. Roy: *Zs. Kr.*, **111**, 185 (1959).
38. Bradley and Grim: *Am. Min.*, **36**, 182 (1951).

TRIDYMITE

TRIDYMITE. *vom Rath* (*Ann. Phys.*, **135**, 437, 1868). Asmanite *Maskelyne* (*Proc. Roy. Soc. London*, **17**, 371, 1868; *Phil. Trans.*, **161**, 361, 1871). Pseudotridymite *Lacroix* (*Bull. soc. min.*, **13**, 162, 1890). Christensenite *Barth and Kvalheim* (*Norske Vidensk. Akad. Oslo*, no. 22, 1944).

High-tridymite (β-tridymite; β_2-tridymite; Upper high-tridymite). Middle-tridymite (β_1-tridymite; Lower high-tridymite). Low-tridymite (α-tridymite). In some early usage the term γ-tridymite has been used for low-tridymite, β-tridymite for middle-tridymite, and α-tridymite for high-tridymite.

In the classical interpretation,[1] tridymite is the stable form of silica at temperatures between 870° and 1470°. At higher temperatures it transforms into cristobalite, and at lower temperatures into high-quartz. These conversions are of the reconstructive type and are extremely sluggish. The conversions can be accelerated by mineralizing or fluxing agents such as alkali oxides or halides and sodium tungstate. High-tridymite can be heated over 1470° to its melting point (1670°), and can be supercooled below 870°. On cooling, it undergoes two displacive or high-low inversions, at about 163° to middle-tridymite, which on further cooling inverts at about 117° to low-tridymite. Tridymite also can crystallize directly as a metastable form at temperatures below 870°. The two low-temperature inversions are both subject to super-cooling and, further, show a significant variation[2] in the temperatures at which they occur, depending on the content of other elements in solid solution and on the amount of structural disorder present (see further under *Cristobalite*). The inversion between the high and middle forms has been observed as low as 125°, and that between the middle and low forms at 100°. Disordered material containing three-layer cristobalite sequences may show an additional inversion near 240° (cristobalite). Some tridymite containing a relatively large amount of material in solid solution shows only a single inversion, between the low and high forms directly.[3] Because of the evidence for extensive solid solution in both natural and synthetic tridymite, the question has been raised[4] whether this is essential for the existence of the phase and if tridymite actually appears as a phase in the pure one-component system SiO_2.

HIGH-TRIDYMITE

STRUCTURE

Hexagonal; dihexagonal dipyramidal—$6/m \, 2/m \, 2/m$. Space group $C6/mmc$. Unit cell contents Si_4O_8.

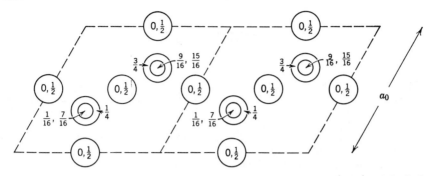

FIG. 105. Atomic arrangement in high-tridymite, projected on (0001). Small circles silicon, large circles oxygen.

UNIT CELL DIMENSIONS

a_0 in Å	c_0 in Å	Temp.	Material	Ref.
5.04	8.24	200°	Natural	5

The crystal structure[5] of high-tridymite is known only approximately. The atoms are in the following special positions of $C6/mmc$:

Si: (f) $\pm$ $(\frac{1}{3}, \frac{2}{3}, u;$ $\frac{2}{3}, \frac{1}{3}, u + \frac{1}{2})$ with $u = 0.44$

O: (c) $\pm$ $(\frac{1}{3}, \frac{2}{3}, \frac{1}{4})$

O: (g) $\frac{1}{2}, 0, \frac{1}{2};$ $0, \frac{1}{2}, \frac{1}{2};$ $\frac{1}{2}, \frac{1}{2}, 0;$ $\frac{1}{2}, 0, 0;$ $0, \frac{1}{2}, 0;$ $\frac{1}{2}, \frac{1}{2}, \frac{1}{2}.$

The structure, like that of the other polymorphs of SiO_2, is based on a three-dimensional linkage of (SiO_4) tetrahedra by sharing of the oxygens (Fig. 105). In high-tridymite the three-fold axes of the individual tetrahedra are parallel to [0001]. The structure, like that of cristobalite, is relatively open as compared to that of low-quartz. Tridymite and cristobalite bear a polytypic

TABLE 31. ANGLE TABLE FOR HIGH-TRIDYMITE

$a:c = 1:1.653;$ $p_0:r_0 = 1.9088:1$

Forms[6]	ϕ	ρ	M	A_2
c 0001	...	0° 00′	90° 00′	90° 00′
m 10$\bar{1}$0	30° 00′	90 00	60 00	90 00
a 11$\bar{2}$0	0 00	90 00	90 00	60 00
q 10$\bar{1}$6	30 00	17 39	81 17	90 00
o 10$\bar{1}$3	30 00	32 28	74 26	90 00
f 10$\bar{1}$2	30 00	43 39	69 49	90 00
r 30$\bar{3}$4	30 00	55 4	65 48	90 00
p 10$\bar{1}$1	30 00	62 21	63 42	90 00

Less common: l 54$\bar{9}$0, i 32$\bar{5}$0, g 20$\bar{2}$3, z 40$\bar{4}$3, x 81$\bar{9}$8.

relation to each other in the manner of hexagonal and cubic closest packing. Tridymite corresponds to the two-layer sequence $ABAB\ldots$, and (ideal) cristobalite to the three-layer (cubic) sequence $ABCABC\ldots$. Random interstratifications of these sequences may occur (see *Cristobalite* and *Opal*), and also systematic interstratifications (superstructures). The latter are represented among the structural types of low-tridymite.

MORPHOLOGY

The axial ratio of high-tridymite has not been determined by morphological measurement at temperatures in which this form is stable. The existing morphological data have been obtained at room temperature on inversion pseudomorphs of low-tridymite and, in part, on pseudomorphs of quartz after high-tridymite. The thermal deformation at the high-low inversions probably is quite small, so that the measured axial ratio is probably close to the true value at temperatures slightly above the inversion to high-tridymite. X-ray study confirmed the hexagonal symmetry and (morphological) unit originally assigned by vom Rath. Orthorhombic, pseudohexagonal axes also have been employed,[21] and the mineral also has been assigned to triclinic symmetry on morphological grounds.[8] An angle table is given in Table 31.

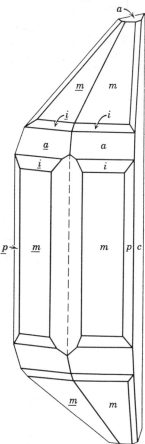

In crystal habit, high-tridymite is in general found as thin plates, flattened on $\{0001\}$ and bounded laterally by $m\{10\bar{1}0\}$; $a\{11\bar{2}0\}$ is usually present only as small modifying faces. The pyramid $o\{10\bar{1}3\}$ is very common, $p\{10\bar{1}1\}$

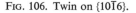

FIG. 106. Twin on $\{10\bar{1}6\}$.

less so. The plates may be extremely thin and fragile; also thick tabular. The crystals usually are below a few millimeters in size, but have been found up to several centimeters across as phenocrysts in igneous rocks. Also found as fanlike groups and spheroidal aggregates; as rosettes and star-like clusters.

Twinning is extremely common.[7] (1) Twin plane $\{10\bar{1}6\}$, both as contact and as penetration twins (Figs. 106–108). Two individuals may occur as wedge-shaped contact twins, giving an acute edge of juncture with $c\underline{c}$ 35°18′ (twice $qc = 17°39′$); similarly as trillings, with the twin angle then 70°36′

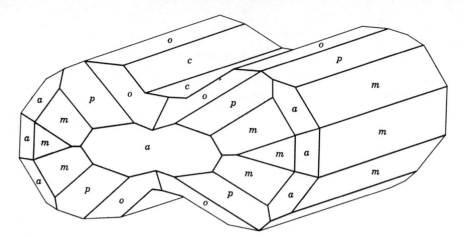

FIG. 107. Trilling on {10$\bar{1}$6}.

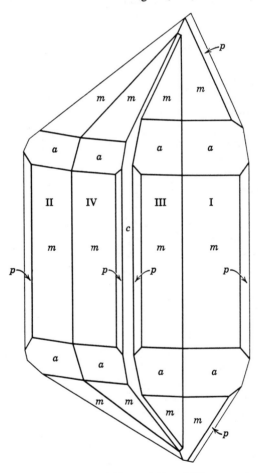

FIG. 108. Twinned crystal. Parts I and III, and II and IV, are each twinned on {10$\bar{1}$6.} Parts I and II are twinned on {30$\bar{3}$4}.

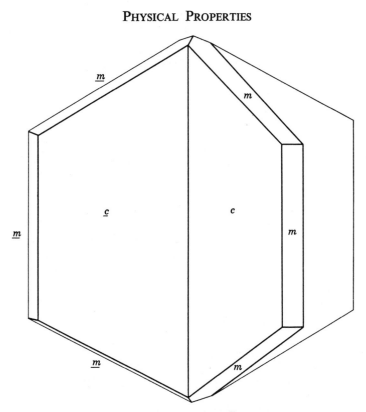

FIG. 109. Twin on {30$\bar{3}$4}.

(approximating to the octahedral angle), or as more complicated groupings. More commonly interpenetrant, as intersecting single-crystals or as intersecting twinned groups, and then with a sheaf-like appearance. Also as one relatively large or distinct individual intersected by other crystals or groups in twinned position. The twins may have deep, narrow re-entrant angles or may be completely joined together. (2) Twin plane {30$\bar{3}$4}, uncommon, often in combination with twinning on {10$\bar{1}$6}. The twin angle $c\underline{c}$ is 69°52′ (also near the octahedral angle). (See Figs. 108 and 109.) The pseudoisometric angular relations of high-tridymite and low-quartz have been remarked.[9]

PHYSICAL PROPERTIES

Optically, uniaxial positive. The optical elongation is negative. The density at 200° is 2.22 (from the known thermal dilation relative to the assumed value 2.262 for low-tridymite at 0°). The density at 200° calculated from the (approximate) cell dimensions measured at that temperature is 2.20.

MIDDLE-TRIDYMITE

Believed to be hexagonal. Uniaxial.

LOW-TRIDYMITE

STRUCTURE

Orthorhombic; crystal class not known. The four different crystals so far investigated by single-crystal x-ray methods are evidently polytypes representing two-layer, four-layer, ten-layer and twenty-layer structures (see tabulation below). High-tridymite corresponds to the two-layer (2H) polytype. Different polytypes have been recognized in different parts of a single-crystal. The thermal behavior of 1 and 2 is not known; on heating, 4 inverted directly from the low to the high form at 127°, while 3 showed the low to middle inversion at 121° and the middle to high inversion at 135°. At a temperature just over the high inversion the material of 3 and 4 showed an orthorhombic pattern of satellitic x-ray reflections which were not reversible with temperature change. A twenty-layer polytype has been synthesized, with inversions at 64°, 113°, and 138°, as have other polytypes of unidentified periodicity.[27]

UNIT CELL DIMENSIONS IN ANGSTROM UNITS

	a_0	b_0	c_0	$4.07 \times$	Z	Material	Ref.
1.	9.98	17.26	8.18	2	32	Poland	63
2.	9.90	17.1	16.3	4	64	Natural	5
3.	9.91	17.18	40.78	10	160	San Cristobal, Mex.	10
4.	9.91	17.18	81.57	20	320	Plumas Co., Calif.	10

X-ray powder diffraction data are given in Table 32. Dimensionally, low-tridymite is pseudohexagonal with $b_0 = a_0\sqrt{3}$. The crystal structure of low-tridymite is not known, but it doubtless is closely related to that of high-tridymite and of middle-tridymite.

MORPHOLOGY

So far as known, all natural crystals of low-tridymite are inversion pseudomorphs after high-tridymite, and the morphology of low-tridymite itself is not known.

TABLE 32. X-RAY POWDER DATA FOR LOW-TRIDYMITE

Synthetic material;[12] filtered Co radiation, in Angstroms.
(Reproductions of the low-angle part of the chart patterns of
tridymite polytypes have been published.[14])

d	I	d	I
4.30	vs	1.874	vvw
4.08	s	1.764	vvw
3.81	s	1.680	vvw
3.25	vw	1.613	vvw
2.96	w	1.581	vvw
2.47	m	1.534	vvw
2.37	vvw	1.519	vvw
2.29	vw	1.387	vvw
2.07	vvw	1.365	vvw
2.03	vvw	1.332	vvw
1.965	vvw	1.294	vvw

v = very strong; s = strong; m = moderate;
vw = very weak; vvw = very, very weak.

PHYSICAL PROPERTIES

Cleavage lacking or not readily observed; a parting or cleavage, perhaps inherited from high-tridymite, has been reported[11] on $\{10\bar{1}0\}$ and $\{0001\}$ in material from meteorites. Fracture conchoidal. Brittle to very brittle. Colorless to white. Transparent. Luster vitreous, sometimes pearly on $\{0001\}$. Hardness 7. Density 2.26 ± 0.01. Not fluorescent in ultraviolet radiation. Blackened by exposure to x-rays.

Optically, biaxial positive. Inversion twinning is present as interpenetrating lamellae and bands or, in basal sections, in sectors resembling the pseudohexagonal twins of aragonite. The extinction also may be patchy, streaky, or jagged. The optic plane has been observed perpendicular to the hexagonal outline of the original high-tridymite crystal, with the acute bisectrix, γ, parallel to the original c-axis. On heating, the plates become isotropic (uniaxial) at the inversion to middle-tridymite. The indices of refraction vary considerably. This variation is a function of compositional variation, but analytical data are lacking to permit a numerical correlation of the optical properties with composition. The indices of refraction apparently increase with increasing substitution of Al, Na, etc. The optical variation in tridymite is much greater than that in low-quartz or in cristobalite. The reported indices of refraction are given in Table 33. Some of the reported values are not identified as to the particular index or indices of refraction that they represent.

Table 33. Indices of Refraction of Low-Tridymite

α	β	γ	Biref.	2V, measured	Material	Ref.
1.468	1.470	1.475	0.007	40°	Meteorite	13
1.469	1.469	1.473	0.004		Synthetic	1
1.470		1.474	0.004		Synthetic	15
1.471	1.472	1.474	0.003	76° 15′	Meteorite	16
1.472		1.477	0.005		Nevada	17
1.473		1.474	0.001		Synthetic	18
1.474		1.478	0.004	30–40°	California	19
1.474	1.475	1.478	0.004		Meteorite	20
1.4773	1.4775	1.4791	0.0018	43°	Mexico	21
1.478	1.479	1.481	0.003	66–90°	California	22
1.479	~1.479	1.481	0.002	72°	Russia	61
1.479	1.480	1.483	0.004		Antarctica	23
	1.469	1.474			Synthetic	60
	1.472–1.476				Japan	24
		1.474			Synthetic	18
	1.475				Natural	65
	1.475				Moravia	25
	1.477	1.479			Japan	60
	1.477				Synthetic	62
	1.478				Mexico	26
	1.478				Synthetic	64
	1.478				New Zealand	26
	1.478–1.480				Japan	24

The following sections refer to both high-tridymite and low-tridymite.

COMPOSITION

Ideally, silicon dioxide, SiO_2. The older analyses of tridymite, all of very doubtful value, together with two recent partial analyses by spectrographic methods, are listed in Table 34.

Much or all natural high-tridymite and also material synthesized from systems containing Al, Na, or other cations is said to be a solid solution. Essentially pure tridymite, however, has been synthesized.[27] The principal mechanism of solid solution is believed to be a substitution of Al for Si coupled with the entrance of alkalies or of other cations into interstitial positions. This is suggested by the derivative structural relations[28] obtaining between certain alkali aluminosilicates and high-tridymite. It also is indicated by some analytical observations. Thus material[23] with relatively high indices of refraction (Table 33, Antarctica), with a single inversion at about 135°, gave optical emission spectra very similar to those afforded by a mixture of 94.8 weight

TABLE 34. CHEMICAL ANALYSES OF TRIDYMITE

	1	2	3	4	5	6	7	8	9	10
SiO_2	96.1	95.5	97.5	96.76	97.43	96.3	95.77	97.84		
Al_2O_3									2.4	2.7
Fe_2O_3	1.9	1.7	1.4	1.98	1.12		3.16	1.65	0.36	0.25
MgO	1.3	1.2	1.1	0.42	1.51	1.1	Tr.	Tr.	0.3	0.4
CaO				0.97	0.58		Tr.	Tr.	0.4	0.2
Na_2O									0.80	0.67
K_2O									0.37	0.75
Rem.	0.66	0.66		1.39		1.6	1.07	1.01	0.263	0.28
Total	99.96	99.06	100.0	101.52	100.64	100.0	100.00	100.50		
d				2.43	2.245					

1,2. Cerro San Cristóbal, Mexico.[30] MgO includes Al_2O_3. Rem. is ign. loss. Stated to contain matrix rock and Fe from mortar. 3. Hungary.[31] 4. Hungary.[32] Rem. is ign. loss 0.34, TiO_2 1.05. Analyst doubted purity of sample. 5. Breitenbach meteorite.[33] 6. Breitenbach meteorite.[34] Rem. is FeO. 7,8. Rittersgrün meteorite.[35] Rem. is ign. loss. 9. Cerro San Cristóbal.[36] Rem. is TiO_2 0.26, MnO 0.0005. Spectrographic analysis. 10. Lyttleton, New Zealand.[36] Rem. is TiO_2 0.28, MnO 0.003. Spectrographic analysis.

per cent quartz and 5.2 per cent nepheline, together with weak Fe and Ca spectra. Another sample, a twenty-layer polytype with a single inversion at 127°, was found[29] to contain Na, Ca, and Al. The two spectrographic analyses cited in Table 34 show amounts of Mg, Ca, Na, and K approximately compensating for the Al and Fe present (in substitution for Si). A specimen from Russia with similar optics contained Na and Al < 1; Ca and Mg $0.0X$, Ba and Mn $0.00X$, and Fe > 1 per cent.[61] A range of compositional variation also is indicated by the observed variation in the optical properties (Table 33) and in the thermal behavior. Additional chemical study is needed.

SOLUBILITY

Insoluble in ordinary acids, but more readily attacked by HF than quartz. Soluble in hot, concentrated solutions of alkalies.

SYNTHESIS

The literature on the crystallization of tridymite from synthetic polycomponent melts and hydrothermal systems is large. Obtained in crystals by fusion of dried precipitated silica in sodium tungstate[37] at about 950–1200°; and by heating 3 parts cristobalite with 1 part sodium tungstate[59] at 1400°. By heating silicic acid, silica glass, or quartz in H_2O at 880–1000° and appropriate pressures.[27] Also by fusion of silica in sodium metaphosphate[38] at about 1400°. By devitrification of soda lime, soda lead, and other glasses; crystals up to 2 cm. in size have been observed.[39] The tridymite obtained from polycomponent systems, especially when containing alkalies and Al, probably is not stoichiometric SiO_2.

ALTERATION

Tridymite very commonly occurs converted to quartz.[40] In many occurrences in igneous rocks, both as euhedral crystals and as phenocrysts, all the tridymite is represented by such pseudomorphs. The pseudomorphs generally are white or milky in color and opaque, with a granular structure. There appears to be a tendency for the quartz to orient relative to the tridymite. This has been described as a parallelism of quartz $\{10\bar{1}1\}$ and $\{01\bar{1}1\}$ with tridymite $\{0001\}$ from optical and x-ray study.[41] Other work[42] showed a tendency for the c-axis of the quartz to be inclined at roughly 60° to the c-axis of the original tridymite. In the latter orientation, quartz $\{0001\}$ or $\{30\bar{3}2\}$ is roughly parallel to tridymite $\{10\bar{1}1\}$ or $\{0001\}$, respectively. Partly resorbed phenocrysts of tridymite up to 17 mm. across in latite at Home, Colorado,[43] are converted to single-crystals of quartz with the 60° relation mentioned.

Tridymite has been observed converted to cristobalite (which see).

OCCURRENCE

Tridymite, originally crystallized in the high form, occurs in relatively siliceous volcanic rocks, especially rhyolite, trachyte, and andesite, and is found less frequently in basaltic rocks. It is found well crystallized in cavities and vesicles associated with cristobalite, quartz, sanidine, augite, hornblende, fayalite, and hematite. Tridymite also occurs in the groundmass of such rocks, usually microscopic but sometimes as large phenocrysts. Tridymite in general is much more abundant than cristobalite in igneous rocks; both minerals are rare in occurrences older than Tertiary. Cristobalite (opal) or fibrous quartz (chalcedony) rather than tridymite is formed when silica is deposited from solutions at relatively low or ordinary temperatures. Tridymite also occurs as a reaction or recrystallization product of siliceous inclusions in igneous rocks,[44] and has been observed[45] formed from quartz in the contact metamorphism of sandstones. It is common in stony meteorites and constitutes several per cent of some falls;[13] cristobalite and quartz are rare in such occurrences.

Originally found in andesite in the Cerro San Cristóbal, near Pachuca, Mexico. Later proved to be widely distributed, at times as a rock-forming mineral. Among the classic occurrences may be mentioned the trachytes of the Siebengebirge, Rhineland, Germany;[46] at San Pietro Montagnon and other places in the Euganean Hills of northern Italy;[47] at Puy Capuchin (Monte Dore) and other localities in the Puy-de-Dôme, Plateau Central, France.[48] In the United States notably in the Tertiary rhyolites and andesites of the San Juan region in southwestern Colorado.[49] Here it is the chief silica mineral in the rhyolites and latites and makes up as much as one-fourth or more of some of the rocks; it is confined to the groundmass, where it is especially common in the more porous, coarsely crystallized parts that crystallized last. Some of the latites contain coarse tridymite phenocrysts easily seen in hand specimens. Tridymite also is an abundant mineral in some of the rhyolite tuffs, where it in part was formed after the tuff was deposited. The total amount of tridymite in the rocks of the San Juan region has been estimated at 350 cubic miles, none of it analyzed. Tridymite also is found associated with cristobalite, fayalite etc., in the lithophysae of the rhyolites, and obsidians in Yellowstone Park, Wyoming;[50] similarly in Iceland,[51] California,[52] and other places. Found with cassiterite, lussatite, and chalcedony in veinlets in rhyolite in Lander County, Neveda.[53] With kaolinite and alunite as an alternation of dacite by acid hot-spring waters at Lassen, California.[19] In the rocks of the Keli region, middle Caucasus, Russia,[54] where a dacite contained 5 per cent of tridymite in vesicles. In the Tertiary igneous complex of Skye and Rhum, Scotland.[55]

HISTORICAL

Tridymite was recognized by vom Rath in 1861 as a new polymorph of silica, in specimens sent to him from the Cerro San Cristóbal, Mexico. Some years later he discovered cristobalite on a personal visit to the locality. The name tridymite was derived from $\tau\rho\iota\delta\nu\mu o\varsigma$, three-fold, in allusion to the common occurrence in trillings. Once described, tridymite was soon identified in a number of other natural occurrences, and it was shown to be identical with synthetic material that had been earlier prepared.[56] The existence of the inversion between low-tridymite and middle-tridymite was recognized[57] optically in 1884, and the additional inversion, between middle-tridymite and high-tridymite, which does not show optically, was recognized by Fenner[1] in 1913. Asmanite, a supposedly new form of silica described from the Breitenbach meteorite very shortly after the discovery of tridymite itself, became accepted as identical with tridymite after numerous investigations.[58] Although type asmanite apparently has not been studied by x-ray methods, tridymite has been identified by x-ray study in many meteorites. The crystal structure of high-tridymite was worked out by x-ray methods (Gibbs, 1927), but the crystal structures of the two low-temperature forms remain unknown. As with cristobalite, the recognition of the relatively open structure of tridymite, and the discovery of alkali aluminosilicates with related structures, led to postulates that the chemical composition of this mineral is variable.

References

1. Fenner: *Am. J. Sci.*, **36**, 331 (1913).
2. Flörke: *Ber. deutsch. Keram. Ges.*, **32**, 369 (1955).
3. Lukesh and Buerger: *Am. Min.*, **27**, 143 (1942); Barth and Kvalheim: *Norsk Vidensk. Ak. Oslo*, no. 22, 1, 1944.
4. Flörke (1955); Eitel: *Bull. Am. Ceram. Soc.*, **36**, 1942 (1957); Holmquist: *Zs. Kr.*, **111**, 71 (1958). See also Hill and Roy: *Trans. Brit. Ceram. Soc.*, **57**, 496 (1958).
5. Gibbs: *Proc. Roy. Soc. London*, **113A**, 351 (1927). Space Group *C62c* or *C6/mmc*, latter assigned from holohedral morphology.
6. Goldschmidt: *Atlas der Krystallformen*, **9**, 1923, p. 2; Billows: *Mem. reale accad. naz. Lincei, Cl. Sci.*, ser. 5, **13**, 506 (1922); Rosický: *Festschr. V. Goldschmidt*, Heidelberg, 1928, p. 229.
7. Summary in Goldschmidt (1923); see also Billows (1922), Rosický (1928), and Boeris: *Atti. soc. Ital. sci. nat. Milan*, **38**, 17 (1899) and *Zs. Kr.*, **34**, 294 (1901).
8. Billows (1922).
9. Mallard: *Bull. soc. min.*, **13**, 161 (1890); Beckenkamp: *Leitfaden der Krist.*, Berlin, 1919, p. 72, and *Zs. Kr,*. **36**, 483 (1902).
10. Lukesh and Buerger (1942).
11. Maskelyne: *Proc. Roy. Soc. London*, **17**, 371 (1868), and *Phil. Trans.*, **161**, 361 (1871); vom Rath: *Ann. Phys.*, Erg. Bd. **6**, 383 (1873).
12. Clarke: *J. Am. Ceram. Soc.*, **29**, 25 (1946). Converted to angstroms from kx units.
13. Foshag: *Am. J. Sci.*, **35**, 374 (1938).
14. Hill and Roy (1958); Flörke (1955).

15. Ivanov: *Trav. inst. pétr. ac. sci. URSS*, no. 9, 23, 1936.

16. Heide: *Cbl. Min.*, 1923A, 69.

17. Knopf: *U.S. Geol. Surv. Bull.* **640G**, 133, 1916.

18. Wilson: *J. Soc. Glass Technol.*, **2**, 186 (1918).

19. Anderson: *Am. Min.*, **20**, 240 (1935).

20. Frondel: priv. comm., 1959, on Indarch (carbonaceous chondrite).

21. Mallard (1890).

22. Durrell: *Am. Min.*, **25**, 501 (1940).

23. Barth and Kvalheim (1944).

24. Kuno: *Bull. Earthquake Res. Inst.*, **11**, Pt. 2, 1933.

25. Rosický (1928).

26. Mason: *Am. Min.*, **38**, 866 (1943).

27. Hill and Roy (1958).

28. See Buerger: *Am. Min.*, **39**, 600 (1954).

29. Lukesh and Buerger (1942). The poor analysis, not published, was made by F. A. Gonyer on a tiny sample (C.F.).

30. vom Rath: *Ann. Phys.*, **135**, 437 (1868).

31. Hoffmann: *Föld. Közl.*, **2**, 71 (1872).

32. Koch: *Min. Mitt.*, **1**, 344 (1878).

33. Maskelyne: *Phil. Trans.*, **161**, 364 (1871).

34. vom Rath: (1873).

35. Winkler: *Nov. Acta Leopold-Carolin. Ak.*, **40**, 339 (1878).

36. Leininger analyst in Mason (1953).

37. Fenner (1913); Quensel: *Cbl. Min.*, 1906, 657, 728; Peyronel: *Zs. Kr.*, **104**, 261 (1942).

38. Rose: *Monatsh. Ak. Wiss. Berlin*, 1867, 129; Schwartz: *Zs. anorg. Chem.*, **76**, 424 (1912),

39. Le Châtelier: *Bull. soc. min.*, **39**, 150 (1916).

40. Mallard (1890); Billows (1922); Hawkes: *Geol. Mag.*, 1916, 205; Boeris (1899); Moehlman: *Am, Min.*, **20**, 808 (1935); Geijer: *Geol. För. Förh.*, **34**, 70 (1913).

41. Mallard (1890); Flörke (1955).

42. Wager, Weedon, and Vincent: *Min. Mag.*, **30**, 263 (1953); Ray: *Am. Min.*, **32**, 643, (1947).

43. Ray (1947).

44. Lacroix: *C. R.*, **223**, 409 (1946); Wells: *Min. Mag.*, **29**, 715 (1951); Staudt: *Cbl. Min.*, 1925A, 47; Weymouth: *Am. J. Sci.*, **16**, 237 (1928).

45. Harker: *Metamorphism*, London, 1950, p. 68; Taylor: *Min. Mag.*, **25**, 544 (1940); Osborne: *J. Roy. Soc. New South Wales*, **82**, 309 (1950).

46. vom Rath (1868) and *Ann. Phys.*, **135**, 449 (1868); Zirkel: *Jb. Min.*, 1870, 825; Laspeyres: *Verhl. Nat. hist. Ver. Rheinl.*, 1900, p. 200.

47. Schuster: *Min. Mitt.*, **1**, 71 (1878); Lasaulx: *Zs. Kr.*, **2**, 252, 267 (1878); Boeris (1899).

48. Lacroix: *Min. de France*, **3**, 1901, p. 165; Gonnard: *Bull. Soc. Min.*, **8**, 310 (1885).

49. Larsen, Irving, Gonyer, and Larsen: *Am. Min.*, **21**, 679 (1936).

50. Foshag: *Proc. U.S. Nat. Mus.*, **68**, Art. 17 (1926).

51. Wright: *Bull. Geol. Soc. Amer.*, **26**, 255 (1915).

52. Wright: *J. Wash. Ac. Sci.*, **6**, 367 (1916).

53. Knopf: *Econ. Geol.*, **11**, 652 (1916).

54. Ustiev: *Trav. inst. petr. ac. sci. URSS*, no. 6, 159, 1934.

55. Black: *Min. Mag.*, **30**, 518 (1954); Wager, Weedon, and Vincent (1953).

56. Rose: *Monatsh. Ak. Wiss. Berlin*, 1867, 140, and 1869, 449; *Ber. deutsch. Chem. Ges.*, **2**, 388 (1869).

57. Merian: *Jb. Min.*, **I**, 193 (1884); Mallard (1890).
58. See summary in Hintze: *Handbuch der Mineralogie*, **1**, 1906, Lfg. 10, p. 1458.
59. Fenner: *J. Soc. Glass Technol.*, **3**, 116 (1919).
60. Kondo and Yamauchi: *J. Soc. Chem. Ind. Japan*, **38**, Suppl., 651 (1935).
61. Zolutukhin: *Chem. Abs.*, **53**, 21464 (1959).
62. Longchambon: *C. R.*, **180**, 1855 (1925).
63. Gajda: *Arch. Min. Warsaw*, **20**, 85 (1957).
64. Clews and Thompson: *J. Chem. Soc. London*, **121**, 1442 (1922).
65. Weymouth (1928).

CRISTOBALITE

CRISTOBALITE. *vom Rath (Jb. Min.*, **I**, 198, 1887). Christobalite, Crystobalite, etc. *wrong orthogr.* High-cristobalite (β-cristobalite), Low-cristobalite (α-cristobalite) *Mallard (Bull. soc. min.*, **13**, 161, 1890); *Fenner (Am. J. Sci.*, **36**, 354, 1913). Metacristobalite (high form).

Cristobalite has two polymorphs: high-cristobalite, an isometric high-temperature form stable from 1470° to the melting point at 1728°, at atmospheric pressure, but existing metastably down to a displacive inversion at about 268°, where it converts to low-cristobalite; and low-cristobalite, a tetragonal metastable form that exists at lower temperatures. The conversion from high-cristobalite to tridymite at 1470° is of the reconstructive type and is extremely sluggish in the absence of fluxes. The high-low inversion temperature at 268° varies over a wide range, depending on the amount of structural disorder present and on the chemical composition of the sample.[1] Highly ordered high-cristobalite (3C) with the ideal composition SiO_2 inverts[2] at about 268°. With increasing disorder and probably with an increasing amount of material in solid solution the inversion temperature may be as low as 130° or less. Often it has been found in the range 175–250°. The inversion is subject to marked supercooling. The amount of supercooling apparently tends to be greater in highly ordered material, amounting to 30° or more, giving the actual high-low inversion much below 268° on cooling (but at 268° on heating). In highly disordered material an additional inversion may be observed, and the dilatometric curves tend to round off.

HIGH-CRISTOBALITE

STRUCTURE

Isometric; tetartoidal—2 3 (?)[4] Space group $P2_13$ (?). Unit cell contents Si_8O_{16}.

UNIT CELL DIMENSIONS AT DIFFERENT TEMPERATURES

Temp.	a_0 in Å	Ref.	Temp.	a_0 in Å	Ref.
20°	7.09	75	600°	7.1363	75
250	7.1285	75	800	7.1462	75
290	7.109	6	1000	7.1492	75
300	7.1306	75	1100	7.1341	75
400	7.1362	6	1200	7.1483	75
500	7.140	16	1300	7.1473	75

TABLE 35. X-RAY POWDER SPACING DATA FOR HIGH-CRISTOBALITE AND LOW-CRISTOBALITE

High-cristobalite. Cu radiation (λ 1.5405 Å). Synthetic material,[16] at 500°, with $a_0 = 7.16$ Å			Low-cristobalite. Mo radiation (λ 0.7093 Å). Synthetic material,[17] at 17°, with a_0 4.973, c_0 6.95 Å		
d	I	hkl	d	I	khl
4.15	100	111	4.04	100	101
			3.138	12	111
2.92	5	211	2.845	14	102
			2.489	18	200
2.53	80	220	2.468	6	112
			2.342	<1	201
2.17	10	311	2.121	4	211
			2.024	3	202
2.07	30	222	1.932	4	113
			1.874	4	212
1.99	5	320	1.756	1	220
			1.736	1	004
1.793	5	400	1.692	3	203
			1.642	1	104
1.688	5	411	1.612	5	301
			1.604	2	213
1.639	60	331	1.574	1	310, 222
			1.535	2	311
1.469	50	422	1.495	3	302
			1.432	2	312
1.379	20	511	1.423	1	204
			1.401	1	223
1.265	30	440	1.368	1	214
			1.353	1	321
1.209	30	531	1.345	1	303
			1.336	1	105
1.130	20	620	1.301	2	313
			1.282	2	322
1.089	5	533	1.235	<1	224
			1.224	<1	401
1.029	5	444	1.207	1	410
			1.1842	2	323
1.000	10	711	1.1762	1	215
			1.1659	1	314
0.956	10	642	1.1556	<1	331
			1.1112	1	420
0.929	10	731	1.0989	3	421

Linear thermal expansion:[5] 8.53×10^{-6} Å per degree C. Indexed x-ray powder spacing data are given in Table 35.

The crystal structure of high-cristobalite was first determined by Wyckoff.[6] This structure, based on the space group $Fd3m$, had a linear arrangement of the Si-O-Si bonds, with 8 silicons in $0, 0, 0$; $\frac{1}{4}, \frac{1}{4}, \frac{1}{4}$; etc., and 16 oxygens in $\frac{1}{8}, \frac{1}{8}, \frac{1}{8}$; $\frac{1}{8}, \frac{3}{8}, \frac{3}{8}$; $\frac{3}{8}, \frac{1}{8}, \frac{3}{8}$; $\frac{3}{8}, \frac{3}{8}, \frac{1}{8}$; etc. A slightly deformed version of this structure, based on the space group $P2_13$, was later reported[7] with non-linear Si-O-Si bonds (Fig. 110). The nearest Si-O distances range between 1.58 and 1.69 Å, average 1.63 Å. In this structure the atomic positions are: 8 Si and 4 O in three sets of $(4f)$: u, u, u; $u + \frac{1}{2}, \frac{1}{2} - u, \bar{u}$; $\frac{1}{2} - u, \bar{u}, u + \frac{1}{2}$; $\bar{u}, u + \frac{1}{2}, \frac{1}{2} - u$, with u (Si, I) $= 0.255$, u(Si, II) $= -0.008$, u(O) $= 0.125$. The remaining 12 oxygens are in general positions (b) with $x = y = 0.660$, $z = 0.062$.

The crystal structure of high-cristobalite is based on a relatively open three-dimensional linkage of (SiO$_4$) tetrahedra, with sharing of all oxygens. The crystal structures of cristobalite and tridymite are approximately related in the manner of hexagonal and cubic closest packing of the oxygens, and a further analogy extends to quartz. The structure of ideal cristobalite is of the three-layer type, 3C, with the stacking sequence $ABCABC\ldots$ along [111] directions. This structure is taken by pure material crystallized at relatively high temperatures. At lower temperatures an increasing amount of random (or systematic) interstratification of two-layer, $ABAB\ldots$ stacking sequences of tridymite may take place (see further under *Opal*). The one-dimensional disordering[8] is accompanied by variations in the temperature of the high-low inversion in the thermal response of the material, and in the x-ray diffraction effects, but not by large changes in the density or indices of refraction.

An oriented overgrowth of cristobalite with twinned tridymite has been observed,[9] with cristobalite [110] (111) parallel to tridymite [11$\bar{2}$0] (0001).

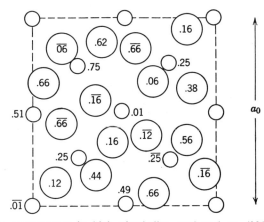

FIG. 110. Atomic arrangement in high-cristobalite, projected on (001). Small circles silicon, large circles oxygen.

PROPERTIES

The calculated density at 500° (from a_0 7.13 Å) is 2.20. Cleavage apparently lacking. Isotropic.

LOW-CRISTOBALITE

STRUCTURE

Tetragonal; tetragonal trapezohedral—4 2 2 (?).[10] Space group $P4_12_1$ (?). Unit cell contents Si_4O_8. Indexed x-ray powder spacing data are given in Table 35.

UNIT CELL DIMENSIONS AT DIFFERENT TEMPERATURES

a_0 in Å	c_0 in Å	Temp.	Ref.
4.97	6.93	Room	10
4.9715	6.9193	22°	11
4.9733	6.9262	30°	12
4.973	6.95	27°	13
4.970	6.910	Room	14

The crystal structure of low-cristobalite is a slight distortion of that of high-cristobalite (Fig. 111). The tetragonal cell of low-cristobalite is pseudo-cubic, with c_0 (6.9 Å) and the prism diagonal $\sqrt{2}a_0$ (7.0 Å) corresponding to

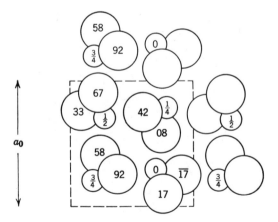

FIG. 111. Atomic arrangement in low-cristobalite, projected on (001). Small circles silicon, large circles oxygen.

the unit cell of high-cristobalite (a_0 7.1). The atomic positions based on $P4_12_1$ are:

Si: (a) $u\,u\,o$; $\bar{u}, \bar{u}, \tfrac{1}{2}$; $\tfrac{1}{2} - u, u + \tfrac{1}{2}, \tfrac{1}{4}$; $u + \tfrac{1}{2}, \tfrac{1}{2} - u, \tfrac{3}{4}$.

O: (b) $x\,y\,z$; $\bar{x}, \bar{y}, z + \tfrac{1}{2}$; $\tfrac{1}{2} - y, x + \tfrac{1}{2}, z + \tfrac{1}{4}$; $y + \tfrac{1}{2}, \tfrac{1}{2} - x, z + \tfrac{3}{4}$; $y\,x\,\bar{z}$; $\bar{y}, \bar{x}, \tfrac{1}{2} - z$; $\tfrac{1}{2} - x, y + \tfrac{1}{2}, \tfrac{1}{4} - z$; $x + \tfrac{1}{2}, \tfrac{1}{2} - y, \tfrac{3}{4} - z$;

with $u = 0.30$, $x = 0.245$, $y = 0.10$, $z = 0.175$

PHYSICAL AND OPTICAL PROPERTIES

Color white or milky white. No apparent cleavage. Brittle. Hardness $6\tfrac{1}{2}$. Density 2.32 (calc. from a_0 4.973, c_0 6.95 Å); 2.33 $\pm$ 0.01 (meas.; H_2O at room temperature). The measured density often is relatively low, down to 2.2, especially in material prepared by calcination of flint or chalcedony, because of small particle size and porosity. Not fluorescent in ultraviolet radiation.

Optically, uniaxial negative ($-$). The indices of refraction of natural and synthetic material in general are close to ω 1.484, ϵ 1.487 in white light at room temperature (Table 36). In some instances it is uncertain whether the single

TABLE 36. INDICES OF REFRACTION OF LOW-CRISTOBALITE

ω	ϵ	Mean Index (?)	Material	Ref.
1.4841	1.4862		Ellora, India	21
1.484	1.486		Synthetic	13
1.484	1.487		Synthetic	22
1.484	1.487		Silica brick	23
1.484	1.487		Fe slag	24
1.484	1.487		Synthetic	38
		1.487	Synthetic	25, 26
		1.485	Yellowstone	27
		1.485	Idaho	28
		1.485	Synthetic	28, 29, 30, 31, 32, 76
		1.485	New Zealand	33
		1.4858	Wisconsin	34
		1.485	Hawaii	35
		1.485	Washington	36
		1.484	Bohemia	39
		1.484	Synthetic	37, 40
		1.483	California	41
		1.483	Synthetic	40
		1.482	Natural	37

value reported represents ω or a mean value. The reported measurements were made by the immersion method, for which the precision obtained is generally ±0.002 to ±0.003 unless special care is taken. On this basis the data do not show a significant variation. Inverted crystals tend to be so complexly twinned or fine grained that it is difficult to measure other than a value averaging the actual orientations present in the grain as viewed. Very low values for the mean index may be obtained on porous fine-grained material.[18] The birefringence is uniformly low, in the range 0.002–0.003. In some instances the observed birefringence is almost zero, perhaps an effect of overlap in closely twinned aggregates. On heating, becomes isotropic (high-cristobalite). The few measurements of the indices of refraction that diverge considerably from the data of Table 36 have higher birefringence and seem suspect (ω 1.4824, ϵ 1.4876, Caucasus;[19] 1.482, 1.489, Japan[49]).

The inversion from high- to low-cristobalite may be accompanied by the development of twinning. Inverted single-crystals of high-cristobalite may show an internal structure of birefringent sectors or lamellae, or a plaid-like pattern, usually in three mutually perpendicular interpenetrating sets.[20] In (100) sections through inverted octahedra the lamellae extinguish parallel to the original a-axes, and in (111) sections to the trace of the a-axes.

The following sections refer to both high-cristobalite and low-cristobalite.

MORPHOLOGY

Cristobalite occurs both massive, as submicrocrystalline aggregates constituting opaline crusts and cavity fillings (see *Opal*) or with a microscopic fibrous structure (see variety *Lussatite*), and as distinct crystals. The crystallized material, formed at relatively high temperatures as high-cristobalite, has in all instances inverted to low-cristobalite or has been converted to low-quartz. The natural crystals generally are small, under 1 mm. in size. The habit is usually octahedral, rarely cubical,[42] and the forms {110} and {331} also have been observed.[43] The octahedra commonly are more or less skeletalized, with perfect corners and edges but with sunken faces, or grade into dendritic or branching skeletal growths. The skeletal growths are often elongated along [100] axes, and have octahedral terminations. Also found as spherulitic growths, ranging up to several inches in size, with a radial fibrous structure. High-cristobalite occurs twinned on (111), either as simple twins flattened parallel to the composition surface, or as repeated groupings. The high-cristobalite produced in synthetic systems, by the devitrification of glass, etc., typically develops as skeletal octahedra or as spherulitic, fern-like or dendritic growths and as serrated needles or blades.

COMPOSITION

Ideally, silicon dioxide, SiO_2. Essentially stoichiometric cristobalite has been synthesized,[2] but most or all natural material, and synthetic material crystallized from systems containing cations such as Al and alkalies in addition to Si, has been said to represent a solid solution. Although chemical analyses of crystallized cristobalite are lacking, aside from the original approximate analysis of 1887, theoretical crystallochemical considerations and the well-known existence of silicates and other oxysalts that are structural derivatives of cristobalite suggest that cristobalite can or does undergo compositional variation through the substitution of Al for Si with valence compensation effected by the coupled entrance of alkalies or other cations into interstitial positions. The very numerous chemical analyses of opal that have been reported outwardly support this belief, but the analyses are unsatisfactory in that they include an unknown and sometimes undoubtedly large amount of non-essential material present in mechanical admixture or in an adsorbed or capillary state. The entrance of H_2O or H_3O into interstitial positions may be a factor, especially in opal and other material formed at low temperatures, and the substitution of (OH) for oxygen, already indicated in quartz and in some opal, is an added possibility.

The original and only analysis of natural cristobalite, made on 81 mg., reported[9] SiO_2 91.0, Fe_2O_3 and a little Al_2O_3 6.2, total 97.2 per cent; Ca was sought but not found and the ignition loss was stated to be less than 1 per cent. Although this analysis has been cited to support the idea that cristobalite is a solid solution, the analyst remarked, "Die kleine Menge von Eisenoxyd und Thonerde rührt ohne Zweifel vom Gesteine (reich an Flitterchen von Eisenglanz) her, welches nur unvollkommen von den oktaëdern getrennt werden konnte." Most of the analyses of natural tridymite cited in this general connection also have been uncritically interpreted.

The measured values of the indices of refraction of natural and synthetic cristobalite hardly show a significant variation (Table 36). Tridymite, on the other hand, shows a wide variation in optical properties. It is surprising that the chemical composition, and the relation of the unit cell dimensions, indices of refraction, and other properties thereto, of these important substances should receive so little factual study.

SOLUBILITY

Coarsely crystallized (low-) cristobalite is insoluble in acids, except HF, but is attacked by fused alkalies. The submicrocrystalline, hydrous, opaline types (opal) are easily soluble in alkaline solutions. On the solubility in water see further under *Quartz*.

SYNTHESIS

Obtained as crystals, usually dendritic or skeletal, by the devitrification of silica glass; the rate of crystallization, extremely slow below about 1200°, increases with increasing temperature and complete devitrification may be reached in an hour or two at 1600°. The rate is accelerated by water vapor. The nature of the devitrification product is influenced by the presence of cations present in the silica glass;[45] tridymite is favored by Li, Na, and K, and cristobalite by Mg, Ca, Ba, and Pb. At low temperatures, below about 400°, in the presence of alkaline water or water vapor, silica glass recrystallizes to quartz. Also formed by calcining quartz; the fine-grained fibrous varieties chalcedony and flint convert more rapidly than does coarsely granular quartz, but the rate becomes appreciable only over about 1200°. Obtained in crystals[44] by fusing dried silica gel or "precipitated silica" in $NaPO_3$ at 700–800°.

VARIETIES

Two main varieties of (low-) cristobalite are distinguished: lussatite, with a fibrous structure visible to the unaided eye or under low magnification; and opal, with a dense, often glassy appearance. The division is arbitrary, since the two types are gradational. Opal is treated in a separate section (page 287).

LUSSATITE. Lussatite *Mallard* (*Bull. soc. min.*, **13**, 63, 1890; *C. R.*, **110**, 245, 1890). Lussatin *Laves* (*Naturwiss.*, **27**, 706, 1939). Pseudolussatin *Braitsch* (*Heidelberg. Beitr. Min. Pet.*, **5**, 331, 1957). Menilite, pt. Opal, pt.

Fibrous varieties of cristobalite are of wide occurrence in association with opal and chalcedony in near-surface geologic environments and in low-temperature hydrothermal occurrences. Such material earlier has been classed with chalcedony, which it closely resembles, and with tridymite, but x-ray study[46] has shown that it is identical with (low-) cristobalite. Specific gravity 2.0–2.1, in the range of that of opal and lower than that of chalcedony. Color white, gray, bluish gray, yellowish, brownish, etc. Translucent to opaque. Found as crusts and botryoidal aggregates; small-stalactitic; spherulitic; as layers with chalcedony in agates, etc.; also in opal and opalized wood. The fibers commonly are elongated[47] on [1$\bar{1}$0], [10$\bar{1}$], [01$\bar{1}$] and show positive optical elongation; also prismatic or fibrous on [111] (?) with negative optical elongation, or indistinctly lamellar on (111). The birefringence is lower than in chalcedony. Lussatite from Weitendorf, Austria,[52] has ω about 1.440. Coarsely fibrous to columnar cristobalite commonly occurs as spherulites in rhyolite, obsidian, and other igneous rocks but has not been described in the present connections. A fibrous cristobalite also has been observed in a blast-furnace slag.[49]

Some birefringent opal without optical evidence of fibrous structure shows a fiber texture on x-ray study. Fibrous cristobalite (lussatite) may undergo a phase conversion[50] in nature to fibrous quartz (chalcedony); a mutual orientation may be preserved, with cristobalite fibers elongated on [110] converting to chalcedony elongated on [11$\bar{2}$0], or on [111] to [0001]. It has been suggested that the fiber orientation in lussatite and opal has been controlled by stresses set up during the crystallization of an initial gel.[51] The chemical composition of lussatite presumably is similar to that of opal, perhaps lower in water. An analysis[48] of lussatite (?) from Bojanovic, Czecho-slovakia, gave SiO_2 92.60, $(Fe, Al)_2O_3$ 3.36, CaO 0.47, MgO 1.13, ign. loss 2.72, total 100.28, and material from another locality gave[61] SiO_2 90.64, ign. loss 2.62.

Lussatite was originally described from the bitumen veins of Lussat and nearby places in Puy-de-Dôme, France; also from the Faroe Islands, Corn-wall, and Tresztyan, Hungary. In serpentine at localities in western Moravia;[53] at Weitendorf, Styria, Austria;[54] in Bohemia; etc.

Lussatite and chalcedony are here employed as inclusive names for the fibrous varieties of cristobalite and quartz, respectively. A number of addi-tional names, for which there seems no real need, have been used according to the optical or crystallographic elongation of fibers; see tabulation below.

Chalcedony (fibrous quartz)			Lussatite (fibrous cristobalite)		
	Elongation			Elongation	
(Chalcedony, common type)	[11$\bar{2}$0]	−	(Lussatite, original)	[110]	+
Quartzin	[0001]	+	Lussatin	[111]?	−
Pseudoquartzin	[0001]	+	Pseudolussatin	[111]?	−
Lutecin	[$\bar{2}$113]?	+	Menilite, etc.	[?]	?
Pseudochalcedony	[?]				
Beekite, etc	[?]	?			

A special name has not been suggested for chalcedony elongated on [10$\bar{1}$0](-).

Additional names have been suggested on other grounds for fibrous varieties of quartz and types of opal in which the fiber characteristics have not been described.

OCCURRENCE

Cristobalite has two main types of occurrence in nature. (1) As opal, or opaline silica, constituting massive, submicrocrystalline aggregates formed at relatively low temperatures and containing a large amount of non-essential water. It occurs as crusts, veinlets, cavity-fillings, and concretionary masses, and also as biochemical deposits and siliceous sinters of various types. In many occurrences it probably has formed directly as more or less disordered low-cristobalite (opal-cristobalite). The properties and geological occurrence of this type of cristobalite are described in a separate section (see *Opal*). (2). As coarsely crystallized material, often euhedral, or as thick-fibrous spherulites, formed at relatively high temperatures and usually in igneous rocks. This material has initially crystallized as high-cristobalite. Some selected localities are given below.

Euhedral cristobalite occurs widely in the vesicles of igneous rocks, particularly in andesites, trachytes, and the lithophysae of rhyolites; typical associated minerals are tridymite, quartz, sanidine, augite, fayalite, and magnetite. Also common in microcrystalline form or as spherulitic growths in the groundmass of rhyolites and other acidic lavas. It also occurs abundantly as spherulites up to an inch or more in diameter in obsidian. The cristobalite is here often associated with sanidine and tridymite.[27] Cristobalite was originally found as octahedra and spinel twins up to 4 mm. in size in andesite on the Cerro San Cristóbal near Pachuca, Mexico. It is widespread in the Tertiary lavas of the western United States, as in the San Juan region of Colorado;[55] in Yellowstone Park, Wyoming;[56] in the Columbia River basalts, and elsewhere in cavities in basaltic rocks; in andesite at Crater Lake, Oregon.[57] Found at numerous occurrences in the Tertiary and Quaternary basalts, dacites, and andesites of the Caucasus and Transcaucasia, Russia;[58] well crystallized in cavities in andesite at Nezdenice, Bohemia;[39] in inclusions in basalt and other rock types, as at Mayen and Niedermendig, Germany;[59] as twinned crystals in rhyolite at Sarospatak, Hungary;[60] in the trachyandesites of Puy de Clierzou, France, amounting to as much as 10 per cent of the rock;[61] as crystals in gas cavities and idiomorphic in the groundmass of an auganite dike at the Kosaka mine, Japan;[62] in various Quaternary lavas in Kamchatka.[64] Dull white spherulites, internally cracked or cavernous and ranging up to an inch in size, occur embedded in lustrous black obsidian at Glass Mountain, Little Lake, Inyo County, California. Near Tokatoka, New Zealand,[67] an andesite dike intruded into limestone has produced a contact zone 4–6 in. thick composed of rounded grains about 0.01 mm. in diameter of hydrogrossular in a matrix of cristobalite, the latter comprising about 60 per cent of the rock.

The cristobalite (and tridymite) of some occurrences in igneous rocks may be completely converted to low-quartz.[65] At Ellora, Hyderabad, India, sharply crystallized octahedra and spinel twins of cristobalite occur perched on fibers of mordenite in cavities in basalt; some of the crystals are converted to randomly oriented aggregates of quartz grains.[66] At this occurrence the temperature of formation must have been relatively low, perhaps 300° or below, and indeed in all its occurrences in volcanic rocks the mineral must have formed at temperatures and pressures at which it was metastable. The opaline varieties of cristobalite, together with chalcedony, in part a conversion product of earlier formed cristobalite, also occur in cavities in rhyolite and other volcanic rocks, but more generally as a product of postconsolidation hydrothermal activity. The microcrystalline and spherulitic cristobalite in the groundmass of dacites and other volcanic rocks has been ascribed in some instances to the alteration of the glassy base of the rock by heated waters.

Cristobalite also has been observed as a recrystallization product of argillaceous sandstone at the contact with basalt in the Transcaucasus, Russia.[68] Pseudomorphs of cristobalite after tridymite have been reported.[69] Distorted octahedra or twins of cristobalite may simulate tridymite in habit; an instance of this kind, which might be a paramorph of cristobalite after tridymite, has been described[72] from basalt near Eschwege, Germany. Cristobalite occurs with cassiterite in veinlets in rhyolitic rocks at Guanajuato, Mexico,[70] and spherulitic lussatite has been found with wood tin and tridymite in Lander County, Nevada.[71] The quartz grains adhering to or embedded in the surface of fulgurites in sand may be partly or completely converted to cristobalite.

HISTORICAL

Cristobalite was found personally by vom Rath in 1884, while making a field examination of the occurrence of tridymite on the Cerro San Cristóbal, Mexico, from whence specimens had been sent to him some years earlier. The question of the separate identities of cristobalite and tridymite remained unsettled for some years. The high-low inversion was discovered by Mallard in 1890 by heating specimens in oil under the microscope. The wide variation in the temperature of this inversion as a function of the previous history of the sample was recognized by Fenner in 1913. The analyses of the crystal structures of high- and low-cristobalite in the period 1925–1935 by Wyckoff, Barth, and Nieuwenkamp, in hand with the earlier determinations of the structures of tridymite and quartz, opened the way to an interpretation of the metastable existence of the high-temperature forms and of the structural and compositional relations obtaining between them and various silicates. The basic control was recognized as the infilling by cations of the interstices in the relatively open structures found for the high-temperature forms, in compensation for the substitution of Al^3 for Si^4 in the tetrahedral framework (Buerger, 1935, 1954).[73] X-ray studies in 1932 and later years also proved that opal, long thought to be amorphous, was a submicrocrystalline aggregate of disordered crystallites of low-cristobalite.

MELANOPHLOGITE *Lasaulx* (*Jb. Min.*, **175**, 250, 627, 1876; 513, 1879). Leukophlogite *Bombicci* (*Acc. sci. ist. Bologna*, **22**, 1891). Sulfuricin (?) *Guyard* (*Bull. soc. chim.*, **22**, 61, 1874); *Brezina* (*Min. Mitt.*, **243**, 1876).

A problematic,[74] pseudomorphous occurrence of opaline silica, chalcedony, and quartz after cubes of an unidentified mineral, perhaps high-cristobalite or fluorite. Found rarely as coatings on crystals of sulfur, celestite, and calcite at Racalmuto, Caltanisetta, and other localities in the sulfur deposits of Sicily. The cubes, usually up to 1 mm. in size, are externally coated by opaline or chalcedonic material, or by minute quartz crystals. Internally, the cubes may be isotropic (opaline) or fibrous, or show a division into six weakly birefringent uniaxial sectors subjacent to the cube faces. Color is usually yellowish to brownish with a strong vitreous luster, also colorless to white. On ignition the opaline material turns black or bluish black (hence the name from $\mu\acute{\epsilon}\lambda\alpha\varsigma$, *black*, and $\phi\lambda\acute{\epsilon}\gamma\epsilon\sigma\theta\alpha\acute{\iota}$, *to be burned*). Analyses show the presence of 5–7 per cent SO_3, and organic material also may be present. A similar material containing about 12 per cent SO_3 occurs as a solfataric alteration of basalt on Kilauea;[63] n 1.46, blackens on heating.

References

1. Flörke: *Ber. deutsch. keram. Ges.*, **32**, 369 (1955); Eitel: *Bull. Am. Ceram. Soc.*, **36**, 142 (1957).
2. Hill and Roy: *J. Am. Ceram. Soc.*, **41**, 532 (1958); Krisement and Trömel: *Zs. Naturforsch.*, **14A**, 912 (1959).
3. Nieuwenkamp: *Zs. Kr.*, **96**, 454 (1937); Barth: *Am. J. Sci.*, **23**, 350 (1932) and **24**, 102 (1932); Wyckoff: *Zs. Kr.*, **62**, 189 (1925); Barth and Posnjak: *Zs. Kr.*, **81**, 376 (1932); Buerger: *Zs. Kr.*, **90**, 186 (1935).
4. Crystal class from x-ray study of Barth (1932); Wyckoff (1925) gave the holohedral space group *Fd3m*. Morphological evidence of the crystal class is lacking.
5. Lukesh: *Am. Min.*, **27**, 226 (1942).
6. Wyckoff (1925).
7. Barth (1932).
8. Flörke (1955); Eitel (1957); Laves: *Naturwiss.*, **27**, 705 (1939).
9. vom Rath: *Jb. Min.*, I, 198 (1887).
10. Crystal class indicated by the x-ray study of Nieuwenkamp: *Zs. Kr.*, **92**, 82 (1935).
11. Jay: *Min. Mag.*, **27**, 54 (1944).
12. Tokuda: *Nippon Kagaku Zasshi*, **79**, 1063 (1958).
13. Swanson and Tatge: *Nat. Bur. Std. Circ. 539*, **1**, 39 (1953).
14. Rotter: *Sborn. Stát. geol. úst. Českoslov. rep.*, **16**, 401 (1949).
15. Nieuwenkamp (1935).
16. Barth and Posnjak (1932).
17. Swanson and Tatge (1953) with summary of earlier work; see also van Valkenburg and Buie: *Am. Min.*, **30**, 526 (1945); Gruner: *Econ. Geol.*, **35**, 867 (1940); Hurlbut: *Am. Min.*, **21**, 727 (1936); Inuzuka: *J. Geol. Soc. Japan*, **47**, 367 (1940); Michel-Lévy: *C. R.*, **226**, 1455 (1948).
18. See Weymouth and Williamson: *Min. Mag,*, **29**, 573 (1951).
19. Ustiev: *Trav. inst. pét. ac. sci. URSS*, no. 6, 159, 1934.
20. Mallard: *Bull. soc. min.*, **13**, 161 (1890); Fenner: *Am. J. Sci.*, **36**, 354 (1913); Bauer: *Jb. Min.*, I, 199 (1887); Holmquist: *Geol. För. Förh.*, **33**, 245 (1911); Rosický: *Festschr. V. Goldschmidt*, Heidelberg, 1928, p. 229; van Valkenburg and Buie (1945).
21. van Valkenburg and Buie (1945).
22. Fenner (1913).
23. Ivanov: *Trav. inst. pét. ac. sci. URSS*, no. 9, 23, 1936.

24. Lapin: *Trav. inst. Pét. ac. sci. URSS*, no. 13, 247, 1938.
25. Weil: *C. R.*, **181**, 423 (1925).
26. Longchambon: *C. R.*, **180**, 1855 (1925).
27. Rogers: *Am. Min.*, **6**, 4, 60 (1921).
28. Weymouth: *Am. J. Sci.*, **16**, 237 (1928).
29. Ramdohr: *Cbl. Min.*, 33, 1920.
30. Bowen and Anderson: *Am. J. Sci.*, **37**, 487 (1914).
31. Riecke and Endell: *Min. Mitt.*, **31**, 501 (1912).
32. Weymouth and Williamson (1951).
33. Hutton: *Trans. Roy. Soc. New Zealand*, **74**, 125 (1944).
34. Hawley and Beavan: *Am. Min.*, **19**, 493 (1934).
35. Dunham: *Am. Min.*, **18**, 369 (1933).
36. Shannon: *Proc. U.S. Nat. Mus.*, **62**, Art. 12 (1923).
37. Kondo and Yamauchi: *J. Soc. Chem. Ind. Japan*, **38**, Suppl., 651 (1935).
38. Anderson: *Am. J. Sci.*, **39**, 418 (1915).
39. Rosický (1928).
40. Washburn and Navias: *J. Am. Ceram. Soc.*, **5**, 584 (1922).
41. Rogers: *Am. J. Sci.*, **45**, 222 (1918) and *J. Geol.*, **30**, 211 (1922).
42. Murdoch: *Am. Min.*, **27**, 500 (1942).
43. Rosický (1928); the form {331} is inconsistent with tetartoidal symmetry.
44. Peyronel: *Zs. Kr.*, **95**, 274 (1936).
45. Rieck and Stevels: *J. Soc. Glass Technol.*, **35**, 284 (1951).
46. Braitsch: *Heidelberg. Beitr. Min. Pet.*, **5**, 331 (1957); Laves (1939); Novák: *Spisy vyd. přiro. fak. Masarykovy Univ.*, no. 153, 1932, and *Bull. soc. min.*, **70**, 288 (1948); Flörke: *Jb. Min., Monatsh.*, 1955, 217.
47. See x-ray studies of Braitsch (1957) and Laves (1939).
48. Kovař: Abstr. in *Zs. Kr.*, **37**, 500 (1903).
49. Pehrman: *Bull. Geol. Inst. Upsala*, **32**, 475 (1948).
50. Braitsch (1957).
51. Laves (1939).
52. Neuwirth: *Min. Mitt.*, **3**, 32 (1952).
53. Slavík: *Cbl. Min.*, 690, 1901: Novák: (1932).
54. Neuwirth (1952); Braitsch (1957).
55. Larsen, Irving, Gonyer, and Larsen: *Am. Min.*, **21**, 679 (1936).
56. Howard: *Am. Min.*, **24**, 485 (1939).
57. Dutton: *Am. Min.*, **22**, 804 (1937).
58. Beljankin and Petrov: *Bull. ac. sci. USSR, Ser. Geol.*, 1936, 303.
59. Lacroix: *Bull. soc. min.*, **14**, 185 (1891); Gaubert: *Bull. soc. min.*, **27**, 242 (1904).
60. Lengyel: *Föld. Közl.*, **67**, 309 (1937).
61. Bentor: *C. R.*, **213**, 211, 289 (1941).
62. Ohashi: *J. Geol. Soc. Tokyo*, **43**, 453 (1936).
63. Payne and Mau: *J. Geol.*, **54**, 345 (1946).
64. Vlodanetz: *Beljankin Jubilee Vol., ac. sci. URSS*, 359, 1946.
65. See Moehlman: *Am. Min.*, **20**, 808 (1935).
66. van Valkenburg and Buie (1945).
67. Mason: Am. Min., **42**, 379 (1957).
68. Beljankin and Petrov: *Am. Min.*, **23**, 153 (1938).
69. Rogers: *Am. J. Sci.*, **45**, 222 (1918); Beljankin: *C. R. ac. sci. URSS*, **21**, 249 (1938); Bentor (1941); Vlodavetz (1946).
70. Switzer and Foshag: *Am. Min.*, **40**, 64 (1955).
71. Knopf: *U.S. Geol. Surv. Bull. 640G*, 133, 1916 and *Econ. Geol.*, **11**, 652 (1916).
72. Ramdohr (1920).

73. Buerger: *Am. Min.*, **39**, 600 (1954) and *Zs. Kr.*, **90**, 186 (1935).
74. Summary of older literature in Hintze: *Handbuch der Mineralogie*, **1**, Berlin, 1906, Lfg. 10, p. 1538; see also Glisczynski and Stoicovici: *Zs. Kr.*, **99**, 238 (1938), and Flörke: *Zs. Kr.*, **112**, 126 (1959).
75. Büssem, Bluth, and Grochtmann: *Ber. deutsch. keram. Ges.*, **16**, 381 (1935).
76. Cole: *J. Am. Ceram. Soc.*, **18**, 149 (1935).
77. Patzak: *Jb. Min.*, *Monatsh.*, 1956, 101.

OPAL

Opal, considered a separate entity among gems and minerals since times of antiquity, on the basis of its outward characters, became classed in most eighteenth century mineralogical works as a type of *silex* or *kiesel*. It was usually placed in the category of translucent fine-grained types of silica then given by many the general name agate, in distinction to the opaque types classed under jasper, and now known as chalcedony or chalcedonic silica. In modern works opal has been classed as a separate species. In general it has been characterized as an amorphous mineral gel. In recent years, however, x-ray diffraction study has shown that opal is not amorphous but is a submicrocrystalline aggregate of crystallites of cristobalite, containing much nonessential water. Opal is here considered as a variety of cristobalite, standing to that species much in the same relation that chalcedony does to quartz. It is described, however, as if it were a species.

OPAL. Upala *Sanskrit*. Opalus, Paederos *Pliny* (*Nat. Hist.*, **37**, 21, 23, A.D. 77). Quartz résinite *Haüy* (*Traité de min.*, **2**, 1801). Hyalite (Hyalit, Hialit, Müller's Glass) *Werner* (Hoffmann: *Min.*, **2a**, 134, 1812; Karsten: *Min. Tab.*, 22, 1800). Glasopal *Hausmann* (*Handb. Min.*, 424, 1813). Menilite *de Saussure* (Delamétherie: *Théor. de la terre*, **2**, 169, 1797; Delarbre and Quinquet: *J. phys.*, **31**, 219, 1787). Leberopal *Karsten* (*Min. Tab.*, 24, 1800). Jasp-Opal *Karsten* (*Min. Tab.*, 26, 1808). Eisenopal *Hausmann* (*Handb. Min.*, **428**, 1813). Semi-Opal, Halb-Opal *Werner* (Emmerling: *Min.*, **1**, 248, 1793; Hoffmann: *Min.*, **2a**, 134, 1812). Wood-Opal, Holz-Opal *Germ.* Cacholong *Mongolian name* (? from *kaschtschilon*, beautiful stone). Mother-of-pearl-opal, Pearlmutter-Opal *Karsten* (*Min. Tab.*, 1808).

Jenzschite *Dana* (*Syst. Min.*, 201, 1868; 194, 1892). Passyite *Marchand* (*Ann. chim. phys.*, **1**, 393, 1874). Grossouvréite *Meunier* (*Bull. soc. géol. France*, **2**, 250, 1902). Granuline pt. (?) *Scacchi* (*Rend. acc. Napoli*, **21**, 176, 1882). Forcherite *Aichhorn* (*Wien. Ztg. Abendbl.*, July 11, 1860; Maly: *J. prakt. Chem.*, **86**, 501, 1862). Neslite *Leymérie* (*Stat. min. de l'Aube*, 116, 1846).

Sinters and Tripoli; Diatomite. Fiorite, Siliceous Sinter, Kieselsinter *Santi* (*Crell's Ann.*, **2**, 589, 1796; Thomson: *J. phys.*, **39**, 407, 1791, and **1**, 108, 1796; Pfaff: *Crell's Ann.*, **2**, 589, 1796); Kieselsinter *Karsten* (*Min. Tab.*, 24, 1808). Kieseltuff *Klaproth* (*Beitr.*, **2**, 109, 1796); Geysirite *Delamétherie* (*Leçons Min.*, 1812; Damour: *Bull. soc. géol. France*, 157, 1848). Geyserite. Michaelite *Webster* (*Am. J. Sci.*, **3**, 391, 1821). Pealite *Endlich* (*U.S. Geol. Surv.*, *6th Ann. Rpt.*, 1873, 153; *Am. J. Sci.*, **6**, 66, 1873). Viandite *Goldsmith* (*U.S. Geol. Surv.*, *12th Ann. Rpt.*, Pt. 2, 407, 1883). Tripoli; Bergmehl (pt.), Kieselmehl, Kieselguhr (*Germ.*). Tripoli slate, Polishing slate; Tripelschiefer, Saugkiesel, etc., (*Germ.*). Randannite *Salvétat* (*Ann. chim. phys.*, **24**, 348, 1848; Lacroix: *Bull. soc. min.*, **39**, 85, 1916); Ceyssatite *Gonnard* (*Min. Puy-de-Dôme*, 1876, 14); Lassolatite *Gonnard* (1876). Vierzonite *Grossouvre* (*Bull. soc. géol. France*, **1**, 431, 1901). Floatstone, Schwimmstein *Germ.*, Silex nectique *Haüy* (*Traité de min.*, **2**, 1801).

MODE OF AGGREGATION

Opal occurs as crusts with a botryoidal, globular, reniform, or ropy surface; small-stalactitic to coralloidal; as concretionary masses with a tuberose or irregular shape. Commonly as a cavity-filling and as veinlets. The siliceous sinters and geyserites are porous to firm in texture; sometimes filamentous, fibroid, or in leathery or sponge-like forms; in cauliflower-like forms; scaly-massive to compact massive. Also massive, rock-forming, poorly compacted (diatomites) to hard and dense.

X-RAY STUDIES

The earliest[1] x-ray powder diffraction studies of opal indicated that the substance was amorphous, as earlier thought on physical and optical grounds. Later work[2] by superior experimental techniques, however, has shown that opal is in general a crystalline aggregate, composed of submicroscopic crystallites of a substance, here called opal-cristobalite, with a more or less disordered internal structure. The large and variable content of non-essential water characteristic of the aggregate is primarily a consequence of the extremely small particle size, or large internal surface, and the manner of formation in nature. In the older literature the crystalline substance present usually was called high-cristobalite, in the ideal sense, but recent x-ray and thermal studies[3] have led to a characterization as a disordered type of low-cristobalite. In this, the one-dimensional disorder present can be broadly described in terms of a layer-like linkage of (SiO_4) tetrahedra, approximately 4.1 Å thick, which can be recognized in the structure of the three classical polymorphs of SiO_2. In quartz these layers are arranged parallel to $(10\bar{1}1)$, in cristobalite parallel to (111), and in tridymite parallel to (0001). The idealized structure of tridymite consists of two geometrically distinguishable kinds of alternating layers, A and B, with the stacking sequence $ABAB\ldots$ The idealized structure of cristobalite, which is taken at high temperatures of formation, is based on the stacking sequence $ABCABC\ldots$ At lower temperatures of formation, and in the presence of alkalies and other cations as noted beyond, an interstratification of both two-layer and three-layer sequences may take place. The interstratification may be periodic, giving rise to superstructures (polytypes), or, more commonly, may be random. In opal the three-layer sequences are dominant (opal-cristobalite). With increasing disorder the x-ray pattern becomes diffuse with the appearance of the stronger diffraction lines of low-tridymite. The high-low inversion temperature also decreases, and the dilatometric curves of highly disordered material show less sharp volume changes.[4]

The disordering is caused by the entrance of cations such as Al, Ca, Mg, and alkalies and probably also of H_2O and (OH) into solid solution. Opal-cristobalite thus is not pure SiO_2 but is a stuffed derivative thereof.

In addition to the diffraction lines of opal-cristobalite, x-ray patterns of opal and especially of opaline rocks may show distinct lines of quartz, chiefly representing chalcedonic material, and of feldspar, clay minerals, and other admixed material. Some opals, especially material of biochemical origin, give faint and very diffuse x-ray patterns. A low, extremely broad peak centering at about $21°$ 2θ is characteristic. The availability of cations that can stabilize cristobalite may be a factor, in addition to the temperature of formation.

Submicrocrystalline opal-cristobalite also forms in artificial silica gels[5] on ageing at ordinary temperatures and more rapidly on hydrothermal treatment. At higher temperatures in water or superheated steam the gel may convert to quartz, depending on the temperature and pressure, with cristobalite appearing first as a metastable, transient phase. The conversion is sensitive to the presence of alkalies and other cations in the wet gel. When heated in air, silica gels or dried, powdery $SiO_2 \cdot nH_2O$ lose water at low temperatures and in the range 800–1400° crystallize rapidly to cristobalite, tridymite, or high-quartz,[6] depending on the amount and kind of alkalies and other cations present.

PHYSICAL PROPERTIES

In its purest form, opal is white to colorless and transparent, commonly milky white or bluish white; also, through admixture of pigmenting material, ranges from yellow and yellowish brown to brown, reddish brown, orange, green, and blue, but usually in pale shades; also gray to black. Opal is sometimes found colored red by disseminated cinnabar[7] or orange-yellow by orpiment; frequently brown or reddish brown by iron oxide. It sometimes shows a rich internal play of colors by reflected light (precious opal). Streak white. Three different color mechanisms occur in opal: pigmentation by finely divided foreign material; interference of light, giving the color play of precious opal; and scattering of light, causing a pale bluish tint in reflected light and a brownish tint in transmitted light. The third mechanism is dominant in the turbid, milky types of opal. Two or all three mechanisms can coexist in a given specimen, but intensive pigmentation by dispersed, opaque foreign material may so reduce the optical transmission that observation of scattering or interference effects may be precluded. In the transparent and subtransparent types of precious opal interference of light is the dominant coloring mechanism, although a weak body-color caused by a foreign pigment also may be apparent. In the hyalite variety of opal, pigmentation, interference, and scattering effects are lacking. Opaque to transparent; generally subtranslucent to subtransparent.

Fracture conchoidal in varying degree; also flat-conchoidal to smoothly undulatory or broadly ribbed, in some types splintery to irregular. Moderately brittle, occasionally markedly so. May spontaneously crack by sudden change of temperature and, more generally, by partial dehydration. Density variable, depending primarily on the water content and on porosity and cracks; usually 1.99–2.25, but ranging down to about 1.8 in sinters and porous material. The density increases with absorption of water, markedly in the case of hydrophane and other very porous, dehydrated types (see *Dehydration*). Hardness $5\frac{1}{2}$–$6\frac{1}{2}$. Some opal, especially hyalite, shows a greenish yellow fluorescence in ultraviolet radiation because of traces of the uranyl ion.[58] Not changed in color by irradiation with x-rays.[73]

OPTICAL PROPERTIES

Opal usually is isotropic, but may show weak anomalous birefringence because of strain and is then uniaxial negative. The index of refraction of different specimens varies with the water content and usually is in the range 1.435–1.455 (Tables 37, 38, and 39). These values are much below the mean index of refraction of coarsely crystallized low-cristobalite, 1.485. The index of refraction of individual specimens increases as additional water is taken up and decreases below the original value as water is removed to leave voids (see *Dehydration*). Minute fibers or bundles of chalcedony are sometimes seen in opal, especially in opaline cherts and other rock-forming types. Fibrous cristobalite, lussatite, sometimes occurs associated with or admixed with opal, but most opal that yields an x-ray diffraction pattern of low-cristobalite does not show such fibers. Microscopic crystals of tridymite were said[51] to be common in opal, probably through confusion with lussatite or chalcedony.

The origin of the color play in precious opal is not well understood. The display of colors is seen in reflected light, not in transmitted light, or very weakly, and the individual colors often are of considerable spectral purity and strength. A spectral line width of about 20 Å has been measured for some red colors. The color play does not appear in monochromatic illumination. Although the colors are caused by the interference of light, the nature of the structure that gives rise to the effect is problematic.[8] It may be an open, regularly spaced grid-work of crystallites of cristobalite. Some specimens of precious opal, however, give an x-ray powder pattern, whereas others do not. Thin lamellae with an index of refraction different from that of the adjoining material, and systems of cracks or fissures, perhaps filled with opal of slightly different index of refraction, also have been postulated. The high saturation of the colors requires that the structure be periodic over a considerable thickness. Strains set up during the drying of the initial gel and of the opal may be a factor.

TABLE 37. INDEX OF REFRACTION OF OPAL

Locality	n, Original	n, after Immersion in H_2O	Ref.
Hyalite, Waltsch	1.458		9
Hyalite, Waltsch	1.455		10
Hyalite, Waltsch	1.4374		10
Hyalite, Mexico	1.451		12
Hyalite, North Carolina	1.418–1.425		16
Hyalite, Guanajuato	1.456–1.458		17
Hyalite, California	1.420		18
Hyalite, Austria	1.457		56
Hyalite, Austria	1.437		56
Hydrophane	1.406	1.446	10
Hydrophane	1.266	1.406	10
Hydrophane	1.387	1.439	10
Hydrophane	1.368	1.443	13
Hydrophane	1.2290	1.3961	15
Hydrophane	1.398	1.4344	15
Tabasheer	1.119	1.364	10
Tabasheer	1.111		14
Opal, plant cells	1.430–1.452		78
Diatomite	1.440–1.448		20
Diatomite	1.441		19
Geyserite, Iceland	1.428–1.445		
Opal, milky	1.442		10
Opal, milky	1.4536		9
Opal, milky, Austria	1.466		56
Opal, milky, Austria	1.439		56
Opal, milky, Austria	1.431		56
Opal, white, Austria	1.473		56
Opal, bluish, Austria	1.439		56
Opal, yellow, Austria	1.444		56
Opal, yellow, Austria	1.458		56
Opal, Brazil	1.428–1.436		55
Opal, Styria	1.431–1.473		56
Opal, Italy	1.440–1.446		57
Opal, brownish, Queretaro	1.418–1.422		16
Opal, wood	1.415–1.423		16
Opal, yellow, Virgin Valley	1.414		16
Opal, milky, Camp Verde, Ariz.	1.448		16
Opal, fire, Guatemala	1.450		10
Opal, Guatemala	1.446		10
Opal, colorless, Guatemala	1.442		10
Opal, fire, Mexico	1.440		11
Opal, fire, Mexico	1.433		12

TABLE 38. SOME CORRELATED PROPERTIES OF OPAL

Locality	Specific Gravity	Index of Refraction	Water Content	Ref.
Opaline chert, California	1.983	1.445	7.34	21
Opaline chert, California	1.992	1.444	7.08	21
Wood opal, California	2.068	1.448	5.96	21
Hyalite	2.028	1.444	8.35	21
Hyalite, Honduras	2.111	1.456	4.93	21
Wood opal, California	2.098	1.451	5.05	21
Opal, in rhyolite	2.009	1.441	8.70	21
Opaline chert, California	2.080	1.448	6.3	21
Opaline chert, Nevada	2.067	1.451	5.34	21
Opaline chert, California	2.016	1.448	4.75	21
Opal, Hungary	2.096	1.4531	6.33	22
Hyalite, Hungary	2.036	1.4465	8.97	22
Hyalite, Japan	2.139	1.4567	4.71	22
Hyalite, Bohemia	2.160	1.4592	3.55	22
Opal, fire, Zimapan	2.008	1.4410	9.16	22
Opal, green, Moravia	2.038	1.4445	7.40	22
Opal, yellowish, Moravia	2.074	1.4491	7.05	22
Opal, milky, Moravia	2.046	1.4528	5.36	22
Opal, milky, Moravia	2.075	1.4499	6.27	22
Opal, milky, Moravia	2.116	1.4525	5.30	22
Opal, milky, Moravia	2.098	1.4518	5.36	22
Opal, milky, Moravia	2.056	1.4499	6.05	22
Opal, milky, Moravia	2.070	1.4478	6.17	22
Opal, milky, Moravia	2.055	1.4496	5.58	22
Opal, milky, Moravia	2.122	1.4501	5.25	22
Opal, milky, Moravia	2.104	1.4512	4.76	22
Opal, milky, Moravia	2.025	1.4425	8.36	22
Hyalite, Waltsch		1.4590	3.4	23
Hyalite, Germany		1.4562	4.84	23
Opal, fire		1.4514	8.5	23
Opal, Hungary		1.4459	9.73	23

TABLE 39. VARIATION IN INDEX OF REFRACTION OF PRECIOUS OPAL[24]

Locality	Original Index (white light)	After Immersion in H_2O for 48 hr.	After Heating at 74°C. for 48 hr.
Hungary: girasol, bluish	1.433	1.451	1.421
Querétaro: fire opal, red brown	1.441	1.443	1.432
Virgin Valley: transparent, pale brown	1.447	1.448	1.442
White Cliffs: milky, green and blue color play	1.456	1.457	1.449

The Virgin Valley and Querétaro material gave strong, broad x-ray powder lines of cristobalite; the other material, a extremely faint and diffuse pattern.

CHEMICAL COMPOSITION

Opal is hydrous silica, $SiO_2 \cdot nH_2O$. The water content is quite variable, ranging up to 20 per cent or more in some natural material, but commonly is in the range from 4 to 9 per cent. A small part of the water may be held structurally in interstitial positions in the cristobalite crystallites that compose the material. The great bulk of the water, however, is adsorbed or capillary and is readily lost on strong desiccation or heating or, in part, by exposure to a dry atmosphere (see *Dehydration*).

Over 100 chemical analyses of opal and sinters have been reported,[25] and a selection is cited in Table 40. Aside from SiO_2 (which on an anhydrous basis generally ranges well over 90 per cent) and H_2O, the principal constituents are Al_2O_3, Fe_2O_3, CaO, MgO, and alkalies. The role of these constituents is not well established, but in many instances they must be ascribed in part or entirely to admixed clay, coprecipitated gels of $Al_2O_3 \cdot nH_2O$ and $Fe_2O_3 \cdot nH_2O$, and other physical impurities. The analyses are in general similar to those of flint, chert, and chalcedony. In part, especially with regard to the Al and alkalies, these constituents may represent material held in solid solution in cristobalite, and are responsible for the (metastable) crystallization of this phase, but there is no proof of this plausible supposition.

The content of Al usually ranges from a few tenths of a per cent of Al_2O_3 up to a few per cent, with Fe_2O_3 ranging to somewhat higher values, 5–7 per cent, and in red or brownish opals, evidently admixed with iron oxide, up to large amounts. Although CaO and MgO vary widely, they generally are present below 1 per cent or are lacking. Alkalies, not always reported in analyses of opal, are generally in the range of a few tenths to 1 weight per cent. There is no correlation of the alkalies with the amount of Al present. Organic material is found in some opals, such as those of Virgin Valley, Nevada, and SO_3 may be present (see *Melanophlogite*). An opal incrustation on sodalite-syenite from the Kola Peninsula, Russia,[59] contained much Na and F, apparently as admixed NaF. Opal and siliceous sinter has been found[60] to have a relatively high ratio of $Si^{28} : Si^{30}$ (30.09). An opal pseudomorph after gypsum has been found[61] to contain 0.0X per cent Ge, and up to 0.03 per cent Ge has been found in opal in fumaroles.[79]

Opal is readily and completely soluble in hot, strongly alkaline solutions.[26] An insoluble residue of admixed clay, chalcedonic silica, sand grains, iron oxide, etc., may be left. Hyalite is less rapidly attacked. Opal also may be partly or completely decomposed by hot, concentrated HCl, affording a gelatinous residue on evaporation, and it is more readily soluble in HF than is chalcedony or powdered quartz. The more rapid attack on opal by acids and alkalies is caused by the high porosity and relatively small particle size of this

TABLE 40. ANALYSES OF OPAL

	1	2	3	4	5	6	7	8	9	10
SiO_2	85.80	91.89	88.73	92.31	89.55	86.54	92.67	96.48	43.60	96.46
Al_2O_3	3.22	1.40	0.99	0.36	0.49	1.73	0.80	Tr.	Tr.	0.08
Fe_2O_3	1.85			}	}		0.14	0.30	52.16	0.09
CaO	0.96		0.49	0.22	0.63	0.55	0.05			
MgO	1.08	0.92	1.48	0.18	0.57	0.74	0.18			Tr.
Na_2O	Not det.								0.08	
K_2O	Not det.		0.34				0.75			
H_2O	6.95	5.84	7.97	5.31	8.03	9.40	5.45	3.33	3.75	3.26
Rem								Tr.	0.37	
Total	99.86	100.05	100.00	98.46	99.27	98.96	100.04	100.11	99.96	99.92
d		2.07		2.198	2.01	1.97		2.13	3.18	2.06

1. Precious opal. Virgin Valley, Nev. (?).[27] 2. Fire opal. Rákos, Hungary.[30] 3. Fire opal. Washington County, Ga.[28] 4. White opal. Faroe Islands.[29] 5. Hyalite. Elba.[31] 6. Milk opal. Elba.[31] 7. Siliceous sinter. Steamboat Springs, Nev.[32] 8. White opal. Nikolaevsky mine, Altai, Russia.[54] With trace SO_3: 9. Red opal. Nikolaevsky mine, Altai, Russia.[54] Rem. is SO_3, Na_2O tr. 10. Opal. Yugoslavia.[62] $H_2O(+)$ 1.34, $H_2O(-)$ 1.92.

material. This behavior has led to the belief that chalcedony, which is soluble in hot alkalies to a certain extent, contains opaline silica as an interstitial material (see further under *Chalcedony*).

DEHYDRATION

The variation in the water content of different opal specimens as found in nature presumably is connected with both initial differences during formation and with dehydration during later history. It has been shown that the dehydration of artificial silica gels is accompanied by a decrease in bulk volume up to a more or less well-defined point, beyond which further loss of water is accompanied by the development of voids. Whether or not a natural opal has formed in this way, it is found that natural material in general has reached a point in which the internal structure has become so established that artificial dehydration over short periods of time is accompanied by the development of voids with little or no change in bulk volume.

In general, the lower the content of water in different specimens of untreated natural material, the higher the specific gravity and the index of refraction. This is shown[21] by the data of Table 38. On the other hand, if an individual specimen is treated under laboratory conditions, it is found that dehydration is accompanied by a decrease in the specific gravity and the index of refraction. An opposite change is produced by hydration. Since natural material in general contains some voids, at least in specimens preserved under museum conditions, hydration by immersion in water may increase the content of water, and also the specific gravity and index of refraction, over the initial values of the specimen.

The water is lost continuously on dehydration, and definite hydrates are not present (as has been claimed[34]). A very small part of the water may be held to high temperatures,[35] perhaps representing H_2O in interstitial solid solution or (OH) in substitution for O in the (SiO_4) tetrahedra. Bound (OH) has been identified in some opal by infrared absorption study.[17] The rate at which water is lost under laboratory conditions varies considerably in different specimens.[36]

Dehydration of large pieces usually is attended by cracking, and on large loss of water initially transparent or translucent material becomes white, opaque, and porous. Desiccated, porous natural material readily imbibes water, in instances up to 50 per cent or more of its volume.[33] Hyalite has been observed to convert to quartz under shear at high confining pressures.[47] The hydrophane variety of opal, in its natural state white or yellowish and virtually opaque, by the absorption of water becomes almost transparent,[33] and some specimens acquire the play of color of precious opal. Organic liquids also are absorbed,[15] as are index-liquids and wax, and ammonia.[63] Cacholong opal also is very porous and adheres to the tongue but does not become transparent on immersion in water.

SYNTHESIS

Hard, opal-like masses of hydrous silica, sometimes transparent, or showing a color play, have been obtained experimentally by reaction of fluosilicic acid with glass,[37] by precipitation of sodium or potassium silicate or ethyl silicate solutions in various ways,[38] by the decomposition of silicon chloride or silicon fluoride,[39] and in other ways.

VARIETIES

The varietal nomenclature of opal, like that of the fine-grained varieties of quartz, has been greatly extended by trivial terms that refer to variations in color, gross structure, state of aggregation, and other secondary characters. A few of these terms have been preserved in the synonymy, which see. Other names or terms of more specific reference that are in general use are mentioned below.

Precious Opal; Noble Opal. Exhibits a play of brilliant colors, variously red, orange, green, or blue, appearing as spangles, wavy or flame-like bands, sheets, etc., usually set in a translucent to subtransparent matrix of a milky white or other body-color. Black opal has a black or other very dark body-color with a play of color. Harlequin opal has a variegated appearance, with a mosaic-like pattern of color in rounded, angular, or roughly rectangular patches of about equal size. Pin-fire opal exhibits closely spaced specks or pinpoints of color. The terms matrix opal and mother-of-opal are applied to matrix material containing closely spaced veinlets or specks of opal, sometimes cut as cabochons or plates.

Fire Opal. A type of precious opal with a dominantly red or orange play of color usually set in a relatively transparent background with a pale yellowish, to yellowish red, orange, or brownish red body-color. The name is also improperly extended to transparent or highly translucent opals with an orange-red, brownish red, or red body-color that do not show a play of color. Chiefly from Mexico.

Girasol or Girasol Opal. A kind of precious opal, relatively transparent, with a rather uniform bluish or reddish floating or wavy type of internal light.

Common Opal. In general, opal without a play of color; also opal without a body-color, degree of translucency or markings to make it of value as an ornamental material. Includes milk opal, hyalite, opaque to semi-translucent opal of various ordinary colors, wood opal, rock-forming opaline silica, etc.

Hyalite. A hyaline type, colorless and clear as glass. Often occurs as crusts with a botryoidal, globular, or reinform surface, also stalactitic. Sometimes faintly tinted, usually blue, green, greenish yellow, or yellowish, and passing into translucent milky or white material. Hyalite tends to have a higher index of refraction and lower water content than most opal. Named from ὕαλος, *glass*.

Hydrophane. A white or light-colored translucent to opaque type of opal that becomes virtually transparent when placed in water.

Tabasheer or Tabaschir. A milky white opaline silica deposited within the joints of bamboo. Density 0.5–0.6, with n about 1.12. Dried material very strongly absorbs water and becomes transparent.[40]

Cacholong. An opaque, white to yellowish type of opal with a mother-of-pearl luster; rather porous, but does not become transparent in water.

Milk Opal. A translucent to opaque type of common opal with a milk white, pale bluish white, or greenish white color.

Wood Opal. Common opal, generally yellowish or brownish in color, sometimes dark brown to brownish black, found as the petrifying material of wood and often preserving the details of the woody structure.

Moss Opal. Contains dendritic or other imitative inclusions, in appearance and origin similar to those in moss agate.

OCCURRENCE

Silica formed at or near the surface of the Earth's crust by the action of biochemical processes, which are dominant in the deposition of silica from sea water and lakes, and by inorganic precipitation either directly from solution or by chemical reaction at essentially ordinary temperatures and pressures, is in general deposited in a form variously called opal, opaline silica, or "amorphous silica." These types of silica chiefly consist of submicroscopic crystallites of cristobalite with a more or less disordered structure and are believed to contain significant amounts of H_2O, Al, alkalies, and other elements in interstitial solid solution (see further under *Cristobalite*). The cristobalite is metastable relative to quartz in these surficial geological environments and tends to convert to the stable form, quartz, or to microfibrous varieties thereof, with time and more rapidly with increasing temperature.[72] The chemical and structural nature of the cristobalite and the composition and pH of the aqueous solutions with which it is in contact are other factors in this conversion. It may be noted that the solubility of silica in natural surface and near-surface waters at ordinary temperatures and pressures is primarily determined against this type of cristobalite as the solid phase (giving solubilities of about 120 p.p.m.) rather than against quartz (about 7 p.p.m.).

The great bulk of the opaline silica found in nature is constituted of sedimentary accumulations of the tests of silica-secreting organisms such as diatoms, radiolaria, and sponges. These may form very extensive geologic formations, the diatomaceous rocks of the Monterey formation in California, e.g., reaching a thickness of many thousands of feet. The opaline silica of such deposits may be reconstituted during diagenesis or low-grade metamorphism into chalcedonic or microgranular quartz, forming chert, sedimentary jaspers, flint nodules, etc. Opaline silica also is secreted by some forms of terrestrial plant life, especially by the Gramineae. Concretions of opaline cristobalite have been found in the urinary tract of animals.[78]

Opaline silica also is formed inorganically in a variety of ways. These include chemical weathering processes, direct deposition from hot springs and from low-temperature hypogene solutions at shallow depths, and deposition from the meteoric circulation. Among the important sources of dissolved silica in deep natural waters not of igneous origin are opaline organic remains in sediments, and the chemical alteration, with release of silica (desilication), of glassy material and of silicates in volcanic ashes and tuffs. Cristobalite, presumably present as disseminated opaline silica, is a characteristic constituent, together with montmorillonite, of bentonite.[50] Opal forms by the solfataric alteration[66] of basaltic and other rocks, the acid solutions leaching the alkalies, alkaline earths, Al, and Fe and leaving a siliceous residue. Opal has been found as pseudomorphs after augite on Vesuvius.[76] A deposit of kaolin containing cristobalite and alunite formed by the hydrothermal alteration of volcanic tuffs occurs in west Texas.[75] Kaolin with alunite, sulfur, and cristobalite occurs around fumaroles in andesite tuff in the Ibusuki caldera, Kyushu, Japan.[80] The siliceous sinters and geyserites formed about the orifices of hot springs and geysers may be largely opaline and formed both by precipitation from the siliceous waters, through the action of organisms, and in part, especially about fumaroles and along the deeper channels of hot springs, by the chemical alteration of the wall rocks. The opaline silica formed during weathering also is chiefly produced by the chemical breakdown of silicates. Tripoli is a residual deposit, consisting of friable, very fine-grained chalcedonic and opaline silica, formed by the weathering of siliceous limestones and calcareous cherts.

The varieties of opaline silica of mineralogical rather than geological interest, including precious opal, fire opal, milk opal, hyalite, and the like, are formed principally by inorganic deposition in cavities and veinlets, either in association with igneous activity or in the meteoric circulation. Opal, including precious opal, is found frequently, sometimes in association with zeolites, in igneous flow rocks, especially those of an acidic and alkalic character, such as rhyolites and trachytes, and in their tuffaceous equivalents. It also occurs in basalts and other basic effusive rocks, but agate and other types of chalcedonic silica are more typical of this type of environment. In these occurrences the opal is found in cavities and veinlets as a postconsolidation product formed by hydrothermal activity or, especially in tuffs and ash beds, by the entrance of meteoric waters. Both opal and chalcedony very commonly occur as the petrifying material of wood in volcanic ash, tuffaceous rocks, and sediments. Also as irregular masses, nodules, and veinlets in cherty calcareous rocks and sandstones, often replacing bones, shells, and other fossil material. At some occurrences the formation of the opal appears to be a near-surface effect related to weathering processes under arid conditions in rock types of favorable composition. Opal also is found as an alteration product in serpentine,[71] where it may be associated with magnesite or garnierite. It may occur as a cementing material, as in sandstones,[70] or may form veinlets.

In some types of occurrence, especially those in the cavities of igneous rocks, the silica appears to have been initially deposited in gelatinous form and later to have hardened into opal; the process may continue with conversion into chalcedonic silica. In a generalized hydrothermal sequence, opal forms and remains metastably as such in the lowest range of temperatures, below roughly 100–150°. Chalcedony forms either directly or by relatively rapid conversion from opaline silica at somewhat higher temperatures, and coarsely crystallized quartz is deposited at still higher temperatures. In some instances fibrous cristobalite (lussatite) apparently forms by crystallization of opal and has been observed to convert in turn to chalcedony.[42] Gelatinous silica itself has been found[41] in some igneous and sedimentary occurrences, as in the diabase sills of New Jersey, in the Chalk formation of England, in cavities in flinty and marly rocks in France, in solfataras, and with hydrophane in porphyry at Hubertusberg, Saxony. On exposure to air, as has been obtained with artificial silica gels, it dries out, hardens, and sets to a hyaline mass or whitens and falls to a powder which under the microscope is seen to be opaline.

LOCALITIES

The principal localities for opal of gem quality and a few other localities for material of specimen interest are mentioned beyond. A traditional locality for opal that has supplied fine gem material since Roman times is in the neighborhood of Červenica, notably on Mount Simonka and Mount Libanka, in Saros Comitat in northern Hungary. This was the principal source of precious opal until the development of the Australian occurrences in the latter 1800's. Here precious opal occurs as nests and rounded masses in altered and bleached zones in an andesite associated with much milk opal and other types of common opal. The formation of the opal has been ascribed to ascending hydrothermal solutions. The precious opal largely occurs as thin seams and bands in massive milk opal; also as nodules in cavities. Remarkable gem pieces up to 600 grams in weight have been found. Hungarian opal, often accepted as a standard of comparison, has a rather characteristic appearance, showing a display of brilliant colors, chiefly reds, greens, and violet-blues, that are irregularly distributed as relatively small patches and spangles against a more or less milky or whitish background. The most valued material is relatively translucent, with only a slight milky appearance, and a strong color play. Other European localities for opal yield chiefly the common varieties, which occur abundantly in the basaltic rocks of northern Ireland, the Faroe Islands, and Iceland; at numerous localities in Germany, as in the trachytes of the Siebengebirge; a milk white and bluish white opal in serpentine at Kosemütz in Silesia; in the Puy-de-Dôme, Plateau Central, France.

Fire opal and precious opal are found at various places in the state of Querétaro, Mexico. The opal occurs there[64] as nodular masses and patchy areas in a reddish brown to pinkish rhyolite. Cavities may be only partly filled with opal, often banded as milky, hyaline, and other types, or they may contain loose nodules. Mexican or Querétaro opal varies considerably in detail. The fire opals range in body-color from pale brownish yellow, straw yellow, and reddish brown shades through orange and orange-red to brown, deep brownish red, and red. These opals, which in general are relatively transparent, may show color play, usually red and green. Uniformly colored types are sometimes cut as faceted stones. The reddish and brownish body-tints are caused by iron oxide. Other types of Querétaro gem opal are virtually transparent with a bluish white haze or brilliant spangles of color, or have a milky white appearance with bright flakes of color and resemble Hungarian opals. Fine fire opal also has been obtained at Zimapan, Hidalgo, and was brought to Europe in the early 1800's; an analysis was reported by Klaproth in 1807. The Querétaro opal is said to have been first worked in the latter 1800's. Opal was used for mosaics and other purposes by the Aztecs. Opal, including gem types, is found widely in Honduras, notably near Erandique, as veins and which may show an interlamination of platy masses, or streaks of precious opal and common opal, in trachytic rocks. The deposits have been known since the early nineteenth century.

Numerous important localities for precious opal are known in Australia.[65] In New South Wales, the Lightning Ridge area afforded fine black opals showing a gray to black body-color with a lively play of red, green, and deep blue or purple, or with uniform deep blue to dark purple background play of color and intercalated green and red flames At this place and at White Cliffs opal and precious opal occur as thin veinlets and seams often trending along bedding planes and joints in sandstones and conglomeratic rocks, the Desert Sandstone, of Upper Cretaceous age. The opal workings are shallow. At White Cliffs opal with color play was found as pseudomorphs after large groups of glauberite crystals[43] resembling pineapples; also opalized fossils, including belemnites, brachiopods, gastropods, and animal bones. Many gem localities are known in Queensland in a similar geologic setting. The Queensland opal usually is of a milky or bluish white body-color with flashes of blue and green, and also red. Gem opal also is obtained at Andamooka, Coober

Pedy, and other localities in South Australia and in western Australia. Some of the Australian opal, especially that from White Cliffs, shows a patterned arrangement of the color play; the valued Harlequin opal has a variegated display in more or less regular arrangement of round, angular, or roughly rectangular areas, and the pinpoint opals have minute flecks of color. Much Queensland and other Australian opal is found in relatively thin veinlets or sheets and is cut as flat plates, often backed with other material, rather than *en cabochon.* Australian opal in general shows colored areas of larger size than Hungarian opal, as well as relatively large areas of uniform color, although not as marked as in most Virgin Valley opal; in many instances two colors predominate, usually either red and green or green and blue.

In the United States, common opal and wood opal are very widely distributed. Precious opal has been obtained notably in the Virgin Valley in Humboldt County, Nevada. These deposits, found about 1906, contain opal chiefly as the petrifying material of wood embedded in Tertiary ash and tuff beds with interstratified coarse detrital material. The wood shows all stages of petrifaction by precious or common opal or chalcedony, from partly silicified lignitic material to completely replaced masses. The precious opal usually occurs in cracks or seams in the more or the completely silicified wood, or envelops such material; also as solid masses or limb sections of wood, often having a zonal structure but not preserving the cellular structure of the wood, and possibly representing casts. The Virgin Valley precious opal occurs in pieces of relatively large size, and remarkable specimens have been preserved.[49] Most of the precious opal ranges from nearly colorless or faintly brownish or yellowish with a limpid appearance to brownish black or black, then of diminished transparency. The color play typically is red, often with orange and yellow, and is developed over relatively broad areas; also shades of green and, less commonly, blue and purplish tints which may be disposed as a rather uniform glow. Translucent bluish and white or milky white types of opal sometimes with color play also are found. The Virgin Valley opal is particularly subject to cracking and to surface crazing by dehydration, with internal strains accentuated by zones of varying water content concentric to the outline of the wood sections; this behavior has limited its application as a gem material.

Precious opal has been found in a number of minor occurrences elsewhere in the United States, as near Opal Mountain in San Bernardino County, California; near Whelan in Latah County and in Lemhi and Owyhee counties, Idaho; and in Oregon. Notable localities for opalized wood include Yellowstone Park, Wyoming; localities in Socorro and Sierra counties, New Mexico; Clover Creek, Lincoln County, Idaho; the Latah formation over a wide area in Klickitat, Yakima, and Benton counties, Washington. Opalized termite pellets have been found in fossil wood.[52]

Colorless opal, hyalite, is of very wide occurrence. Found particularly as crusts and stalactitic masses in cavities in basaltic rocks, as at Waltsch, Bohemia; the Puy-de-Dôme, France; numerous localities in the Columbia River basalts of the northwestern United States; Iceland; the Faroe Islands. Also as spherules at Tateyama and Sankyo, Japan;[44] at Querétaro and Guanajuato, Mexico; etc. Hyalite often occurs as crusts and films along fracture and joint surfaces in pegmatite, granites, and other rocks; it often shows a bright lemon-yellow fluorescence in ultraviolet radiation because of traces of uranyl compounds released from accessory minerals during weathering. Fine specimens have been found in Mitchell County, North Carolina. Hyalite also is found associated with precious opal, especially in occurrences in volcanic rocks. Opaline siliceous sinters are found in regions of hot spring and geyser activity, as in Yellowstone National Park, Wyoming; Steamboat Springs, Nevada; Iceland; and New Zealand. Hyaline opal is a product of the decomposition of a Roman cement in the hot springs of Plombières, France.[45] Opal has been found in hollow spherulites in perlite at Hosaka, Iwashiro Province, Japan.[53] Opal and chalcedony occur as hydrothermal alteration products of pitchstone and volcanic ash on Specimen Mountain, Rocky Mountain National Park, Colorado.[67] Opal containing dendritic enclosures occurs as nodules in limestone in Morrill County, Nebraska.[68] Reniform, tuberose, and other concretionary forms of opal (menilite, pt.),[69] usually of grayish or grayish brown appearance, or black, and dull in luster, are found in argillaceous limestone and marls, as at Limagne and Aurillac, Plateau Central, and in the Paris basin, France.

HISTORICAL

The name opal is believed to have derived from the Sanskrit *upala, precious stone* or *gem*. Opal is said to have been obtained in ancient times from India, with material of lesser quality from Egypt, Arabia, and various localities in the Mediterranean region, but from Roman times almost to the twentieth century the principal source of supply was Hungary. In Byzantine time Constantinople was the center of distribution of Hungarian opal. Opal (*opalus, paederos*) was the favorite gem of the Romans. Pliny termed precious opal of incomparable beauty and elegance, and relates the story of Nonius, a Roman senator who gave up house and country rather than yield a ring containing an opal the size of a hazelnut to the demand of Marcus Antonius. According to a modern authority, the finest opals were then worth over $20,000 a carat in present-day exchange. The high rank of the precious opal was long continued. B. de Boodt (1647) acknowledged it the most beautiful of gems, and Dutens (1779), who gave the same opinion, said that the finest opals were valued as high as diamonds. Ure (1853)[81] said, "In modern times fine opals of moderate bulk have been frequently sold at the price of diamonds of equal size: the Turks being particularly fond of them."

The popularity of opals decreased during the nineteenth century, partly because of a foolish superstition that they were unlucky (said to have been started by a romantic novel, *Ann of Geierstein, or, The Maiden of the Mist,* written in 1829 by Walter Scott, in which an opal talisman appears). The traditionally high value of opal, especially the milky precious opals from Hungary, also declined in this period with the appearance of abundant fine material, particularly black opal, from Australian deposits. Black opals and opals with a patterned display of color, such as Harlequin opals, are today among the most highly valued types. Black opal from an unstated source was known in Europe in the eighteenth century. In fact, as opposed to superstition, the principal ill fortune associated with opals is their tendency to crack by exposure to a dry atmosphere or to undergo diminution of the color play and translucency, either by dehydration or by contact with grease and oil. Sudden temperature changes, e.g., during lapidary operations, should be avoided. The tendency to crack varies considerably in material from different localities, and even from a given locality, and is said to be least marked in Hungarian opal and more marked in that from Central America, Mexico, and especially the Virgin Valley, Nevada.

Girasol, from the Italian *gira-re, to turn,* and *sole, the sun,* is a name of relatively recent origin. It was early applied to precious stones that display fire when turned in sunlight, and also to fire opal, but now generally refers to opal with a uniform, diffuse bluish or reddish play of color, usually in a slightly milky background.

ALTERATION

Opal occurs very commonly as a replacement of wood, as already noted, and has been found as pseudomorphs after glauberite,[43] gypsum,[46] calcite, aragonite, apophyllite,[74] siderite, and apatite.

References

1. Rinne: *Zs. Kr.,* **60,** 55 (1924); Lehmann: *Zs. Kr.,* **59,** 455 (1924); Kerr; *Econ. Geol.,* **19,** 1 (1924); Baier: *Zs. Kr.,* **81,** 219 (1932).
2. Levin and Ott: *Zs. Kr.,* **85,** 305 (1933) and *J. Am. Chem. Soc.,* **54,** 828 (1932); Greig: *J. Am. Chem. Soc.,* **54,** 2846 (1932); Dwyer and Mellor: *Proc. Roy. Soc. New South Wales,* **68,** 47 (1934); Büssem, Bluth, and Grochtmann: *Ber. deutsch. keram. Ges.,* **16,** 381 (1935); Inuzuka: *J. Geol. Soc. Japan,* **47,** 367 (1940); Taliaferro: *Am. J. Sci.,* **30,** 450 (1935); Sudo: *Science (Japan),* **19,** 40 (1949); Cirilli and Giannone: *Rend. accad. Sci. Napoli,* **11,** 1940–41; Minato: *J. Min. Soc. Japan,* **1,** 67 (1935); Flörke: *Jb. Min., Monatsh.,* **10,** 217 (1955), and *Fortschr. Min.,* **37,** 73 (1959); Bushinskii and Frank-Kamenetskii: *C. R. acad. sci. USSR,* **96,** 817 (1954); Neuwirth: *Min. Mitt.,* **3,** 32 (1952); Braitsch: *Heidelberg. Beitr. Min. Pet.,* **5,** 331 (1957); Swineford and Franks: *Soc. Econ. Paleont. Min. Spec. Publ.* 7, 111, 1959; Raman and Jayaraman *Proc. Indian Ac. Soc.,* **38B,** 343 (1953); and **38A,** 101 (1953); Miller: *Rocks and Min.,* **21,** 276 (1946); Fenoglio and Senero: *Atti reale accad. sci. Torino,* **78,** 265 (1943); Vasileo and Veselovsky: *J. Phys. Chem. USSR,* **7,** 918 (1936).

3. Flörke: (1959) and (1955); *Ber. deutsch. keram. Ges.*, **32**, 369 (1955); Eitel: *Bull. Am. Ceram. Soc.*, **36**, 142 (1957); Nieuwenkamp and Laves: *Zs. Kr.*, **90**, 377 (1935); Jagodzinski and Laves: *Schweiz. Min. Pet. Mitt.*, **28**, 456 (1948).
4. Flörke (1955).
5. Krejci and Ott: *J. Phys. Chem.*, **35**, 2061 (1931); Gillingham: *Econ. Geol.*, **43**, 242 (1948); Flörke: *Fortschr. Min.*, **32**, 33 (1954); Wyart: *C. R.*, **220**, 830 (1945); Dubrovo: *C. R. acad. sci. USSR*, **23**, 50 (1939); Dwyer and Mellor: *J. Roy. Soc. New South Wales*, **67**, 420 (1933).
6. Rieck and Stevels: *J. Soc. Glass Technol.*, **35**, 284 (1951).
7. Cf. Ross: *U.S. Geol. Surv. Bull.* 1042–D, 88, 1956; Pollock: *Trans. Inst. Min. Met. Eng. Tech. Publ. 1735*, 1944.
8. Baier: *Zs. Kr.*, **81**, 183 (1931); Behrends: *Sitzber. Ak. Wiss. Wien*, **64**, 1 (1871); Brewster: *Edinburgh Phil. J.*, **36**, 385 (1845); Gürich: *Jb. Min.*, Beil. Bd. **14**, 472 (1901); Reusch: *Ann. Phys.*, **124**, 431 (1865); Kleefeld: *Jb. Min.*, **II**, 146 (1895); Raman and Jayaraman: (1953) and (1953).
9. Zimanyi: *Zs. Kr.*, **22**, 327 (1893).
10. Des Cloizeaux: *Min.*, 23, 1862.
11. Ites: *Zs. Kr.*, **41**, 303 (1905).
12. Sinkankas: *Gemstones of North America*, New York, 1959, p. 108.
13. Reusch (1865).
14. Brewster: *Edinburgh J. Sci.*, 285, 1828.
15. Stscheglayev: *Ann. Phys.*, **64**, 325 (1898).
16. Levin and Ott, (1933).
17. Keller and Pickett: *Am. Min.*, **34**, 855 (1949).
18. Swartzlow and Keller: *J. Geol.*, **45**, 101 (1937).
19. Lupander: *Geol. För. Förh.*, 521, 1934.
20. Bushinskii and Frank-Kamenetskii: (1954).
21. Taliaferro (1935).
22. Kokta: *Rozpravy České Ak.*, ser. 11, **40**, no. 21 (1931).
23. Rinne: *Jb. Min.*, Beil. Bd. **39**, 388 (1914); and *Keram. Rundschau*, **37**, 772 (1929).
24. Frondel: priv. comm., 1959.
25. Hintze: *Handbuch der Mineralogie*, Berlin, 1906, **1**, Lfg. 10, p. 1535; Doelter: *Handbuch der Mineralchemie*, **2**, 1914, Pt. 1, p. 148; Simpson: *J. Roy. Soc. West. Australia*, **16**, 25 (1930); Zarter: *Lotos*, **85**, 22 (1937); Croce: *Jb. Min.*, **I**, 148 (1934).
26. Rammelsberg: *Ann. Phys.*, **112**, 177 (1861); Leitmeier: *Zbl. Min.*, 632, 1908; Lunge and Millberg: *Zs. angew. Chem.*, 426, 1897; Winkler: *Abhl. Leop.-Carol. Ak. Naturfor. Halle*, **40**, 331 (1878).
27. Foster: *Rocks and Min.*, **17**, 100 (1942).
28. Brush cited in Dana: *Syst. Min.*, 152, 1854.
29. Forchhammer: *Ann. Phys.*, **35**, 331 (1835).
30. Loczka: *Föld. Közl.*, **21**, 375 (1881).
31. d'Achiardi: *Atti soc. toscana sci. nat. Pisa*, **11**, 114 (1899).
32. Woodward cited in Weed: *Am. J. Sci.*, **42**, 168 (1891).
33. Kunz: *Am. J. Sci.*, **34**, 479 (1887); Church: *Min. Mag.*, **8**, 181 (1889); Blasius: *Zs. Kr.*, **14**, 258 (1888); Judd: *Nature*, **35**, 488 (1887); Hintze: *Zs. Kr.*, **13**, 392 (1887).
34. Hannay: *Min. Mag.*, **1**, 106 (1877); Damour: *Bull. soc. géol. France*, **5**, 157 (1848).
35. d'Achiardi (1899); Doelter: *Handb. Min.*, **2**, Pt. 1, 248, 1914; Tammann: *Zs. phys. Chem.*, **27**, 323 (1898).
36. Taliaferro (1935); ref. 1.
37. Bertrand: *Bull. soc. min.*, **3**, 57 (1880); Cesàro: *Bull. acad. roy. Belg.*, **26**, 721 (1893).

38. Meunier: *C. R.*, **112**, 953 (1891); Monier: *C. R.*, **85**, 1053 (1887); Frémy: *C. R.*, **72**, 702 (1871); Ebelmen: *C. R.*, **21**, 502, 1527 (1845) and **26**, 854 (1848); Becquerel: *C. R.*, **67**, 1081 (1868).

39. Langlois: *Ann. chim. phys.*, **52**, 331 (1858); Merz: *J. prakt. Chem.*, **99**, 177 (1866).

40. Judd: *Nature*, **35**, 489 (1887); Mallet: *Zs. Kr.*, **6**, 96 (1882); Bütschli: *Abh. Geol. Wiss. Göttingen, Math. Nat. Kl.*, 149, 1908; Macie: *Phil. Trans.*, **81**, 368 (1791).

41. Jannetaz: *Bull. soc. géol. France*, **18**, 637 (1861); Jukes-Browne: *Quart. J. Geol. Soc.*, **45**, 403 (1889); Hinde: *Phil. Trans.*, Pt. 2, 403 (1885); Edge: *Min. Mag.*, **19**, 11 (1920); Taliaferro: (1935); Kormilitsyn: *Mem. soc. russe min.*, **80**, 269 (1951); Spezia: *Atti reale accad. sci. Torino*, **34**, 705, 1899; Levings: *Trans. Inst. Min. Met.*, **21**, 478 (1911); Naboko and Silnichenko: *Geokhimiya*, 1957, 253; Furcron: *Georgia Min. Newsletter*, **7**, 122 (1954).

42. Braitsch: *Heidelberg. Beitr. Min. Pet.*, **5**, 354 (1957); Lengyel: *Föld. Közl.*, **66**, 129, 278 (1936).

43. Anderson and Jevons: *Rec. Australian Mus.*, **6**, Pt. 1, 31 (1905); Raggatt: *J. Roy. Soc. New South Wales*, **71**, 336 (1938).

44. Wada: *Min. Japan*, **1**, 11 (1905) and **2**, 60 (1906).

45. Daubrée: *Ann. Mines*, **12**, 294 (1857) and **13**, 245 (1858).

46. Lacroix: *Nouv. arch. Muséum Paris*, **9**, 201 (1897).

47. Larsen and Bridgman: *Am. J. Sci.*, **36**, 81 (1938).

48. Raman and Jayaraman: *Proc. Indian Acad. Sci.*, **38B**, 343 (1953).

49. Foshag: *Rocks and Min.*, **8**, 9, 13, 16 (1933); Sinkankas (1959), p. 116, Dake: *The Mineralogist*, **9**, 7, 22 (1941).

50. Gruner: *Am. Min.*, **25**, 587 (1940) and *Econ. Geol.*, **35**, 867 (1940); Endell, Wilm, and Hofmann: *Zs. angew. Chem.*, **47**, 540 (1934); Flörke (1955).

51. Rose: *Monatsh. Ak. Wiss. Berlin*, 1869, 461.

52. Rogers: *Am. J. Sci.*, **36**, 389 (1938).

53. Otsuki: *Beitr. Min. Japan*, no. 5, 274, 1915.

54. Pilipenko: *Minerology of the western Altai*, *Bull. Imp. Tomsk Univ.* **63**, 1915 (*Min. Abs.*, **2**, 111, 1923).

55. Mason and Greenberg: *Arkiv Min. Geol.*, Stockholm, **1**, 519 (1954).

56. Neuwirth (1952).

57. Fenoglio and Senero (1943).

58. Haberlandt and Herneggar: *Sitzber. Ak. Wiss. Wien, Math. Nat. Kl.*, Abt. 11a, **7**, 359 (1947); Iwase: *Bull. Chem. Soc. Japan*, **11**, 377 (1936); Smith and Parsons: *Am. Min.*, **23**, 515 (1938); and others.

59. Gerasimovsky: *Fersman Mem. Vol.*, *Ac. Sci. USSR*, 1946, 15.

60. Allenby: *Geochim. Cosmochim. Acta*, **5**, 40 (1954).

61. Zák: *Sborník ústřed. ústavu geol.*, **18**, 641 (1951).

62. Lorković: *Min. Abs.*, **2**, 430 (1925).

63. Sameshmia: *Bull. Chem. Soc. Japan*, **6**, 165 (1931).

64. Foshag: *Gems and Gemology*, Los Angeles, 1953, **7**, p. 278, and ref. 12.

65. Croll: *Bull. Bur. Min. Res. Geol. Geophys.*, *Australia 17*, 1950; David: *Geol. of Australia*, London, 1950, Pt. III (3), p. 379.

66. See Payne and Mau: *J. Geol.*, **54**, 345 (1946); Macdonald: *Am. J. Sci.*, **242**, 496 (1944); Lacroix: *Bull. soc. min.*, **30**, 219 (1907); Shimada and Dozono: *Nippon Kagaku Zasshi*, **78**, 1661 (1957); Anderson: *Am. Min.*, **20**, 240 (1935).

67. Wahlstrom: *Am. Min.*, **26**, 551 (1941).

68. Mitchell: *Rocks and Min.*, **22**, 1115 (1947).

69. Lacroix: *Min. de France*, Paris, 1901, **3**, p. 332.

70. Storz: *Die sekundäre authigene Kieselsaüre*, Berlin, 1931, Pt. 2, p. 184.

71. Barvíř: Abstr. in *Zs. Kr.*, **31**, 525 (1899).
72. See Doelter: (1914), pp. 176–182; see also ref. 42 and refs. on origin of agate (p. 215); Slavík: *Věst. Král. České Spol. Nauk*, no. 16, 1942, and no. 16, 1946; Tolman: *Econ. Geol.*, **22**, 407 (1927); Hoss: *Heidelberg. Beitr. Min. Pet.*, **6**, 59 (1957); White, Brannock, and Murata: *Econ. Geol.*, Fiftieth Ann. Vol., Pt. 1, 99 (1955).
73. Pough: *Am. Min.*, **32**, 38 (1947).
74. Bailey: *Am. Min.*, **6**, 565 (1941); Scheit: *Min. Mitt.*, **29**, 263 (1910).
75. Shurtz: *J. Geol.*, **59**, 60 (1951).
76. Abbolito: *Per. min. Roma*, **19**, 117 (1950).
77. Baker: *Australian J. Botany*, **8**, 69 (1960).
78. Baker, Jones, and Milne: *Australian J. Agricult. Res.*, **12**, 473 (1961).
79. Kimura *et al.*: *J. Chem. Soc. Japan, Pure Chem. Sect.*, **73**, 589 (1952).
80. Muraoka: *Bull. Geol. Surv. Japan*, **2**, 74 (1951).
81. Ure: *Dictionary of Arts, Manufact. and Mines*, New York, **1**, 1853, art. on "Lapidary."

KEATITE

Keatite is tetragonal, crystallizing in the tetragonal-trapezohedral (4 2 2) class as indicated by the enantiomorphous space groups $P4_12_1$ and $P4_32_1$ found by x-ray study. Morphological data are lacking. The substance is known only as a microcrystalline synthetic product,[1] predominantly forming square platelets up to about 0.05 mm. in size. The plates, presumably flattened on (001), are optically negative with ω 1.522 and ϵ 1.513. The specific gravity is 2.50 (meas.), 2.50 (calc.). Observations on the cleavage and hardness are lacking.

TABLE 41. X-RAY POWDER SPACING DATA FOR KEATITE[1]

vw = very weak; vvw = extremely weak

hkl	d obs., in Å	Relative Intensity	hkl	d obs., in Å	Relative Intensity
100	7.46	vvw	303	1.879	$\frac{1}{2}$
101	5.64	$\frac{1}{2}$	400	1.864	1
110	5.28	$\frac{1}{2}$	...	1.667	$\frac{1}{2}$
111	4.50	2	...	1.636	vw
102	3.72	7	...	1.589	vw
201	3.42	10	...	1.562	$\frac{1}{2}$
112	3.33	2	...	1.489	$\frac{1}{2}$
211	3.11	2	...	1.441	vw
113	2.516	$\frac{1}{2}$	...	1.412	$\frac{1}{2}$
222	2.246	$\frac{1}{2}$	...	1.389	vw
213	2.174	$\frac{1}{2}$	...	1.366	vw
004	2.148	$\frac{1}{2}$	...	1.321	vw
312	2.067	$\frac{1}{2}$	...	1.246	vw
114	1.988	vw			

The unit cell dimensions of keatite are a_0 7.456, c_0 8.604 Å (powder diffractometer method, with quartz as internal standard). The cell contains $Si_{12}O_{24}$. The x-ray powder spacing data are given in Table 41. The crystal structure[2] of keatite is based on (SiO_4) tetrahedra sharing corners, with 8 silicons and 24 oxygens in general positions of 8(b) and arranged into four-fold spirals parallel to c. The spirals are linked together by silicon atoms, in special positions of 4(a), located on diagonal two-fold axes, that bond to an oxygen atom in each of four adjacent spirals. The atomic coordinates are given in Table 42, and a projection of the structure on (001) is shown in Fig. 112. Beta-spodumene, $LiAlSi_2O_6$, is structurally related to keatite.[4]

TABLE 42. COORDINATES OF ATOMS IN KEATITE

Atom	Position	Coordinates x	y	z
Si_1	8(b)	0.326	0.120	0.248
Si_2	4(a)	0.410	...	...
O_1	8(b)	0.445	0.132	0.400
O_2	8(b)	0.117	0.123	0.296
O_3	8(b)	0.344	0.297	0.143

Keatite has been synthesized by heating commercial dried silica ("analytical reagent silicic acid") in water containing an alkali, added as the carbonate or hydroxide of Li, Na, or K or as Na_2WO_4. The concentration of alkali is small (e.g., 1 ml. of $0.01N$ NaOH solution per 20 ml. H_2O contained in the

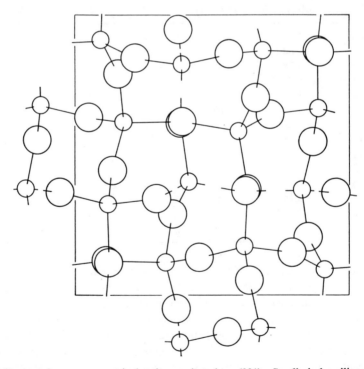

FIG. 112. Atomic arrangement in keatite, projected on (001). Small circles silicon, large circles oxygen.

autoclave), and the ratio of SiO_2 to H_2O is not critical. Keatite, together with small amounts of quartz or cristobalite, was obtained over the entire investigated range from 380° to 585°C. and 5000 to 18,000 p.s.i. Silica gel formed by the hydrolysis of tetraethyl orthosilicate also has been used successfully. The hydrothermal crystallization of amorphous silicic acid at temperatures in the range 325–440° and pressures from 15,000 to 59,000 p.s.i. generally proceeds with the metastable formation first of cristobalite, then of keatite, and finally of the stable phase quartz.[6] In this system increasing pressure favors the attainment of equilibrium more than does increasing temperature.

Keatite is soluble in cold HF. Heating in air at 1620° for 3 hours completely converted microcrystalline keatite to cristobalite, with little if any change when heated at 1100° for 37 hours. Conversion to the other known polymorphs of SiO_2 is easily effected by heating in the presence of Na_2WO_4 in the temperature range of stability of these forms. Keatite may be identical with one of the unidentified polymorphs earlier reported[3] as intermediate phases in the conversion of silicic acid to cristobalite by heat. It is not known to occur in nature. Named[5] after Paul P. Keat, who discovered the phase.

References

1. Keat: *Science*, **120**, 328 (1954).
2. Shropshire, Keat, and Vaughan: *Zs. Kr.*, **112**, 409 (1959).
3. Endell: *Kolloid Zs.*, **111**, 19 (1948).
4. Skinner and Evans: *Am. J. Sci.*, **258A**, 312 (1960).
5. Sosman: *Science*, **119**, 738 (1954).
6. Carr and Fyfe: *Am. Min.*, **43**, 908 (1958).

COESITE

COESITE. Silica C, Coesite *Sosman* (*Science*, **119**, 738, 1954).

STRUCTURE

Dimensionally, coesite is precisely pseudohexagonal around the b-axis, with $a_0 = b_0$ and $\beta = 120°$, but the true symmetry is monoclinic, in the prismatic class, as shown by x-ray single-crystal photographs. Thus a 0-level (010) Weissenberg or precession photograph shows three central lattice lines 60° apart with identical spacings, corresponding to the directions of the a-axis, the c-axis, and the bisector of the ac angle, but variations in the intensities of the reflections and missing reflections show that the b-axis is in fact not an axis of six-fold symmetry.

UNIT CELL DIMENSIONS

Setting:	Second ($C2/c$)	Second ($C2/c$)	First ($B2/b$)
a_0 in Å	7.23	7.16	7.17
b_0 in Å	12.52	12.39	7.17
c_0 in Å	7.23	7.16	12.38
β	120°	120°	120°
$a_0:b_0:c_0$	0.577:1:0.577	0.578:1:0.578	1:1:1.726
Z	$Si_{16}O_{32}$	$Si_{16}O_{32}$	$Si_{16}O_{32}$
Spec. grav., calc.	2.85	2.93	2.93
Ref.	2	4	3

The unit cell here used[1] is based on the C-face-centered cell in the conventional orientation of a monoclinic crystal with b horizontal. The space group is then $C2/c$ (on the basis of not wholly satisfactory evidence for holohedry) or is Cc. In the setting with the unique axis vertical, by interchange of b and c, the space group can be described as either $I2/a$, $A2/a$, or $B2/b$; the third description has been employed in the structural investigation.[3]

The x-ray powder spacing data[4] for synthetic coesite are given in Table 43. The indexing of the powder lines cannot be given uniquely in most instances because of the pseudosymmetry. The indices given are for planes which appear from single-crystal data to contribute most to the observed intensity.

The crystal structure of coesite[3] is based on (SiO_4) tetrahedra joined into a three-dimensional network of a type different from those of other SiO_2 polymorphs but resembling that of the feldspars. The network is composed of two

310

TABLE 43. X-RAY POWDER DATA FOR COESITE[4]

Synthetic material, with a_0 7.16 Å, b_0 12.39, c_0 7.16;
filtered copper radiation (?), in angstrom units

hkl	d, obs.	d, calc.	I, chart
020	6.217	6.198	5
021	4.40	4.383	5
130, 111	3.432	3.439	50
002, 040, $\bar{2}$21	3.098	3.099	100
220	2.77	2.772	15
131	2.68	2.705	15
201, $\bar{2}$41	2.350	2.343	5
112, 150	2.303	2.302	10
240, 22$\bar{3}$	2.195	2.191	10
151, 310, 132	2.034	2.038	10
330	1.846	1.848	10
$\bar{2}$61	1.789	1.789	10
260, 222	1.716	1.717	15
113, $\bar{3}$52, 17$\bar{1}$	1.711	1.703	10

kinds of rings of four tetrahedra, one parallel to (010) and the other approximately parallel to (001). Each ring has eight external connections by sharing of oxygens with other tetrahedra, to give diagonal chains criss-crossing on alternate levels parallel to (001). The chains on the same level are not connected but are joined by the rings of glide-equivalent chains on the levels above and below. The coordinates of the atoms in coesite are given in Table 44.

TABLE 44. ATOMIC COORDINATES IN COESITE

Atom	x	y	z
Si_1	0.1403	0.0735	0.1084
Si_2	0.5063	0.5388	0.1576
O_1	0	0	0
O_2	$\frac{1}{2}$	$\frac{3}{4}$	0.1166
O_3	0.2694	0.9405	0.1256
O_4	0.3080	0.3293	0.1030
O_5	0.0123	0.4726	0.2122

MORPHOLOGY

Synthetic crystals sometimes have a simple gypsum-like habit, flattened on {010} and elongated on the c-axis with well-developed faces of {130} and terminal faces of {001}, {011}, {$\bar{1}$01}, {$\bar{1}\bar{1}$1}, {$\bar{1}$11}, and other forms. The faces of the individual terminal forms usually are not developed in full complement or are of unequal size. Most crystals do not have a conspicuous zonal development and are variable in habit, with an unequal development of the equivalent faces of individual forms. The axial orientation and the identity of the principal faces may not be apparent at sight. Almost always {010} is present, usually as the dominant form, and one or more faces of {130} are ordinarily present. Some habits are shown in Figs. 113 and 114.

TABLE 45. ANGLE TABLE FOR COESITE

Monoclinic, prismatic—$2/m$

$a:b:c = 0.577:1:0.577;$ $\beta\ 120°;$ $p_0:q_0:r_0 = 1:0.5:1;$ $p_0' = 1.1547,$

$q_0' = 0.5773,$ $x_0' = 0.5773$

Forms	ϕ	ρ	ϕ_2	$\rho_2 = B$
c 001	90° 00′	30° 00′	60° 00′	90° 00′
b 010	0 00	90 00	...	0 00
i 130	33 42	90 00	0 00	33 42
m 110	63 26	90 00	0 00	53 26
w 011	45 00	39 15	60 00	63 26
v 021	26 34	52 14	60 00	45 00
D $\bar{1}$01	−90 00	30 00	120 00	90 00
r 111	71 34	61 18	10 20	73 54
R $\bar{1}$11	−45 00	39 15	120 00	73 54

Since $\beta = 120°$, pairs of faces such as (001) and ($\bar{1}$01) or (011) and ($\bar{1}$11) have the same rho angle and cannot be distinguished morphologically. Similarly, equant plates flattened on (010) may have a hexagonal outline (but under the microscope extinguish asymmetrically). A choice between the two

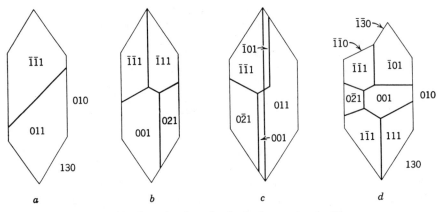

FIG. 113. Crystals of coesite, in the face-centered setting.

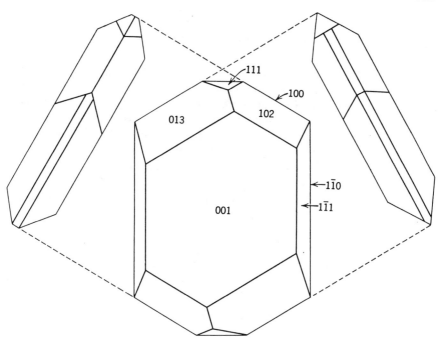

FIG. 114. Crystal of coesite, in the body-centered setting.

a-axis directions can be effected by x-ray single-crystal photographs; both settings have $\beta = 120°$, but one affords an end-face-centered cell (here taken) and the other a body-centered cell. Table 45 is an angle table[1] for coesite.

TWINNING

 Twins on {100} and {021} have been observed by x-ray single-crystal study.[2] In the {021} law the twin plane is exactly at 45° to b^* and c^* since $d(001) = d(020)$. Neither twin law can be recognized by morphological measurements because of the precise hexagonal pseudosymmetry which produces exact parallelism of unlike planes in the twinned parts. All the faces on a twinned crystal can be indexed on a single lattice. Similarly, there is no doubling of reflections on single-crystal x-ray photographs. The twinning can be identified, however, from x-ray *n*-level photographs of the reciprocal lattice in the case of the {100} law, and from appropriately oriented O-level photographs for the {021} law, all of which show reflections inconsistent with an untwinned lattice. Re-entrant angles have been observed in twins on {021}, with (010) of the twinned individuals at 90° to each other.

PHYSICAL PROPERTIES

Observations on cleavage are lacking. Colorless and transparent. Luster vitreous. The hardness of coesite on the Mohs scale is not known, but it is greater than that of quartz and appears to be about $7\frac{1}{2}$. Hardness values[5] obtained by the Knoop method give values slightly less than those of spinel and topaz. The specific gravity of synthetic material has been given[5] as 3.01 and, in a later study,[6] as 2.93 ± 0.02. The latter value is in better agreement with those calculated from the measured unit cell dimensions and the number of formula units, 16, indicated by the equivalent positions available in the space group. The higher value, 3.01, may perhaps be real and caused by undetermined material in solid solution.

OPTICAL PROPERTIES

Colorless. Biaxial, positive.

INDICES OF REFRACTION (WHITE LIGHT)

α	β	γ	$2V$	Ref.
1.599	1.604	54°	5
1.593	1.597	64°	9
1.594	1.597	61°	10

Data obtained on synthetic material. Optical orientation not known. Natural and synthetic material usually is so fine grained that only the mean index of refraction can be obtained (~1.595 − 1.60).

COMPOSITION

Silicon dioxide, SiO_2. Complete quantitative analyses are lacking; but chemical tests,[5] such as complete volatilization on heating with ammonium bifluoride and the transformation without loss of weight into silica glass and cristobalite when heated at 1700°, indicate that the substance is at least essentially SiO_2. A spectrographic analysis[7] of natural material from Meteor Crater containing coesite with some quartz afforded over 99 per cent SiO_2 and less than 1 per cent of other cations (identity not stated). The main mechanism of compositional variation that could be expected in coesite would be the substitution of Al (and possibly of Fe^3 at elevated temperatures) for Si with valence compensation effected by the entrance interstitially of alkalies or H_3O. The substitution of $(OH)_4$ for (SiO_4) is a possibility.

Coesite is nearly insoluble in HF at room temperature, but is rapidly dissolved by fused ammonium bifluoride.

OCCURRENCE

Coesite was first discovered in nature[7] in 1960 at Meteor Crater, Arizona, although it had been known as a synthetic compound since 1953. Meteor Crater is a bowl-shaped depression about 4000 ft. wide and 570 ft. deep, formed by a giant meteorite that was disintegrated and largely vaporized by the impact. The floor of the crater is underlain by talus, alluvial deposits, and Pleistocene lake beds resting on a lens-like body of brecciated material up to 600 ft. thick. The country rocks include a white fine-grained saccharoidal quartz sandstone, the Coconino sandstone of Permian age, that locally has been fused to vesicular silica glass. Coesite occurs abundantly in sheared and compressed areas of this sandstone as a fine-grained nearly isotropic matrix in which the fractured quartz grains are embedded. It also is a subordinate constituent of sandstone that has largely been converted to glass. Cristobalite with n 1.483 also occurs in the silica glass. .The coesite presumably formed under an impact shockwave in excess of 20 kilobars and, from the association with silica glass, at elevated temperatures. Coesite also has been reported[8] in pumaceous tuff and stressed granite near the rim of the Rieskessel meteoritic impact caldera in Bavaria, Germany. Also found at the Wabar meteor crater, near Al Hadida, Arabia,[16] and in fossil meteorite craters[17] at Kentland, Newton County, Indiana, and near Sinking Springs, Ohio.

The experimental data on the pressure-temperature equilibrium curve for quartz-coesite and the estimated temperature gradients within the Earth indicate that quartz should convert to coesite at depths below 60–100 km. Coesite might then be expected to occur in rocks containing free silica that have formed at such depths, providing that the coesite has not been destroyed by later metamorphism or alteration.

SYNTHESIS[11]

Originally obtained by heating a mixture of sodium metasilicate and diammonium phosphate to temperatures between 500° and 800° at 35,000 atm. for periods of about 15 hours. The reaction rate becomes very slow below about 500°. Other fluxes that have been employed successfully in place of sodium phosphate include boric acid, ammonium chloride, ammonium vanadate, and potassium fluoborate, and potassium silicate can be substituted for sodium metasilicate. The presence of alkalies is undesirable because of the possibility of formation of alkali silicates, and probably the most satisfactory starting material is "silicic acid," $SiO_2 \cdot nH_2O$. This material gave 100 per cent conversion to coesite at 580° and 30,000 bars in 1 hour, and the rate of conversion increases with increasing temperature in the coesite field. Coesite forms in 1 hour or less from dry quartz at temperatures in excess of 1200° and at sufficiently high pressure, but at lower temperatures some H_2O must be present for the reaction to run in a reasonable length of time. Coesite also has been obtained by the reaction of elemental Si with $AgCO_3$, SiO_2 and Ag being formed, and by the direct conversion of powdered flint, silica glass, quartz, tridymite, and cristobalite. Euhedral crystals up to 1 mm. in size have been obtained.

The equation representing the equilibrium curve separating the stability fields of quartz and coesite has been determined[12] as $P = 22.5T + 9500$ and as[13] $P = 19.5 + 0.0112T$, where P is in bars and T is in degrees C., with coesite stable in the high-pressure region. Coesite exists metastably apparently indefinitely at ordinary temperatures and pressures. It readily converts to quartz when treated at relatively high temperatures and pressures in the quartz stability field. A polymorph of BeF_2 isostructural with coesite has been synthesized.[14]

NAME

After Loring Coes Jr., who synthesized the substance in the research laboratory of The Norton Company, manufacturers of abrasives, Worcester, Mass.

References

1. Ramsdell: *Am. Min.*, **40**, 975 (1955) by Weissenberg method.
2. Axial ratio from x-ray cell on the synthetic crystals of Coes (*Science*, **118**, 131, 1953). The symmetry was originally stated by Coes to be probably triclinic.
3. Zoltai and Buerger: *Zs. Kr.*, **111**, 129 (1959).
4. Dachille and Roy: *Zs. Kr.*, **111**, 451 (1959). Powder data also are given by Coes (1953), Boyd and England: *J. Geophys. Res.*, **65**, 749 (1960) and Khitarov, Slutskiy, and Arsenyeva: *Geokhim.*, **8**, 666 (1957).
5. Coes (1953).
6. Dachille and Roy (1959).
7. Chao, Shoemaker, and Madsen: *Science*, **132**, 220 (1960).
8. Pecora: *GeoTimes*, **5**, no. 2, 16 (1960).
9. Boyd and England (1960).
10. Khitarov, Slutskiy, and Arsenyeva (1957).
11. Coes (1953); Griggs and Kennedy: *Am. J. Sci.*, **254**, 722 (1956); MacDonald: *Am. J. Sci.*, **254**, 749 (1956); Dachille and Roy (1959); Khitarov *et al.* (1957); Boyd and England (1960).
12. MacDonald (1956).
13. Boyd and England (1960).
14. Dachille and Roy (1959).
15. Rogers: *Am. J. Sci.*, **19**, 195 (1930).
16. Chao, Fahey, and Littler: *Science*, **133**, 882 (1961).
17. Cohen, Bunch, and Reid: *Science*, **134**, 1624 (1961).

STISHOVITE

STISHOVITE. Unnamed polymorph. *Stishov and Popova* (*Geokhimiya*, no. 10, 837–839, 1961). Stishovite *Chao, Fahey, Littler, and Milton* (*J. Geophys. Res.*, **67**, no. 1, 1962).

Stishovite is tetragonal, crystallizing in the ditetragonal dipyramidal class ($4/m\ 2/m\ 2/m$) from the isostructural relation to rutile indicated by x-ray powder diffraction study.[1] The space group is the $P4/mnm$. Unit cell dimensions:[2] a_0 4.1790 Å, c_0 2.6649 (±0.0004). Unit cell contents Si_2O_4. Indexed x-ray powder spacing data are given in Table 46. Stishovite is the first mineral to be found in which Si occurs in octahedral rather than tetrahedral coordination with oxygen.

The hardness as measured by a micro-indentation method is 2080 kg. per mm.[2] parallel to the elongation and 1700 kg. per mm.[2] perpendicular thereto. Observations on the cleavage are lacking. Specific gravity 4.35 (meas. on synthetic material), 4.28 (calc. from the cited cell dimensions of unanalyzed natural material). Stishovite is much denser than is coesite (2.93) or quartz (2.65). Colorless and transparent. Optically uniaxial, positive: ω 1.799, ϵ 1.826 (±0.002) on synthetic material:[1] elongation positive.

Composition SiO_2. Chemical analyses of natural material are lacking. In HF, less soluble than coesite and much less soluble than quartz. Stishovite converts to cristobalite when heated in air at 900°C. It was first synthesized in 1961 by two Russian investigators, S. M. Stishov and S. V. Popova, at 1200–1400° and a pressure reported to be above 160 kilobars (the actual pressure may have been about 30 per cent less because of calibration errors[3]). Acicular crystals and laths were obtained up to about 0.5 mm. in length. Later identified[2] as extremely fine grains, mostly of submicron size, associated with coesite and silica glass in sandstone at Meteor Crater, Arizona. The formation of the stishovite and coesite is attributed to high transient shock pressure and high temperature accompanying the impact of a very large meteorite.

References

1. Stishov and Popova: *Geokhimiya*, no. 10, 837–839 (1961).
2. Chao, Fahey, Littler, and Milton: *J. Geophys. Res.*, **67**, no. 1 (1962).
3. Kennedy and LaMori: *Progress in Very High Pressure Research*, New York, 1961, pp. 304–313.

TABLE 46. X-RAY POWDER DATA FOR STISHOVITE

Film recording in 114 mm diameter camera, copper radiation, nickel filter. Calculated spacings from a_0 4.1790 Å, c_0 2.6649.

d (calc.)	hkl	Natural[2]		Synthetic[1]	
		d (means)	I	d (meas.)	I
2.955	110	2.959	100	2.95	100
2.247	101	2.246	18	2.24	30
2.089	200			2.09	1
1.979	111	1.981	35	1.98	42
1.869	210	1.870	13	1.87	21
1.530	211	1.530	50	1.53	72
1.478	220	1.478	18	1.476	30
1.332	002	1.333	9	1.326	15
1.322	310	1.322	4		
1.292	221	1.291	1	1.293	2
1.234	301	1.235	25	1.234	42
1.215	112	1.215	9	1.215	11
1.184	311	1.185	2	1.184	4
1.159	320	1.159	7		
1.123	202	1.123	1		
1.085	212	1.084	1	1.086	3
1.063	321	1.062	2	1.063	4
1.045	400	1.045	2	1.044	5
1.014	410	1.013	1	1.014	5
0.9895	222	0.9900	6	0.987	15
0.9850	330	0.9850	2		
0.9473	411	$0.9475\alpha_1$	4	0.947	11
		$0.9470\alpha_2$	2		
0.9383	312	$0.9382\alpha_1$	3	0.940	7
		$0.9375\alpha_2$	2		
0.9345	420				
0.9239	331				
0.8818	421	$0.8824\alpha_1$	0.7	0.881	3
		$0.8813\alpha_2$	0.3		
0.8689	103	$0.8691\alpha_1$	2	0.876	3
		$0.8683\alpha_2$	1		
0.8507	113	$0.8506\alpha_1$	1	0.851	2
		$0.8501\alpha_2$	0.7		
0.8358	430				
0.8221	402	$0.8220\alpha_1$	2	0.823	11
		$0.8215\alpha_2$	1		
0.8196	510	$0.8194\alpha_1$	2	$0.819\alpha_1$	15
		$0.8193\alpha_2$	1	$0.819\alpha_2$	7
0.8067	412	$0.8065\alpha_1$	1	$0.8065\alpha_1$	9
		$0.8065\alpha_2$	0.7	$0.8065\alpha_2$	4
0.8023	213	$0.8023\alpha_1$	6	$0.8023\alpha_1$	21
		$0.8022\alpha_2$	2	$0.8023\alpha_2$	11
0.7975	501, 431	$0.7975\alpha_1$	4	$0.7972\alpha_1$	21
		$0.7973\alpha_2$	1	$0.7971\alpha_2$	11
0.7921	332	$0.7920\alpha_1$	4	$0.7919\alpha_1$	21
		$0.7919\alpha_2$	2	$0.7919\alpha_2$	11

NATURAL HIGHLY SILICEOUS GLASSES

The natural glasses that closely approach silica glass in composition are of two main types: fulgurites, formed by lightning striking and melting quartz sand or rock; and meteoritic glass, formed by the impact of meteorites on quartz sand or sandstone. The ash glass formed by the combustion of vegetal material is somewhat lower in silica content but also may be included here. A summary account of these materials and of synthetic silica glass is given beyond.

A number of additional types of natural glasses are known. The igneous glasses, such as obsidian, represent rock magmas that have been cooled sufficiently rapidly to prevent crystallization. These rocks range in composition up to roughly 80 per cent SiO_2. Tektites are a group of natural glasses found as small masses with a globular, tear-like, or other fluidal shape, often modified by weathering, that occur in many parts of the world. They include the moldavites of Central Europe, the australites of Australia, the billitonites of the East Indies, and the shonites of Sweden, among other types. The SiO_2 content of tektites generally is in the range from 68 to 82 per cent SiO_2, and is lower than that of most types of fulgurites and meteoritic glass. The bediasites of Texas and some other tektites contain lenticular grains or drawnout threads of silica glass.[1] The specific gravity of tektites generally falls in the range 2.30–2.46, and the index of refraction between 1.48 and 1.52. The color is dark green to greenish or brownish black or black. The origin of tektites is problematic but is usually considered to be extra-terrestrial. Glass also occurs as an accessory constituent in some meteorites, especially the stony types. It occurs in the fused crust, or internally as an interstitial material or as microscopic inclusions in enstatite, tridymite, or other constituents. Fused feldspar glass (maskelynite) occurs in some meteorites, and a glass with a relatively low index of refraction that may be highly siliceous has been noted. Mylonitic veins and zones resembling glass (pseudotachylite) are found both in meteorites and in terrestrial rocks.

SYNTHETIC SILICA GLASS

Silica glass is colorless and transparent. It is white or milky in color and translucent to opaque when it contains fine pores or bubbles that have been entrapped in the viscous melt, and is then silky in luster when drawn into rods. Hardness about 7, comparable to that of quartz. Density 2.203 ± 0.001 at 0°. The index of refraction for Na (589.29 mμ) is 1.4585, and in white light about 1.462. The index of refraction increases with increasing content of the common impurities Al, Fe, Mg, Ca, and alkalies. Silica glass is relatively transparent in the near and middle ultraviolet, although less so than quartz. It is virtually opaque at a wavelength of 193 mμ and less. Ideally, silica glass is optically isotropic and is often so found, but preparations that have been mechanically worked into rods or other shapes, or that have undergone particular heat treatment, may show a granular or ribbon-like birefringent structure. The effect is distinct from the strain birefringence commonly present. Silica glass is not optically active.

Silica glass is colored smoky brown to brownish violet by irradiation[44] with x-rays and with the radiations from radioactive substances. The color is rapidly discharged with an accompanying bright bluish luminescence at temperatures above a few hundred degrees. The depth of the irradiation color varies in different preparations, and sometimes in different parts of a single preparation, as in streaks or bands, presumably depending on the content of Al or other impurities. Silica glasses prepared from natural quartz crystals that vary in the depth of color acquired under irradiation show the same relative response.[2] Silica glass is triboluminescent. It may show a faint green fluorescence under x-ray or electron excitation, but it does not fluoresce visibly in ultraviolet radiation.

The x-ray diffraction pattern of silica glass shows a single broad diffraction peak corresponding to a spacing of about 4.32 Å. The structure of silica glass is a random three-dimensional network in which each silicon is tetrahedrally coordinated by four oxygens and each oxygen is shared by two silicons, the two bonds to an oxygen being roughly diametrically opposite. The orientation of one tetrahedral group with respect to a neighboring group about the connecting Si-O-Si bond is random. Each atom thus has a definite number of nearest neighbors at a definite distance, giving rise to the observed diffraction effect, but there is no periodicity in three dimensions and the substance is non-crystalline.[3]

Silica glass does not have a sharp melting point but becomes less viscous with increasing temperature. It can be worked readily only in an oxyhydrogen flame. Below 1710°, the melting point of cristobalite, silica glass is thermodynamically unstable. Theoretically, silica glass should devitrify, if quenched to a temperature below 1710°, to the crystalline polymorph of silica stable at that temperature, affording tridymite, e.g., in the range from 870° to 1470°, but cristobalite usually is afforded as the stable or metastable phase at all temperatures below 1710°. The rate of devitrification is inappreciable at temperatures below about 1000° and is rapid above about 1200°. The devitrification to cristobalite begins at the surface of the glass and takes place relatively rapidly in the milky or opaque types of fused silica that contain many fine pores and bubbles. The crystallization is markedly accelerated by water vapor.

FULGURITES

The name fulgurite (from the Latin *fulgur*, *lightning*) is applied to the vitreous tubes and crusts formed when lightning strikes and locally melts sands or the bare surfaces of rocks. The best-known fulgurites are from quartz sands, and take the form of tubes up to a half-inch or slightly more in diameter. The tubes course downward from the surface of the ground, generally decreasing in diameter and branching as they descend, and usually extend for several feet. Examples 20–30 ft. in length are fairly common, and some range up to 60 ft. or more. There may be local bulbous or knobby enlargements in the tube, which correspond to relatively porous or impure layers in the sand, and the tubes may be deflected or caused to bifurcate by pebbles in their path. The tubes sometimes terminate in sack-like enlargements, usually crumpled or flattened. Platy or rugose isolated masses of glass may occur at the end of the path, or the further course of the lightning may be indicated by a mealy consistency of the sand, as if the grains were only superficially fused and loosely cemented together.

The outer surfaces are rough, with adhering sand, and warty, spiny, or thread-like excrescences may be developed as if droplets of fused silica had been projected outward. Tiny perforations with rounded edges may be present in the walls of the tube. The tubes often are crudely corrugated or ridged parallel to their length or have a folded appearance. They have been likened to a vegetable stalk much contracted by drying. Only two ridges may be present, giving a wing-like appearance to the cross-section, or three or four ridges may be present, yielding a rough cross-like or I-beam appearance in cross-section. The ridges may be somewhat flattened, resembling flanges;[4] the middle plane of the flanges or ridges may have a relatively dark color.[5] In a number of instances the corrugations have a tendency for a spiral arrangement, although usually distinct only over short lengths of the fulgurite, and the direction of the spiral apparently always is right-handed[6] (see section on *Symmetry*, under *Quartz*, for convention as to hand). The axis of the tube itself generally is irregular, with random bends and crooks, but a spiral tube (also right-handed) has been described.[7] The origin of the surface corrugations and ridges is problematic. For the most part they seem to be primary features, although the collapse of the tubes while still viscous may be a factor.

The inner opening of the tubes may be nearly round but more generally is either crudely elliptical or has a jagged or toothed appearance with three, four, or five corners and indented sides as seen in cross-section. The shape of the opening may vary along the length of the fulgurite. The inside surface of the tubes is smooth and glassy, sometimes with a silvery luster, and in some specimens resembles an applied glaze which locally may show evidence of flowage. Glassy threads may extend completely across the opening as if the tube had been pinched together while still viscous and then had been again distended. Blister-like bubbles, some exploded leaving rounded edges, may be present on the inner walls. The walls usually are about 0.5–2 mm. thick, but may be paper thin, and there is no close relation between tube diameter and wall thickness. The toothed cross-section and flanges may have arisen through the ruptural deformation and ultimate collapse of an expanding tube of silica glass, analogous to the pyritohedral symmetry of some agate geodes.

The central hole and the vesicles in the glass are generally attributed to the

TABLE 47. INDEX OF REFRACTION AND DENSITY OF FULGURITES

n (white)	Specific Gravity	Ref.
1.462	2.203	Silica glass
1.461	2.20	15
1.462	...	41
1.462	1.05–1.25	14
1.465	2.21	39
1.457	...	14
...	2.197, 2.07	11
1.458	2.1	28

thermal expansion of air entrapped in the sand, and the flattening to the pressure of the surrounding sand while the glass was plastic. The expansion of water vapor was originally suggested by Watts in 1790. A supposed mechanical action of the lightning in forcing the sand apart, the open walls then being fused, also has been suggested. A contributing factor may be SiO_2 vaporized by the heat of the lightning, followed by collapse of the tube under the near vacuum produced when this gas condenses at a high temperature while the glass is still relatively fluid. The boiling point of SiO_2 is 2950° at 1 atm.,[8] and the vapor pressure at 1900° is about 19 mm.[9] The black or dark gray color of the glass in some specimens may be atomically dispersed elemental Si, since SiO_2 is readily reduced[10] to Si at very high temperatures by H, C, and CO. Silicon melts at 1420° and boils at 2355° at 1 atm.

In color, fulgurites are externally grayish white, yellowish brown, tan, or brown, sometimes dark gray to black. The surface also may be mottled or spotted black or brown. A local brown color may be produced where the tube transects an iron-rich layer in the sand, and the tube may here be distended because of the greater fluidity of the glass. The glass itself is colorless or faintly greenish or yellowish in small grains. It is isotropic, and the reported values of the index of refraction are close to that of pure silica glass (Table 47). Chemically, the glass ranges between about 90 and 99 per cent SiO_2. The principal impurities are ferric iron, to which the yellowish color is owing, and Al together with alkalies, Ca, and Mg (Table 48). A number of partial analyses also have been reported[12] with the following percentages of SiO_2: 91.23, 92.6, 94.26, 95.91, 96.44, 97.3. In some instances at least the silica content of the fulgurite is higher than that of the adjacent sand.[11]

TABLE 48. ANALYSES OF FULGURITES

	1	2	3	4	5	6	7	8
SiO_2	93.8	99.0	93.4	88.46	91.66	90.2	97.3	92.6
Fe_2O_3	⎱ 3.8	0.3	Tr.	1.16	⎱ 6.69	0.7	⎱ 1.1	⎱ 3.2
Al_2O_3	⎰	0.7	5	6.69	⎰	0.9	⎰	⎰
CaO	0.6			0.17	0.38	0.1	2.0	3.5
MgO				0.17	0.12	0.5		
Na_2O				0.01	0.77	0.6	Tr.	Tr.
K_2O				2.68	0.73	0.5		
Rem.					0.33	6.5		
Total	98.2	100.0	98.4	99.80	100.68	100.0	100.4	99.3

1. Germany.[42]   2. New Jersey.[14]   3. Australia.[40]   4. Western Australia.[39]   5. Illinois.[11]   Rem. is ign. loss.   6. Holland.[13]   Rem. is insol. 0.9, carbonaceous matter 5.6.   7,8. Germany.[11]

The glass contains numerous bubbles, which tend to be smaller and more numerous in the outer parts of the wall and in the ridges. The bubbles may be elongated and radially arranged. Inclusions of quartz grains are common in the peripheral parts. These grains generally are white and opaque, or turbid, and are minutely cracked, presumably because of thermal shock and passage through the 573° inversion. Adherent quartz grains may be fused on the inner side. Cristobalite pseudomorphs or rims of cristobalite around quartz grains may be present. Evidence of devitrification of the glass itself almost never is observed. Minute crystallites of undetermined nature have been seen in a few instances.[13]

Inclusions of mullite (?), presumably due to the local dissociation and crystallization of included clayey material before the Al diffused into the body of the glass, have been reported.[14] The specific gravity of the clear glass is about 2.2 (Table 47); much lower values are obtained in vesicular material, and higher values are caused by the presence of admixed Fe, quartz grains, etc. Sand fulgurites may be quite fragile because of bubbles and crack systems radiating out from included grains of quartz.

Fulgurites are common.[15]   A general condition for their formation is the presence of a dry dielectric such as quartz sand overlying a water-containing stratum or the water table. They are especially abundant or are relatively easily observed in sand dunes, as in the Sahara,[16] and it is estimated that upwards of 2000 fulgurites occur in an area of about 8 sq. mi. in the Kalahari Desert.[17] Numerous examples are known from commercial sand pits in the Atlantic Coastal Plain[18] and in other regions. Several fulgurites may be found in a small area and may represent separate branches of a single lightning stroke. At Olkusz, Silesia, 26 fulgurites, mostly associated in groups, were found in an area of 200 by 100 ft.[19] Fossil fulgurites are known.[20] Undoubted fulgurites were found as early as 1706 in Germany, and they were probably known to the ancients.[21] Their origin was recognized in 1790 by Withering,[22] who described a fulgurite found under a tree that had been struck by lightning. The fulgurite was found while the foundation was being dug for a memorial tablet warning the passerby against the danger of taking refuge under a tree during a thunderstorm. Numerous experiments have been described in which fulgurites have been produced artifically by passing an electric discharge through sand or other materials,[23] and they have been produced when broken high-tension power lines have fallen upon sandy soils. A class of objects of similar appearance produced by the silicification of plant roots and by concretionary processes in sands has been termed pseudofulgurites.[24] They are usually composed of calcium carbonate, limonite, or chalcedonic silica.

Fulgurites of the second class, which may be termed rock fulgurites in distinction to the sand fulgurites just described, are formed when lightning strikes the bare surface of rocks.[25] This type of fulgurite forms thin glassy crusts, with which may be associated short tubes or perforations lined with glass in the rock. The holes may show features indicating that molten glass welled out of them onto the surface. Glasses of this type may be relatively low in silica, depending on the nature of the fused rock. Described instances include fused granite, andesite, hypersthene basalt, mica schist, hornblende schist, glaucophane schist, serpentine, and other rock types. These glasses are relatively fluid, and devitrification phenomena may be observed.[13] The color, index of refraction, and specific gravity of the glass vary widely with composition. Rock fulgurites are common on the high peaks of mountain ranges, and may be so abundant, as on the summit of Little Ararat, Armenia,[26] as to give the outcrops a glassy, perforated appearance. Fulgurites of this type were first recognized in 1786 by Saussure, who described vitreous droplets and bubbles on granite from Mt. Blanc, Switzerland. He duplicated the effect experimentally, using an electric discharge from Leyden jars.[27]

METEORITIC SILICA GLASS

Natural glasses are known which have undoubtedly been produced by the fusion of silicious terrestrial materials in the intense heat developed when large meteorites have struck the Earth's surface. The composition of these glasses is primarily determined by the composition of the bedrock or sand upon which the meteorite fell, being virtually pure silica glass in the case of some desert sands, but it may be modified by the inclusion of vaporized material from the meteorite itself. These glasses are extremely stable both chemically and physically and in some instances apparently have been preserved after the craters in which they were formed were removed by erosion. Other less stable glasses formed similarly perhaps have been destroyed by weathering. The name lechatelierite, after the French chemist Henri Le Châtelier (1850–1936), has been proposed[28] for the natural glasses, chiefly including fulgurites and meteoritic glass, that closely approach silica glass in composition.

At Meteor Crater, Arizona, the first described[29] occurrence of meteoritic glass, the glass occurs on the rims and at the bottom of the crater, a depression about 4000 ft. in diameter and 570 ft. deep, as colorless to white masses and slabs. The glass is highly vesicular to porous, and is isotropic with n about 1.460. The specific gravity cannot be measured accurately because of the presence of minute bubbles, and low values of about 2.1 have been obtained. Thin sections of this glass, as of other meteoritic glasses, may show flow structures. The glass sometimes contains unmelted grains of quartz, and there is a gradation through wholly melted material, which may retain somewhat the structure of the original sandstone, to specimens with most of the quartz grains still intact but with some interstitial glass. The unmelted quartz grains may show rhombohedral parting, perhaps acquired by thermal shock at the 573° inversion. A little cristobalite (coesite?) is present in some of the glass, apparently as the result of devitrification. The glass contains about 98.6 per cent SiO_2 and, with the Libyan Desert glass described beyond, approaches more closely to pure silica glass than do other known occurrences of natural glass aside from fulgurites.

At Wabar, Arabia,[30] silica glass occurs very abundantly as cindery masses strewn over the ground, and the walls of the craters appear to be built wholly of this material. Pieces of the original iron meteorite, a medium octahedrite that was largely disintegrated and vaporized by the impact, are found around the craters. The masses of glass seem to be individual bombs ejected from the craters by explosions. The larger bombs are rough and cavernous, often showing ropy forms, and the surface is dull and dark gray or black in color. They consist mainly of a bluish gray or brownish glass full of bubbles of all sizes up to 6 cm. across. Smaller bombs consist internally of a very cellular snow-white glass, resembling pumice. On the outside they consist of black glass, often only a thin skin, with a glazed surface and almost free from bubbles. The glass contains numerous small black magnetic spheres, ranging in size from about 0.14 mm. in diameter down to dust, that are composed of α-iron (Fe 91.2, Ni 8.8 per cent). The metal spheres were formed by condensation from the vaporized meteorite, droplets raining down into the pool of molten and boiling silica in the crater on the desert floor. The glass is isotropic. Analyses of the glass, together with measured values of the index of refraction and specific gravity, are given in Table 49. The high content of Fe and Ni in the dark glass, in about the same ratio as in the metal, is due to the inclusion of vaporized metal from the meteorite.

The silica glass at the craters at Henbury, Central Australia,[30] was produced by fusion of a ferruginous sandstone. Fragments of the original meteorite, a medium octahedrite, are abundant. The glass is dull brown to black in

TABLE 49. ANALYSES OF METEORITIC GLASS

	1	2	3	4	5
SiO_2	98.20	92.88	68.88	86.34	89.81
TiO_2	0.23	0.12	3.64		
Al_2O_3	0.70	2.64	5.60		
Fe_2O_3	0.53	0.23	8.46		
FeO	0.24	0.53	7.92		
MnO	...	0.01	0.05		
NiO	0.02	...	0.28		
MgO	0.01	0.47	2.03		
CaO	0.30	1.46	2.51		
Na_2O	0.33	0.42	0.03		
K_2O	0.02	1.61	1.43		
H_2O^+	0.03	0.32	0.03		
H_2O^-	0.03	0.11	0.05		
Total	100.64	100.81	100.91		
d	2.206	2.10	2.31	2.296	2.284
n	1.4624	1.468	1.545		

1. Libyan Desert.[33] 2. Wabar, Arabia.[30] With SrO 0.01, P_2O_5 tr. White type.
3. Henbury, Central Australia.[30] Black type. 4,5. Mt. Darwin, Tasmania.[46]

color, cellular, with a glazed and ropy or pimpled surface. A few metallic spheres are present. In composition, the glass is relatively low in silica and high in iron (Table 49).

Silica glass not associated with known craters or with meteorites but presumed to have formed by the impact of large meteorites has been found at Mount Darwin on the west coast of Tasmania[31] and in the Libyan Desert.[32] The Darwin glass occurs in amounts of thousands of tons as vesicular and slaggy masses in glacial deposits. It contains small metallic spheres. The glass is isotropic, with a slightly variable index of refraction and specific gravity. The composition is given in Table 49. The Libyan silica glass occurs as pieces and lumps up to 16 lb. in weight, in part pitted and worn by wind-blown sand, on the surface in valleys between parallel ranges of very large sand dunes. The glass occurs in a roughly oval area about 130 by 53 km. in size. Some of the glass was worked by prehistoric man. It has a pale greenish yellow color, in part clear and transparent, with bubbles and white spherulites of cristobalite. It is relatively high in silica (Table 49) with correspondingly low values for the index of refraction and specific gravity. The origin of the Darwin and Libyan glasses has been discussed in relation to the possible terrestrial origin of tektites.[33]

ASH GLASS

Slag-like or scoriaceous glasses rather high in SiO_2 may be produced by the combustion of vegetal matter, especially grass and wheat straw. Many of the known examples are from haystack fires (straw-silica)[34] or from the combustion of grain.[45] The ash of gramineacous plants may range upwards of 5 per cent

TABLE 50. ANALYSES OF ASH GLASS

	1	2	3
SiO_2	66.04	57.40	70.11
Al_2O_3	1.55	1.81	0.48
Fe_2O_3	0.59	0.59	0.72
CaO	6.00	8.56	4.94
MgO	3.80	5.56	3.36
K_2O	11.98	13.58	8.76
Na_2O	6.88	8.98	7.97
Carb. matter	2.69	3.16	1.88
Rem.			1.20
Total	99.53	99.64	99.42
n			1.502

1, 2. Ash glass. South Australia.[38] 3. Straw-silica glass. Ramona, Calif.[34] Rem. is TiO_2 0.04, MnO 0.11, H_2O (110°) 0.02, P_2O_5 1.03.

of the dry weight and is typically high in silica. Some species in this family secrete free silica, such as the gel-like tabasheer of the bamboo and the opal of the epidermal cells of oats.

Stones of ash glass range up to 20 or 30 lb. in weight. The color varies from green to smoky gray and black. The vesicles may contain black carbonaceous matter, which also may be dispersed in finely divided form through the glass itself. Flow structures may be seen in thin section, and ovoid gas vesicles are typically present. Analyses of straw-silica glasses are given in Table 50. The relatively high content of alkalies, especially of K_2O, is typical of the inorganic content of grasses and distinguishes these glasses from tektites and sand fulgurites. The natural glass from Macedon, Victoria, apparently also is an ash glass and has been ascribed to sedimentary material fused by bush fires.[35] This material apparently is relatively high in silica and is described as similar to Darwin glass. It has also been suggested[31] that the Macedon glass formed by the impact of a meteorite. An ash glass containing melillite and tridymite resulting from a wheat-straw fire has been mentioned.[36] The clinkers (wood-ash stones)[37] found within the trunks of trees partly burned in forest fires differ in that they essentially consist of potassium-calcium carbonate (fairchildite) and lack glass.

References

1. Barnes: *Univ. Texas Publ. 3945*, 477, 1940.
2. Laemmlein: *Dokl. Ak. Nauk USSR*, **43**, 247 (1944).
3. Warren: *Zs. Kr.*, **86**, 349 (1933); *J. Appl. Phys.*, **8**, 645 (1937).
4. Barbour: *Nebraska State Mus. Bull.* **6**, 45, 1925.
5. Anderson: *Nebraska State Mus. Bull.* **7**, 49, 1925.
6. Anderson (1925); Fenner: *Rec. South Australian Mus.*, **9**, 127 (1949); Bayley: *Am. J. Sci.*, **43**, 327 (1892); Hobbs: *Am. J. Sci.*, **8**, 17 (1899); Gümbel: *Zs. deutsch. geol. Ges.*, **34**, 647 (1882); Petty: *Am. J. Sci.*, **31**, 188 (1936); Wood: *Nature*, **84**, 70 (1910). The twist is right-handed in both hemispheres.
7. Wood (1910).
8. Ruff: *Trans. Am. Electrochem. Soc.*, **68**, 89 (1935).
9. Ruff: *Zs. anorg. Chem.*, **117**, 172 (1921).
10. von Wartenberg: *Zs. anorg. Chem.*, **79**, 71 (1913).
11. Merrill: *Proc. U.S. Nat. Mus.*, **9**, 83 (1886); Fischer: *Jb. Min.*, Beil. Bd. 56, 69 (1927–28).
12. Fischer (1927–28), with literature.
13. Julien: *J. Geol.*, **9**, 673 (1901).
14. Myers and Peck: *Am. Min.*, **10**, 152 (1925); Rogers: *J. Geol.*, **54**, 117 (1946).
15. Leading references: Fischer (1927–28); Barrows: *School of Mines Quart. (Columbia Univ.)*, **31**, 294 (1910); Van Tassel: *Bull. inst. roy. sci. nat. Belg.*, **31**, no. 9 (1955); Wichmann: *Zs. deutsch. geol. Ges.*, **35**, 849 (1883); Merrill (1886); Richardson: *Min. Coll.*, **3**, 131, 145, 161 (1896–97); Chirvinsky: *Bull. int. ac. sci. Bohême*, **23**, 214 (1923).
16. Lacroix: *Bull. serv. mines, Govt. Gen l'Afrique Occid. Franc.*, No. 6, 23, 1942, and *Bull. soc. min.*, **38**, 188 (1915).
17. Lewis: *South African Geograph. J.*, **19**, 50 (1936).
18. Petty (1936).
19. Roemer: *Jb. Min.*, 33, 1876.
20. Barrows (1910), Fischer (1927–28); Van Tassel (1955).
21. Böttiger: *Ann. Phys.*, **72**, 317 (1822).
22. Withering: *Phil. Trans.*, 293, 1790.
23. Butcher: *Proc. Phys. Soc.*, **21**, 254 (1909); Wood (1910); Petty (1936); Rollman: *Ann. Phys.*, **134**, 605 (1868); Fenner (1949).
24. Lacroix (1942); Seelye: *New Zealand J. Sci. Techn.*, **32B**, 1 (1951); see also Hill: *Rocks and Min.* **23**, 802 (1948).
25. Diller: *Am. J. Sci.*, **28**, 252 (1884); Aston and Bonney: *Quart. J. Geol. Soc.*, **52**, 452 (1896); Bonney: *Geol. Mag.*, **6**, 1 (1899); Julien (1901).
26. Abich: *Sitzber. Ak. Wiss. Wien*, **60**, 153 (1870).
27. Saussure: *Voyages dans les Alps*, **4**, 1786, p. 472.
28. Lacroix: *Bull. soc. min.*, **38**, 182 (1915).

29. Rogers: *Am. J. Sci.*, **19**, 195 (1930).
30. Spencer: *Min. Mag.*, **23**, 387 (1933).
31. Spencer: *Min. Mag.*, **25**, 425 (1939); Hills: *Rec. Geol. Surv. Tasmania*, No. 3, 1915; David, Summers, and Ampt: *Proc. Roy. Soc. Victoria*, **39**, 167 (1927).
32. Clayton and Spencer: *Min. Mag.*, **23**, 501 (1934); Spencer (1939).
33. Spencer (1939).
34. Milton and Davidson: *Am. Min.*, **31**, 495 (1946).
35. Baker and Gaskin: *J. Geol.*, **54**, 88 (1946).
36. Dana, J. D.: *Manual of Min.*, twelfth ed., 1898, p. 443, fig. 7.
37. Milton and Axelrod: *Am. Min.*, **32**, 607 (1947).
38. Fenner: *Trans. Roy. Soc. South Australia*, **64**, 305 (1940).
39. Simpson: *J. Roy. Soc. West. Australia*, **17**, 145 (1930–31).
40. Knibbs, Grimshaw, and Curran: *Science*, **7**, 377 (1898).
41. Rogers: *J. Geol.*, **25**, 526 (1917).
42. Geinitz: *Arch. Nat. Meckl.*, **47**, 60 (1893).
43. Harting, cited in Fischer (1927–28).
44. Cf. Levy and Varley: *Proc. Phys. Soc. London*, **68B**, 223 (1955); Hoffman: *Glastech. Ber.*, **14**, 281 (1936).
45. Vélain: *Bull. soc. min.*, **1**, 113 (1878).
46. David, Summers, and Ampt (1927).

INDEX

331